Dynamics of the Earth

T0181152

V. I. Ferronsky • S. V. Ferronsky

Dynamics of the Earth

Theory of the Planet's Motion Based
on Dynamic Equilibrium

 Springer

V. I. Ferronsky
Water Problems Institute of the Russian
Academy of Sciences
Gubkin st. 3, 119333 Moscow
Russia
ferron@aqua.laser.ru

S. V. Ferronsky (deceased)

ISBN 978-94-007-9273-9   ISBN 978-90-481-8723-2 (eBook)
DOI 10.1007/978-90-481-8723-2
Springer Dordrecht Heidelberg London New York

Cover illustration: Real picture of motion of a body A in the force field of a body B. Digits identify succession of turns of the body A moving around body B along the open orbit C.

Cover design: deblik, Berlin

Revised and updated edition of the book in Russian "Dinamika Zemly: Teoria dvizhenija planety na osnovah dinamicheskogo ravnovesia" by V. I. Ferronsky and S. V. Ferronsky, Nauchniy Mir, 2007

Printed on acid free paper.

Springer is part of Springer Science+Business Media (www.springer.com)

To the Memory of
GEORGY NIKOLAEVICH DUBOSHIN
Teacher and Preceptor

Contents

Preface

The book sets forth and builds upon the fundamentals of dynamics of the Earth as a self-gravitating body whose movement is based on its dynamic equilibrium state. The term "self-gravitating body" refers to the Earth capacity for self-generation of the gravitational energy that gives it planetary motion. The idea of applying this dynamic approach appeared after the classical dynamics of the planet's motion based on hydrostatics had failed. It also followed an analysis of the geodetic satellite orbits and discovery of the relationship between the mean (polar) moment of inertia and the gravitation potential of the planet. The dynamical equilibrium of a self-gravitating body, which generates energy by means of interaction of its mass particles, was applied as an alternative to hydrostatics. In order to derive the equations of a dynamical equilibrium state, the volumetric force and volumetric moment were introduced into Newtonian equations of motion. Here the hydrostatic equilibrium state appeared to be the particular case of a gravitating uniform body subjected to an outer force field. It was based on the theory that the basic mode of motion of a self-gravitating Earth is its interactive particle oscillations which represent the main part of the planet's kinetic energy and appear as oscillations of the polar moment of inertia.

In the second part of the twentieth century, continuous study of space by artificial satellites opened a new page in space sciences. It was determined that the ultimate goal of this scientific program should be an answer of the Solar system's origin. At the same time, in order to solve geodetic and geophysical problems, investigation of the near Earth cosmic space was initiated.

The first geodetic satellites for studying dynamic parameters of the planet were launched almost 50 years ago. They gathered vast amounts of data that significantly improved our knowledge of the inner structure and dynamics of the Earth. They made it a real possibility to evaluate experimentally the correctness of basic physical ideas and hypotheses in geophysics, geodesy and geology, and to compare theoretical calculations with observations. Success in this direction was achieved in a short period of time.

On the basis of satellite orbit measurements, the zonal, sectorial and tesseral harmonics of gravitational moments in expansion of the gravitational potential by a spherical function, up to tens, twenties and higher degrees were calculated. The calculations have resulted in an important discovery having far-reaching effects. The obtained results proved the long-held assumption of geophysicists that the Earth does not stay in hydrostatic equilibrium, which, in fact, is the basic principle of the theories of dynamics, figure and inner structure of the planet. The same conclusion was made about the Moon.

This conclusion means that the model used to determine hydrostatic equilibrium of the Earth, which was applied in order to interpret the outer and central force field, does not satisfy the observed dynamic effects of gravitational interaction of mass particles and should be revised. But the state of scientific knowledge of this phenomenon has been found to be not ready to cope with such a situation. The story of the condition of hydrostatic equilibrium of the planet begins with Newton's consideration, in his famous work "*Philosophiae Naturalis Principia Mathematica*", of the Earth oblateness problem. The investigation based on hydrostatics was further developed by French astronomer and mathematician Clairaut. Later on the hypothesis of hydrostatic equilibrium was extended to all celestial bodies including stars. The authority of Newton was always so high that any other theories for solution of the problem in dynamics and celestial body structure were never proposed. But in current times the problem has arisen of the cause of the discrepancy between theory and observation and a movement has appeared to take over this crisis in the study of fundamentals of the Earth sciences. A situation like this happened at the beginning of the twentieth century when the radioactive and roentgen radiation was discovered and the corpuscular-wave nature of light was proved. This was the starting point for development of quantum mechanics. We seem now to have a similar situation with respect to planetary motion.

The conclusion about the absence of hydrostatic equilibrium of the Earth and the Moon was a reason to start our work with this interesting problem, the results of which are presented in this book. We found a still more serious discrepancy related to the Earth hydrostatic equilibrium, which is as follows. It is known that the planet's potential energy is almost 300 times more than the kinetic one represented by the body's rotation. This relation between the potential and kinetic energy contradicts the requirement of the virial theorem according to which the potential energy of a body in the outer uniform force field should be twice as much of the kinetic one.

Considering the Earth's observed potential energy, its angular velocity should be about seventeen times as much as it is. However, the planet has remained for a long time in an equilibrium state. In fact, the Earth appears to have been deprived of its kinetic energy. Some of the other planets, such like Mars, Jupiter, Saturn, Uranus and Neptune, exhibit the same behavior. But for the Mercury, Venus, our Moon and the Sun, the equilibrium states of which are also accepted as hydrostatic, their potential energy exceeds their kinetic energy by 10^4 times. A logical explanation comes to mind that there is some hidden form of motion of the body's interacting mass particles, together with their respective kinetic energy, which has not previously been taken into account. It is known that the hydrostatic equilibrium condition of a body existing in the outer force field satisfies a requirement of the Clausius virial theorem. The same requirement follows also from the Eulerian equations for a liquid-filled uniform sphere. The virial theorem gives an averaged relationship between the potential and kinetic energies of a body. A periodic component of the energy change during the corresponding time interval is accepted as a constant value and eliminated from consideration. From this evidence it is not difficult to guess that the hidden form of motion and the source of needed kinetic energy of the Earth and the planets including the Moon and the Sun might be found in that eliminated periodic component. In the problem considered by Newton, that component was absent because of his concept of the central gravitational force field, the total sum of which is equal to zero.

Taking into account the relationship between the Earth's gravitational moments and the gravitational potential observed by the satellites, we came back to derivation of the virial theorem in classical mechanics and obtained its generalized form of the relationship between the energy and the polar moment of inertia of a body. In doing so, we obtained the equation of dynamical equilibrium of a body in its own force field where the hydrostatic equilibrium is a particular case of a uniform body in its outer force field. The equation establishes a relationship between the potential and kinetic energies of a body by means of energy of oscillation of the polar moment of inertia in the form of the energy conservation law. An analytical expression of the derived new form of the virial theorem is based on Newton's laws of motion and represents a differential equation of the second order, where the variable value is kinetic energy of the body's oscillating polar moment of inertia. In this case the earlier lost kinetic energy is found by taking into account the oscillating collision of the interacting mass particles, the integral effect of which is expressed through oscillation of the polar moment of inertia. That effect fits the relationship between the potential and kinetic energies in the classical virial theorem. At the same time a novel physical conception about gravitation and electromagnetic interaction is discovered and mechanism of the energy generation becomes clear. The nature of the gravity forces as a derivative of the body's inner energy appears to be discovered.

We initiated the study in dynamics of a self-gravitating body, based on dynamic equilibrium, in the seventies. The results were published in a series of papers (Ferronsky et al., 1978–1996; Ferronsky, 2005, 2008, 2009; Ferronsky, 1983, 1984) and in the books "Jacobi Dynamics" (Ferronsky et al., 1987), and "Dynamics of the Earth" (Ferronsky and Ferronsky, 2007). Recently obtained results related to the problem of the Earth's dynamics are presented in this work. We show here that the new effect, which creates dynamics of the Earth, is its own force field. Earlier, the sum of the inner forces and their moments being affected by the outer central force field were considered as equal to zero. We find that the mass forces of interaction being volumetric ones created the inner force field which appears to be the field of power (energy) pressure. That field, according to its definition, cannot be equal to zero. The resultant of the field pressure appears to be a space envelope. The envelope has a spherical shape for a sphere and an elliptic shape for an ellipsoid. It was found that dynamic effects of the body's force field occur in oscillation and rotation of the shells according to Kepler's laws. A body that has a uniform mass density distribution realizes all its kinetic energy of motion in the form of so-called virial oscillations. It was assumed, earlier, that wave properties of this nature, like oscillations for mass particles in mechanics of bodies, are unessential. We found that virial oscillations of a body initiated by the force field of its own interacting mass particles represent the main part of its kinetic energy. Theories based on hydrostatics ignore that energy. But, as it was noted above, in this case the potential energy of the Earth and other celestial bodies by two or more orders exceeds their kinetic energy represented only in the form of axial rotation of the mass. Such an unusual effect has a simple physical explanation. Still in the beginning of the last century French physicist Louis de Broglie expressed an assumption, proved later on, that any micro-particle including electron, proton, atom and molecule, acquires particle-wave properties. The relationship, discovered by the artificial satellites between changes of the Earth's gravitational potential and the moment of inertia, shows that interaction of the planet's masses takes place on their elementary particle levels. It means that the main form of motion of the interacting mass particles is their oscillation. Continuous "trembling" of the planet's gravitational field, detected by satellites as the gravitational moments change, is another fact proving the de Broglie idea and extending it to the gravitational interaction of celestial body masses.

The dynamical approach to solution of the problem under consideration allowed the authors to expand the body's potential energy on its normal, tangential and dissipative components. The differential equations which determine the main body's dynamical parameters, namely its oscillation and rotation, were written. A rigorous solution of the equations was considered on the basis for bodies with spherical and axial symmetry. The solutions of problems relating to rotation, oscillation, obliquity and oblateness of a body's orbit and itself was considered on the basis of the general solution of dynamics of a self-gravitating body in its own force field. It was found that precession and wobbling of the Earth and irregularity of its rotation depends on effects of the polar and equatorial oblateness and the separate rotation of the planet's, the Sun's and the Moon's shells. The outer force field of a body follows rotation of the resultant envelope of the shells, but with some delay because of the finite velocity of the energy propagation in the outer force field. Also the problems of inner structure of the Earth, the nature of the planet's electromagnetic field and mechanism of the energy generation are considered. Methods for studying some practical tasks like orogenesis, earthquakes, volcanism and climate change are discussed. The theory we present is applicable not only to the planets and satellites, but also to the stars, where hydrostatic equilibrium is considered as an equation of state. Finally, the theory opens a way to understand the physics of gravitation as the internal power (energy) pressure which occurs at matter interaction on the level of molecules, atoms and nuclei.

It is our pleasant duty to express sincere gratitude to Professor G.N. Duboshin of the M.V. Lomonosov Moscow State University for his continuous support and encouragement. We also express gratitude to our colleague S.A. Denisik with whom the authors worked for many fruitful years. We are indebted to Academicians A.S. Monin and B.V. Raushenbach, Corresponding Member T.I. Moiseenko, Professors

E.P. Aksenov, V.G. Demin, I.A. Gerasimov, I.S. Zektser for their support and fruit-ful discussions. Finally, we wish to say many thanks to our old colleagues V.A. Polyakov, V.T. Dubinchuk, V.S. Brezgunov, L.S. Vlasova and Yu.A. Karpychev for their cooperation in research. We want especially to thank Mrs. Galina A. Kargina for assistance in preparation the manuscript for this book.

V. I. Ferronsky

Chapter 1
Introduction

This chapter presents a short history of the birth and development of the dynamics of the Earth based on the assumption of an initial hydrostatic equilibrium state.

Dynamics is the branch of mechanics that deals with the problem of a body in motion and the forces applied to create that motion. Newton's three laws created a theoretical basis for dynamics to solve two types of tasks that are inverse to each other. The first was to find the relevant forces by application of known laws of body motion. The second type of task was to find the law of body motion on the basis of those identified acting forces. Newton solved the problem of the first type. He searched for the force that moves the Earth and the planets around the Sun according to Kepler's laws. Newton concluded that the planets and their satellites are moved along their orbits by interaction of centripetal (attractive) and inertial forces. His solution of the problem that led to the idea of "motion of bodies mutually attracting each other with centripetal forces" and other solved problems related to body motion, became the foundation for Newton's formulation of these laws of motion, including the general law of gravitation.

Newton published results of his study in 1686 in his famous work "*Philosophie Naturalis Principia Mathematica*" in Latin. In the third part of the work, under the title "*About the World System*", he presented a solution to the problem "*Determine the ratio of the planet's axis length to the length of its diameters*". This question about the Earth's oblateness with respect to its axis of rotation appears to be the basis for study of the planet's figure. The figure was presented as a rotating-by-inertia spherical body with polar and equatorial channels filled with a uniform liquid in a hydrostatic equilibrium state. Later, the French mathematician and astronomer Clairaut substituted Newton's idea of a uniform liquid in the channels of the rotating body with the concept of a liquid of fluctuating density along its radius. Even later, other researchers presumed the body's liquid to be an elastic and viscous substance. But the conditions of its hydrostatic equilibrium and inertial rotation of the planet have remained up to now as fundamentals of its dynamics as well as of the other celestial bodies.

Let us briefly consider the main steps in the history of dynamics of the Earth and the story of appearance of the hydrostatic equilibrium idea.

V. I. Ferronsky, S. V. Ferronsky, *Dynamics of the Earth,*
DOI 10.1007/978-90-481-8723-2_1, © Springer Science+Business Media B.V. 2010

1.1   Copernican Heliocentric World System

Nicolaus Copernicus (1473–1543), Polish astronomer and mathematician, the author of the heliocentric world system, was the first to come to the conclusion that the Earth moves around the Sun. For many centuries astronomers and sailors had unconditionally adopted the idea of the geocentric world system developed by the famous Greek astronomer Claudius Ptolemaios. In accordance with his conception, the Earth was the center of the system and all other observed celestial bodies rotated around that center. Copernicus started his work by study and improvement of the Ptolemaic system, presented in a 13-volume description entitled "*The Great Mathematical Construction*" with the Arabian name "*Almagest*" which enclosed the Hipparchos' star catalog describing the location of 1 022 stars in an elliptic system of the sky's coordinates. But very soon he discovered that the complicated cycles of motion of the observed planets and other celestial bodies, presented by Ptolemaios in the form of tables, can be described in a simpler way and in more logistic forms of motion around the Sun. In doing so, Copernicus came to understand the true heliocentric law of motion of the Earth and the other planets. For practical use he compiled, like Ptolemaios, tables of the planets' motion along the sky in circular orbits. The tables were very quickly popularized, first of all among navigators. Near the end of his life, in 1543, the Copernican heliocentric system was published in his work "*About Revolution of the Sky Spheres*". In addition to his astronomic study, Copernicus prepared a set of logarithmic tables that are, in content, close to those used in the present day. In so doing he made a notable contribution to trigonometry.

1.2   Galilean Laws of Inertia and Free Fall

After Copernicus died, his study of motion of the Earth and celestial bodies was continued by the Italian astronomer, mathematician and mechanic Galileo Galilei (1564–1642). Still being a student at a medical university, he came to know Aristoteles' physics and found it unconvincing. Galileo left medicine and started to study mechanics and geometry, in particular works of Euclid, Archimedes and Copernicus. In 1589 he wrote "Dialogue on the Motion", where he disproves the Aristotelian conception of the Earth's motion. In that work he does not mention the name Copernicus but he became a convinced and faithful supporter of the Copernican system and used the ideas from it in his own writing. In 1592–1610 Galileo carried out studies on static and dynamic equilibrium in machines by applying for this purpose the virtual displacement principle. At the same time he studied the free fall of kernels with different densities which, as legend says, were thrown from Pisa's tower. From that experience Galileo derived the laws of bodies in free fall, either on a sloping plain or thrown upward at an angle to the horizon, and also their application to isochronism of a pendulum's oscillation. After construction of his first triple, and later on 32 multiple, telescope, Galileo placed it on the San Marco tower and carried out a long series of observations. He discovered here Venus's phases of motion, rotation and spots of the Sun, mountains on the Moon, and the four satellites of Jupiter. These discoveries gave rise to doubts from other naturalists. The findings were called an optical illusion, arguing that they contradicted the Aristotelian postulates. Despite the fact that Galileo was accepted into the Accademia dei Lincei, known as the Rome Academy of "The Lynx-eyes", he was denounced by the Church for his defense of Copernican ideas, which were declared by the Jesuit Congregation to be heretical and his book was included in the list of prohibited readings. Despite this censure, in 1630 Galileo succeeded in publishing the book "*Dialogue about the Two Main World Systems*" where he defended Copernicus' ideas. Soon however, sale of the book was forbidden by demand of the inquisition and the author was forced to publicly deny the Copernican theory and to make repentance. Shortly thereafter, in the protestant countries "*The Dialogue*" was published in Latin.

By his observations and works Galileo founded the basic principles of mechanics like the laws of inertia and free fall of bodies, the principle of relative motion, the law of mechanical energy conservation in pendulum motion and the law of motion summation. Whereas Archimedes formulated the fundamentals of statics, Galileo laid the basis of dynamics. The assumptions of "The Thinker" were: the world is unbounded, matter is perpetual, and the heavenly bodies are similar to the Earth.

1.3   Kepler's Laws of Planets' Orbital Motion

Johannes Kepler (1571–1630), German astronomer and mathematician, became acquainted as a young man with the Copernican theory which predetermined his way of life. In 1600 Kepler moved to Prague in order to work with the well-known astronomer Tycho Brahe who for many years had maintained observation of the motion of the planets, in particular that of Mars. After the tragic death of Brahe in 1601, Kepler obtained copious materials resulting from these observations and undertook an extensive organization and evaluation. As early as 1604 he published a work that showed that illumination from a source decreases according to the inverse square law. In 1609 his work "*New Astronomy*" appeared, where Kepler published results of his evaluation of Brahe's observation of the motion of Mars. The results were presented as his first and second laws. The first of them states that the orbit of the unperturbed planet's motion is a curve of the second order, in one of the focuses of which the Sun is situated. It follows from the second law that the radius vector joining the unperturbed planet's motion with the Sun sweeps out equal areas in equal times. In 1619 Kepler published a new work entitled "*Harmony of the World*", where he describes his third law. In accordance with that law, in unperturbed elliptic motion of two planets, the ratio of the square of their periods of revolution to the cube of the semi-major axes is the same.

After discovery of the Copernican heliocentric system it was assumed that the planets move around the Sun along circular orbits. Kepler, being a convinced

Fig. 1.1 Kepler's problem

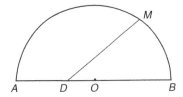

follower of Copernicus, shared the idea. But after analysis of Brahe's data he found that the observational points of the planets' annual motion did not describe circles. The points inscribe the circle but do not form it. However, the results from processing the data of the planets' annual trajectories indicated that they do describe some form of spatial curves. In order to reduce the space coordinates of the planets' motion to mean values and to obtain a plane elliptic figure, Kepler developed a specific method of averaging the observational points based on inscribing polygons into a circle and calculating with infinitesimals. Finally Kepler succeeded in finding a methodology of reducing the data which allows one to obtain an elliptic trajectory and to formulate the first two laws of the planets' motion. In *"New Astronomy"* Kepler presented this method in the form of the following geometric solution of the problem which allows one to find elements of a planet's orbit satisfying his first two laws of motion. Across point D on diameter AB of semicircle $AOBM$ (Fig. 1.1) the straight line DM should be drawn in such a way that it divides the area in the given ratio. The problem was written in the following transcendental equation

$$y - c \sin y = x, \qquad (1.1)$$

which is solved by the given values c and x, when ($|c| < 1$). Here c is the figure's eccentricity which at a value less than unit gives an ellipse and at zero gives a circle. The value x characterizes the scale of averaging taken as a ratio of the semicircle areas formed by the line DM.

Equation (1.1) represents projections of the reduced space coordinates of a body's motion along its orbit on the plane. With the help of this equation astronomers could determine the body's position on a point of the orbit at a given moment of time and solve the reverse problem of determination of the time moment of a body passing through a given point of the orbit. To come back from the projections of the trajectory on the plane to space coordinates in the sky, three angles called the true, mean and eccentric anomalies are used (Fig. 1.2).

Figure 1.2 demonstrates the true anomaly expressed by angle V between the pericenter B of the orbit and the radius vector S of the body, in the direction of the body's motion. In accordance with Kepler's second law the angle V changes in time faster when the body moves in orbit to the pericenter B and its motion is slower in a direction away from the pericenter B.

The mean anomaly is determined by angle M, which lies between the direction to the pericenter and the radius vector of a fictitious point, but is assumed to be moving with constant velocity during which that point passes pericenter B and apocenter A simultaneously with the real body. Thus, while moving from point B to point A the

Fig. 1.2 The true V, mean M and eccentric E anomalies for determination of a body's position on the orbit by Kepler's equation

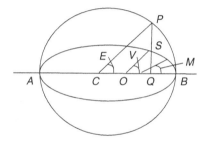

real body precedes the fictitious point, whereas moving from A to B the real body lags behind it.

The eccentric anomaly is expressed by angle E with the point in the center of the orbit and situated between the direction to the pericenter and point P. That point lies on the circle drawn from the geometric center of the orbit and the perpendicular PQ, carried out on the diameter, passes through the point S. The point P plays an auxiliary role to determine the mean M and true V anomalies by formulas

$$E - e\sin E = M, \tag{1.2}$$

$$M = M_0 + v(t - t_0), \tag{1.3}$$

$$\operatorname{tg}\frac{V}{2} = \sqrt{\frac{1+e}{1-e}}\operatorname{tg}\frac{E}{2}, \tag{1.4}$$

where e is the eccentricity of the orbit; M_0 is the mean anomaly at some initial moment of time t_o, which is accepted as an element; v is the mean value of the body's orbit velocity of motion.

It is clear that the meaning of Kepler's problem, represented by Eq. (1.2), is to inscribe into a circle an ellipse, which is the real averaged trajectory of the body's motion on the orbit, by applying the mean velocity value of the motion v and the mean anomaly M_0. Herewith, the inscribed ellipse must touch the circle only in two points of the body's orbit, namely in the perihelion B and aphelion A.

Kepler's first two laws and the equation represent the averaged space picture of a planet's motion over a period of revolution around the Sun. They do not describe small variations of the motion parameters either within each period of revolution or from one period to another. Those variations of the parameters' motion are smoothed by the mean anomaly M, and the Kepler laws and the equation expresses conditions of the hydrostatic equilibrium of the system. Kepler's equation was solved by Newton in his two-body problem in order to find the force which sets the body in motion. Newton's solution was done in the frame-work of Kepler's formulation of the problem, i.e., for the condition of hydrostatic equilibrium of a planet's motion. This remark is important for understanding the logic of his judgment and geometric construction which Newton used for solution of the two-body problem and the problem of the Earth's oblateness. As to the method of averaging of

Kepler's space coordinates by means of infinitesimals, it served to be the ideological base for development of the differential and integral calculus originally initiated by Newton and Leibniz simultaneously, but obviously not without the influence of Kepler's and Huygens works.

1.4   Huygens Laws of Clock Pendulum Motion

Christian Huygens (1629–1695), the Netherlands physicist, mathematician, mechanic and astronomer, was the founder of the wave theory of light and the theory of probability, the author of the first pendulum clock and investigator of the pendulum laws of motion which synchronously follows the Earth's motion. At 22 years of age he published his first work about determination of the arc length of the circle, ellipse and hyperbola. And three years later he wrote a work about the ratio of the circle's length to its diameter, which was called π. Then there was the work "*About calculation of the bones game*", where studies of cycloid, logarithmic and chain lines were undertaken and which became a part of the foundation of the theory of probability. Together with Hook he established the points of freezing and boiling of water. At the same time Huygens actively worked on increasing the luminosity of astronomical telescopes. In 1655, with his own instruments, he discovered the satellite Titan of Saturn, its own rings, the nebulae of the constellation Orion, and the poles of Jupiter and Mars.

Astronomical observations always needed precise and easily calculated measurements of time. In 1657 Huygens designed the first pendulum clock to be driven by a trigger mechanism of motion. In the next year he published a treatise "*The Pendulum Clock*", where his description of the discovery and the study of the pendulum clock motion was presented. It was known that the period of oscillation of a pendulum depends on the amplitude of the oscillation. In order to determine the precise motion of the clock, Huygens developed a construction, astonishing even for modern standards, schematically presented in Fig. 1.3.

Figure 1.3a shows, by dashed lines, barriers having a cycloid configuration, which bounds the swings of elastic filament of the suspended pendulum. The filament from a suspension point O up to some point A sags to both sides of the cycloid.

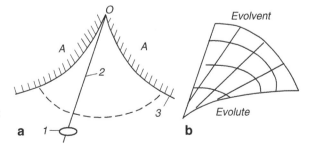

Fig. 1.3 Scheme of the Huygens pendulum clock (**a**), evolvent and evolute (**b**): pendulum (*1*); filament of suspension (*2*); cycloid (*3*)

Below point A the filament is held tight by the weight of the pendulum due to its motion to that point along the cycloid. The pendulum itself during that motion traces the cycloid (shown by dashes). In such a device the period of the pendulum oscillation does not depend on amplitude of the oscillation.

The described Huygens' project was not realized because at that time a more suitable design to solve the problem of synchronizing the oscillations was found. The interest in Huygens' technical idea has been lost and his name is mentioned only in differential geometry in connection with his introduction of the curves known as evolvent and evolute (Fig. 1.3b). In our time Huygens' idea is used for design of geophysical devices like gravitational variometers and gravimeters for measurement of the Earth's gravitational field. Technical solutions for such devices were proposed at the end of the nineteenth century by Hungarian physicist Eötvös. However, Huygens' study in pendulum motion contains much more fruitful, although not realized, ideas.

Recall, that the evolute is a curve which is formed from the locus of the centers of curvature of another plane curve (evolvent). The equation of this curve is a semi-cubic parabola. The evolvent is an unwound form of a curve perpendicular to a family of tangents to the evolute. The meaning of Huygens' idea is the following. First, the relation between an evolute and an evolvent represents a relation between a function and its derivative or between a function and its integral. But these relations exist in the integrated form and are geometrically observable but not in a local form such as in mathematical analysis. Secondly, as it is seen from the drawn family of such unwound curves with different fixed lengths of the pendulum filaments (Fig. 1.3b) in each point of the initial curve, the corresponding evolvent has a peculiarity. And third, an important point for us, the marked peculiarity is always of the same type. It is a semi-cubic parabola like $x^2 = y^3$ or $x = y^{3/2}$. This is a universal law, being a consequence of the simple fact that in each task related to motion we always have some initial conditions inherited by the moving object. For example, in the case of the Huygens' swinging pendulum its suspension filament winds away from the curve at some fixed point.

If one recalls the Kepler laws then it is possible to notice their important properties. The first two laws determine the trajectory of the same body. The third law relates to the family of trajectories traced by different planets of the same Solar System family. This law says that squares of periods of the revolving planets are proportional to cubes of their semi-major axes. It means that on the plane of time-coordinates the above law is expressed by a semi-cubic parabola. And in turn, the above is evidence of the fact that if the motion is considered in the space of time, but not in the space of configurations, then the Kepler laws express a universal law of nature in the integral form. Here constancy of the light velocity plays the role of isochronism of the oscillations.

Huygens applied the design and study of pendulum motion to description of elastic wave propagation, including in anisotropic media (double refraction of light beams in crystals) which he considered in "*The Treatise on the Light*". He discovered here one more effect. Namely, the line of the peculiar points, which was discussed above, determines the edge of the region (this is a space edge according

Fig. 1.4 Huygens' peculiar
points for a family of
evolvents

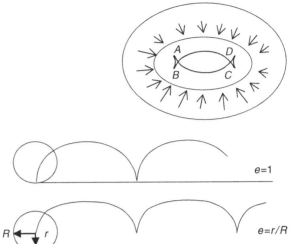

Fig. 1.5 Huygens' solution
of Kepler's equation

to Huygens). The sphere has no such edge. It appears when the waves propagate inside the closing curve. Huygens considered a case with an ellipse (Fig. 1.4).

Here waves propagate with constant velocity to the inner hollow of an ellipse. At the beginning the curve is transferred equidistantly to the ellipse. After that a time comes when the peculiar points A, B, C, D appear. That is not a continuum but a limited number of points. If the considered region is not spatial as Huygens discussed, but it is a space–time region, then the phase transition phenomenon appears that is described by the van der Waals cubic equation.

Finally there is one more important element of analysis proposed by Huygens. In fact, he introduced into physics an integral approach to studying the behavior of a system. An example to explain his approach is a straight-rolling wheel. A point on its rim or on the spoke traces the curve, which gives a solution of the Kepler equation (Fig. 1.5).

Some of Huygen's profound ideas are far from being realized. The laws of the pendulum motion of his clock, which in detail and synchronously follows the Earth's motion, could be physically demonstrated by the appropriate technical implementation not only for gravimetry but also for study of the planet's dynamics. As to the theoretical conclusions, we used them in our previous works and reference will be made below in the book.

1.5   Hooke's Law of Elasticity

Robert Hooke (1635–1703), the great English naturalist, a member and the second Secretary of the London Royal society, scientist and inventor, is founder of the theory of elasticity. Together with Huygens he determined the constant points of boiling and freezing of water. In 1660 Hooke discovered the law of proportionality, within the elastic limit, between a strain and the stress producing it in the body. In fact, the elastic property of a uniform body, being under action of the outer forces, defines hydrostatic equilibrium of this body. Later on, in 1674 in his work *"The attempt to study the Earth motion"* Hooke proposed to develop the respective theory applying three assumptions; (1) all the celestial bodies have an attraction to the center; (2) all the bodies preserve their uniform motion in a straight line up to the moment when some other force will deflect them; (3) the force of attraction is the higher the closer is the body.

1.6 Newton's Model of Hydrostatic Equilibrium of a Uniform Earth

Isaac Newton (1643–1727), the genius intellectual, English mathematician and physicist, is the founder of classical mechanics and astronomy and the author of the gravitation law. His merits and contribution to development of the natural sciences is difficult to be overestimated. The top of Newton's scientific work became the generalization of scientific results of Copernicus, Kepler, Galileo, Huygens, Borelli, Hooke, Galley and other predecessors and contemporaries, all of whose work was presented in his *"Philosophiae Naturalis Principia Mathematica"* and published in 1686. In that book a mathematical (geometric) approach was used for solution of the problems of celestial mechanics and dynamics of the Earth. Later on, an analytical basis for such a purpose was developed by Lagrange, Euler, d'Alembert, Hamilton, Jacobi, Cauchy, Bernoulli and other mathematicians in the seventeenth to nineteenth centuries.

Newton adopted the condition of the Earth's hydrostatic equilibrium state together with the Keplerian laws of motion and his problem. That model of equilibrium comprises the basis for solution of the two-body problem and the problem of the Earth's oblateness.

Newton opens his work with definitions of matter, momentum and innate, applied, centripetal force and with formulation of his three laws of motion. In Book I *"The Motion of Bodies"* the solution of the two-body problem is presented. In Book II *"The Motion of Bodies (in resisting medium)"* the hydrostatics theorems are discussed. And in Book III *"The System of the World"* the solution of the Earth's oblateness problem is considered. Let us recall the original Newton's formulations of the more important principles which we cite and discuss later on in the book. For that purpose we quote from the English translation of Newton's *Principia*, made by Andrew Mott in 1729 (Newton, 1934).

Book I. The Motion of Bodies

Definition I. *The quantity of matter is the measure of the same, arising from its density and bulk conjointly.*

Definition II. *The quantity of motion is the measure of the same, arising from the velocity and quantity of matter conjointly.*

Definition III. *The vis insita, or innate force of matter, is a power of resisting, by which every body, as much as in it lies, continues in its present state, whether it be rest, or moving uniformly forwards in a right line.*

Definition IV. *An impressed force is an action exerted upon a body, in order to change its state, either of rest, or of uniform motion in a right line.*

Definition V. *A centripetal force is that by which bodies are drawn or impelled, or any way tend, towards a point as to a centre.*

Of this sort is gravity, by which bodies tend to center of the earth; magnetism, by which iron tends to the load stone; and that force, whatever it is, by which the planets are continually drown aside from the rectilinear motion, which otherwise they would pursue, and made to revolve in curvilinear orbits. A stone, whiled about in a sling, endeavors to recede from the hand that turns it; and by that endeavor, distends the sling, and that with so much the greater velocity, and as soon as it is let go, flier away. That force which opposes itself to this endeavor, and by which the sling continually draws back the stone towards the hand, and retains in its orbit, because it is directed to the hand as the centre of the orbit, I call the centripetal force. And the same thing is to be understood of all bodies, revolved in any orbit. They all endeavor to recede from the centers of their orbits; and were it not for the opposition of a contrary force which restrains them to, and detains them in their orbits, which I therefore call centripetal, world fly off in right lines, with uniform motion...

The quantity of any centripetal force may be considered as of three kinds: absolute, accelerative, and motive.

Definition VI. *The absolute quantity of a centripetal force is the measure of the same, proportional to the efficacy of the cause that propagates from the centre, through the spaces round about.*

Definition VII. *The accelerating quantity of a centripetal force is the measure of the same, proportional to the velocity which is generates in a given time.*

Definition VIII. *The motive quantity of a centripetal force is the measure of the same, proportional to the motion which is generates in a given time.*

These quantities of forces, we may, for the sake of brevity, call by the names of motive, accelerative, and absolute forces; and for the sake of distinction, consider them with respect to the bodies that tend to the centre of forces towards which they tend; that is to say, I refer the motive force to the body as an endeavor and propensity of the whole towards a centre, arising from the propensities of the several parts taking together; the accelerative force to the place of the body, as a certain power diffused from the centre to all places around to move the bodies that are in them; and the absolute force to the centre, as endued with some cause, without which those motive forces would not be propagated through the space round about; whether that cause be some central body (such as is the magnet in the centre of the magnetic force, or the earth in the centre of the gravity force), or anything else that does not yet appears. For I here design only to give a mathematical notion of those forces, without considering their physical cause and seats...

I likewise call attractions and impulses, in the same sense, accelerative and motive; and use the words attraction, impulse, or propensity of any sort towards a centre, promiscuously, and indifferently, one for another; considering those forces not physically, but mathematically: wherefore the rider is not to imagine that by those words I anywhere take upon me to define the kind, or the manner of any action, the causes or the physical reason thereof, or that I attribute forces, in a true and physical sense, to certain centers (which are only mathematical points); when at any time I happen to speak as attracting, or as endued with attractive powers.

Law I. *Every body continues in its state of rest, or of uniform motion in a right line, unless it is compelled to change that state by forces impressed upon it.*

Law II. *The change of motion is proportional to the motive force impressed; and is made in the direction of the right line in which that force is impressed.*

Law III. *To every action there is always opposite and equal reaction: or, the mutual actions of two bodies upon each other are always equal, and directed to contrary parts.*

Section XI. Motion of Bodies Tending to each other with Centripetal Forces.
Before discussing the problem, essentially Newton notes that "*...I approach to state a theory about the motion of bodies tending to each other with centripetal forces, although to express that physically it should be called more correct as pressure. But we are dealing now with mathematics and in order to be understandable for mathematicians let us leave aside physical discussion and apply the force as its usual name*".

Proposition LVII. Theorem X. *Two bodies attracting each other mutually similar figures about their common centre of gravity, and about other mutually.*

For the distance of the bodies from their common centre of gravity are inversely as the bodies, and therefore in a given ratio to each other; and hence, by composition of ratios, in given ratio the whole distance between the bodies. Now these distances are carried round their common extremity with an uniform angular motion, because lying in the same right line they never change their inclination to each other. But right line that are in a given ratio to each other, and carried round their extremities with an uniform angular motion, describe upon planes, which either rest together with them, or are moved with any motion not angular, figures entirely similar round those extremities. Therefore the figures described by the revolution of those distance are similar.

Proposition LVIII. Theorem XI. *If two bodies attract each other with forces of any kind, and revolve about the common centre of gravity: I say, that, by the same forces, there may be described round either body unmoved a figure similar and equal to the figures which the bodies so moving describe round each other.*

Let the bodies S and P (Fig. 1.6a) revolve about their common centre of gravity C proceeding from S to T, and from P to Q.

Fig. 1.6 The problem of two
bodies mutually attracted

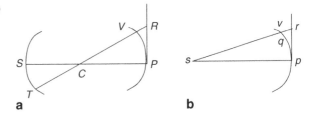

a   **b**

From the given point s (Fig. 1.6b) *let there be continually drown sp and sq equal
and parallel to SP and TQ; and the curve pqv, which the point p described by point
p at its revolution will be equal and similar to the curves which are describes in its
revolution round the fixed point S, will be similar and equal to the curve which the
bodies S and P describes about each other; and therefore, by Theor. XX, similar to
the curves in curves ST and PQV which the same bodies describe about their common centre of gravity C; and that because the proportions of the lines SC, CP, SP
or sp, to each other given.*

Case 1. *The common centre of gravity C (by Cor. IV of The Laws of Motion) is
either at rest, or moves uniformly in a right line. Let us first suppose it at rest, and
in s and p let there be placed two bodies, one immovable in s, the other movable in
p, similar and equal to the bodies S and P. Then let the right lines PR and pr touch
the curves PQ and pq in P and p, and produce CQ and sq to R and r. And because
the figures CPRQ, sprq are similar, RQ will be to tq as CP to sp, and therefore in a
given ratio. Hence if the force with which the body P is attracted towards the body S,
and by consequence towards the intermediate centre C, were to the force with which
the body p is attracted towards the centre s, in the same given ratio, these forces
would in equal times attract the bodies from the tangents PR, rq; and therefore
this last force (tending to s) would make the body p revolve in the curve pqv, which
would become similar to the curve PQV, in which the first force oblique the body P
to revolve; and their revolutions would be completed in the same times. But because
those forces are not to each other in the ratio of CP to sp, but (by reason of the
similarity and equality of the distance SP, sp) mutually equal, the bodies in equal
times will be equally drawn from the tangents; and therefore that the body p may be
attracted through the grater interval rq, there is required a grater time, which will
vary as the square root of the intervals; because, by Lem. X, the space described at
the beginning of the motion are as the square of the times. Suppose, then, the velocity of the body p to be to the velocity of the body P as the square root of the ratio of
the distance sp to distance cp, so that the arcs pq, PQ, which are in a similar proportion to each other, may be described in times that are as the square root to the
distance; and the bodies P, p, always attracted by equal forces, will describe round
the fixed centers C and s similar figures PQV, pqv, the latter of which pqv is similar
and to be figure which the body P describes round the movable body S.*

Case 2. *Suppose now that the common centre of gravity, together with the space in
which the bodies are moved themselves proceeds uniformly in the right line; and (by*

Cor. VI of The Laws of Motion) all the motions in this space will be performed in the same manner as before; and therefore the bodies will describe about each other the same figures as before, which will be therefore similar and equal to the figure pqv.

Corollary I. *Hence two bodies attracting each other with forces proportional to the square of their distance, describe (by Prop. X), both round their common centre of gravity and round each other, conic sections having their focus in the centre about which the figures are described; and conversely, if such figures are described, the centripetal forces are inversely proportional to the square of the distance.*

Corollary II. *And two bodies, whose focuses are inversely proportional to the square of their distance, describe (by Prop. XI, XII, XIII), both round their common centre of gravity, and round each other, conic sections having their focus in the centre about which the figures are described. And conversely, if such figures are described, the centripetal forces are inversely proportional to the square of distance.*

Corollary III. *Any two bodies revolving round their common centre of gravity describe areas proportional to the time, by radii drown both to the centre and to each other.*

Book II. The Motion of Bodies (in Resisting Medium)

Proposition XIX. Theorem XIV. *All the parts of an homogeneous and uniform fluid in any unmoved vessel, and compressed on every side (setting aside the consideration of condensation, gravity, and all the centripetal forces), will be equally pressed on every side, and remain in their places without any motion arising from that pressure.*

Case 1. *Let a fluid be included in the spherical vessel ABC, and uniformly compressed on every side: I say, that no part of it will be moved by that pressure For it and part, other as D, be moved, all such parts at the same distance from the centre on very side must necessarily be moved at the same time by a like motion; because the pressure of them all in similar and equal; and all other motion is excluded that does not come all of them nearer to the centre, contrary to the supposition...*

Proposition XXII. Theorem XVII. *Let the density of any fluid be proportional to the compression, and its parts be attracted downwards by a gravitation inversely proportional to the square of the distances from the centre: I say, that if the distance be taken in harmonic progression, the densities of the fluid at those distances will be in a geometrical progression.*

Book III. System of the World (in Mathematical Treatment)

Proposition II. Theorem II. *That the forces by which the primary planets are continually drawn off from rectilinear motions, and retained in their orbits, tend to the sun; and are inversely as the squares of the distances of the places of those planets from the sun's centre.*

Proposition VII. Theorem VII. *That there is a power of gravity pertaining to all bodies, proportional to the several quantities of matter which they contain.*

Proposition VIII. Theorem VIII. *In two spheres gravitating each towards the other, if the matter in places an all sides round about and equidistant from the centers in similar, the weight of either sphere towards the other will be inversely as the square of the distance between their centers.*

Proposition IX. Theorem IX. *That the force of gravity, considered downwards from the surface of the planets, decreases nearly in the proportion of the distances from the centre of the planets.*

If the matter of the planet were of an uniform density, this proportion would be accurate true. The error, therefore, can be no greater than what may arise from the inequality of the distance.

Proposition X. Theorem X. *That the motions of the planets in the heavens may subsist an exceedingly long time.*

Hypothesis I. *That the centre of the system of the world is immovable.*

Proposition XI. Theorem XI. *That the common centre of gravity of the earth, the sun, and all the planets, is immovable.*

Proposition XII. Theorem XII. *That the sun is agitated by a continual motion, but never recedes far from the common centre of gravity of all the planets.*

Based on the above proofs, Newton considers other versions related to the two-body problem which have became basic principles for celestial and classic mechanics.

In Book III, Proposition XIX Newton considers the problem of the Earth's oblateness as follows:

Proposition XIX. Theorem XIX. *To find the proportion of the axis of a planet to the diameters perpendicular thereto.*

Our countryman, Mr. Norwood, measuring a distance of 905751 feet of London measure between London and York, in 1635, and observing the difference of latitudes to be 2°28', determined the measure of one degree to be 367196 feet of London measure, that is, 57060 Paris toises. M.Picard, measuring an arc of one degree, and 22'55" of the median between Amiens and Malvoisine, found an arc of one degree to be 57060 Paris toises. M.Cassini, the father, measured the distance upon the meridian from the town Collioure in Roussillon to the observatory of Paris; and his son added the distance from the Observatory to the Citadelo of Dunkirk. The whole distance was 486156$^1/_2$ toises and the difference of the latitudes of Collioure and Dunkirk was 8 degrees, and 31'11$^5/_6$". Hence an arc of one degree appears to be 57061 Paris toises. And from these measures arc conclude that the circumference of the earth is 123249600, and its semidiameter 19615800 Paris feet, upon the supposition that the earth is of a spherical figure.

Taking advantage of measurements existing at that time, Newton calculated the ratio of the total gravitation force over the Paris latitude to the centrifugal force over the equator and found that the ratio is equal to 289:1. After that he imagines the Earth in the form of an ellipse of rotation with axis PQ and the channel *ACQqca* (Fig. 1.7).

Fig. 1.7 Newton's problem
of the Earth's oblateness

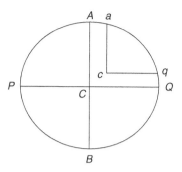

If the channel is filled with water, then its weight in the branch $ACca$ will be related to the water weight in the branch $QCcq$ as 289:288 because of the centrifugal force which decreases the water weight in the last branch by the unit. He found by calculation that if the Earth has a uniform mass of matter and has no motion, and the ratio of its axis PQ to the diameter AB is as 100:101, then the gravity force of the Earth at the point Q relates to the gravity force at the same point of the sphere with radius CQ or CP as 126:125. By the same argument the gravity at point A of a spheroid drawn by revolution around axis AB relates to the gravity in the same point of the sphere drawn from center C with radius AC as 125:126. However, since there is one more perpendicular diameter, then this relation should be 126:125$^{1/2}$. Having multiplied the above ratios, Newton found that the gravity force at point Q relates to the gravity force at point A as 501:500. Because of daily rotation the liquid in the branches should be in equilibrium at a ratio of 505:501. So, the centrifugal force should be equal to 4/505 of the weight. In reality the centrifugal force composes 1/289. Thus, the excess in water height under the action of the centrifugal force in the branch $Acca$ is equal to 1/289 of the height in branch $QCcq$.

After calculation by hydrostatic equilibrium in the channels, Newton obtained that the ratio of the Earth's equatorial diameter to the polar diameter is 230:229, i.e., its oblateness is equal to $(230 - 229)/230 = 1/230$. This result, demonstrating that the Earth's equatorial area is higher than the polar region, was used by Newton for explanation of the observed slower swinging of pendulum clocks on the equator than on the higher latitudes.

At the end of Book III, after discussion of the Moon's motion, the tidal effects and the comets' motion, Newton concludes as follows.

Hitherto we have explained the phenomena of the heavens and our sea by the power of gravity, but have not yet assigned the cause of this power. This is certain, that it must proceed from a cause that penetrates to very centers of the sun, and planets, without suffering the least diminution of its force, that operates not according to the quantity of the surfaces of the particles upon which it acts (as mechanical causes used to do), but according to the quantity of the solid matter which they contain, and propagates its virtue on all sides to immense distances, decreasing always as the inverse square of the distances. Gravitation towards the sun is made up out of the gravitations towards the several particles of which the body of the sun is composed; and in receding from the sun decreases accurately as the inverse square of the distance as far as the orbit of Saturn, as evidently appears from the quiescence of the aphelion of the planets; nay even to the remotest aphelion of the comets, if those 4 aphelions are also quiescent.

But hitherto I have not been able to discover the cause of those properties of gravity from phenomena, and I frame no hypotheses; for whatever is not deduced from the phenomena is to be called an hypothesis; and hypotheses, whether meta-physical or physical, whether of occult qualities or mechanical, have no place in experimental philosophy. In this philosophy particular propositions are inferred from the phenomena, and afterwards rendered general by induction Thus it was that the impenetrability, the mobility, and the impulsive force of bodies, and the laws of motion and of gravitation, were discovered. And to us it is enough that gravity does really exist, and act according to the laws which we have explained, and abundantly serves to account for all the motions of the celestial bodies, and of our sea.

And now we might add something concerning a certain most subtle spirit which pervades and lies in all gross bodies; by the force and action of which spirit the particles of bodies attract one another at near distances, and cohere, if contiguous; and electric bodies operate to greater distances, as well repelling as attracting the neighboring corpuscles; and light is emitted, reflected, refracted, inflected, and heats bodies; and all sensation is excited, and the members of animal bodies move at the command of the solid filaments of the nerves, from the outward organs of sense to the brain, and from brain into the muscles. But these are things that cannot be explained in few wards, nor are we furnished with that sufficiency of experiments which is required to an accurate determination and demonstration of the laws by which this electric and elastic spirit operates.

Lagrange referred to Newton's work as *"the greatest creation of a human intel-lect"*. It was published in England in Latin in 1686, 1713 and 1725 in his life-time and many times later on. We reiterate that the passages above are from the translation by Andrew Mott in 1729 that was printed in 1934.

As it follows from Newton's definition of the centripetal innate forces, his understanding of their meaning and action in nature is very wide. The innate force of matter is the power of resistance. It can develop as the force of a body's resistance due to which it remains at rest or moves with constant velocity. It can develop as a body's resistance (reactive) force to an outer effect and as a pressure when the body faces an obstacle. In modern mechanics this force is understood synonymously as the force of inertia. The resistance force or force of reaction has found its place in the theory of elasticity, and the pressure force is used in hydrodynamics and aerodynamics.

The main meaning of the centripetal force which was introduced by Newton is that each body is attracted to a certain center. He demonstrates this ability of bodies and objects on the Earth to attract to its geometric center by action of the gravity force. Newton distinguishes three kinds of manifestation of the centripetal force, namely absolute, accelerating and moving. The absolute value of this force is a measure of the source power of its action from the center to outer space. The body's attraction to the center and emission of the attraction from the center is demonstrated by Newton in Book III *"The System of the World"*, where in Theorem II he notes that gravity forces from the planets are directed to the Sun. In Theorem IX he says that attraction of the planets themselves goes from their surfaces to the centers. According to Newton's idea the planet's surface is an area of formation of absolute value of the centripetal force from where it emits that force upward and downward.

The accelerating value of the centripetal force by Newton's definition, is a measure proportional to velocity which it develops over a long time. The moving value of the centripetal force is a measure which is proportional to the moment, i.e., to the mass and velocity.

After such a wide spectrum of functions which Newton attributes to the centripetal force, it becomes clear why he was unable to understand its physical meaning and acknowledged: *"But hitherto I have not been able to discover the cause of those properties of gravity from phenomena, and I frame no hypotheses; for whatever is not deduced from the phenomena is to be called an hypotheses; and hypotheses, whether metaphysical or physical, have no place in experimental philosophy. In this philosophy particular propositions are inferred from the phenomena, and afterwards rendered general by induction. Thus it was that the impenetrability, the mobility, and the impulsive force of bodies, and the laws of motion and gravitation, were discovered. And to us it is enough that gravity does really exist, and act according to the laws which we have explained, and abundantly serves to account for all the motions of the celestial bodies, and of our sea"*.

It is worth noting that mathematicians to whom Newton expounded the theory, because of complications in their analytical operations with these forces, introduced the force function to celestial mechanics and analytical dynamics, i.e., energy with its ability to develop pressure. In doing so they practically generalized the physical meaning of the force effects. As to the centripetal forces, later on in Sect. 2.2 of Chap. 2 we shall show that volumetric forces of mass particle interaction in reality generate Newton's physical pressure. which in formulation of practical problems is expressed by the energy. Once more note that Newton, as he said himself, instead of using the correct physical meaning of the concept *"pressure"* gave preference to the concept *"attraction"* as being more understandable to mathematicians.

Newton's problem about the mutual attraction of two bodies, which depict similar trajectories around their common center of gravity and around each other, is based on the geometric solution of Kepler's problem formulated in his first two laws. Newton's solution is founded on his conception of the centripetal and innate forces under which the bodies depict similar trajectories around their common center of gravity and around each other. In celestial mechanics, developed on the basis of Newton's attraction law, the two-body problem is reduced to an analytical problem of one body, the motion of which takes place in the central field of the common mass. Both Newton's geometric theorem and the analytical solution of celestial mechanics are based on the hydrostatic equilibrium state of a body motion due to Kepler's laws. Newton well understood this and expressed it in his hydrostatics laws. But in both cases the two-body problem was solved correctly in the framework of its formulation. The only difference is that by Kepler the planet motion occurs under the action of the Sun's forces whereas Newton shows that this motion results from the mutual attraction of both the Sun and the planet.

In Section V of Book II "*Density and Compression of Fluids: Hydrostatics*" Newton formulates the hydrostatics laws and on their basis in Book III "*The System of the World*" he considers the problem of the Earth's oblateness by applying real values of the measured distances between a number of points in Europe. Applying the found measurements and hydrostatic approach he calculated the Earth's oblateness as being equal to 1/230, where in his consideration the centrifugal force plays the main contraction effect expanding the body along the equator. In fact the task is related to the creation of an ellipsoid of rotation from a sphere by action of the centrifugal force. Here Newton applied his idea that the attraction of the planet itself goes from the surface to its center. In this case the total sum of the centripetal forces and the moments are equal to zero and rotation of the Earth should be inertial. It means that the planet's angular velocity has a constant value.

Inertial rotation of the Earth is accepted a *priori*. There is no physical evidence or other form of justification for this phenomenon. There are also no ideas relative to the mode of a planet's rotation, namely, whether it rotates as a rigid body or there is a differential rotation of separate shells. In modern courses of mechanics there is only analytical proof that in the case when the body exists in the outer field of central forces, then the sum of its inner forces and torques is equal to zero. Thus, it follows that the Earth's rotation should have the mode of a rigid body and the velocity of rotation in time should be constant.

The proof of the conclusion, that if a body occurs in the field of the central forces then the sum of the inner forces and torques is equal to zero, and the moment of momentum has a constant value, is directly related to the Earth's dynamics. Let us see it in a modern presentation (Kittel et al., 1965).

Write the expression of the moment of momentum $\mathbf{L}$ for a mass point m, the location of which is determined by a radius vector r relative to an arbitrarily selected fixed point in the inertial system of coordinates

$$\mathbf{L} \equiv \mathbf{r} \times \mathbf{p} \equiv \mathbf{r} \times m\mathbf{v}, \tag{1.5}$$

where $\mathbf{p}$ is the moment; $\mathbf{v}$ is the velocity.

After differentiation of (1.5) with respect to time, one obtains

$$\frac{d\mathbf{L}}{dt} = \frac{d}{dt}(\mathbf{r} \times \mathbf{p}) = \frac{d\mathbf{r}}{dt} \times \mathbf{p} + \mathbf{r} \times \frac{d\mathbf{p}}{dt}. \tag{1.6}$$

Since vectorial product

$$\frac{d\mathbf{r}}{dt} \times \mathbf{p} = \mathbf{v} \times m\mathbf{v} = 0, \tag{1.7}$$

then taking into account Newton's second law for the inertial reference system, we have

$$\mathbf{r} \times \frac{d\mathbf{p}}{dt} = \mathbf{r} \times \mathbf{F} = \mathbf{N},$$

from which

$$N = \frac{dL}{dt},$$ (1.8)

where N is the torque.

For the central force $F = \hat{r}f(r)$, which acts on the mass point located in the central force field, the torque is equal to

$$N = r \times F = r \times \hat{r}f(r) = 0.$$ (1.9)

Consequently, for the central forces the torque is equal to zero and the moment of momentum L appears to be constant.

In the case when the mass point presents a body composed of n material particles, then moment of momentum L of that system will depend on location of the origin of the reference system. If the reduced vector of the mass center of the system relative to the origin is R_c, then the equation for the moment of momentum L is written as

$$L = \sum_{n=1}^{N} m_n (r_n - R_c) \times v_n + \sum_{n=1}^{N} m_n R_c \times v_n = L_c + R_c \times P,$$ (1.10)

where L_c is the moment of momentum relative to the system's center of the masses; $P = \sum m_n \times v_n$ is the total moment of the system. Here the term $R_c \times P$ expresses the moment of momentum of the mass center and depends on the origin, and the term L_c, on the contrary does not depend on the reference system.

The total torque of the system, which appears as a result of interaction between all the particles, is equal to

$$N = \sum_{n=1}^{N} r_n \times F_n,$$

and the sum of the inner forces is

$$F_i = \sum_{n=1}^{N} F_{ij},$$ (1.11)

here and further the summing is being done at condition $i \neq j$.

The torque of the inner forces is

$$N_{in} = \sum_i r_i \times F_i = \sum_i \sum_j r_i \times F_{ij}.$$ (1.12)

Since

$$\sum_i \sum_j r_i \times F_{ij} = \sum_i \sum_j r_j \times F_{ji},$$ (1.13)

then the torque of inner forces can be presented in the form

$$\mathbf{N}_{in} = \frac{1}{2} \sum_i \sum_j \left(\mathbf{r}_i \times \mathbf{F}_{ij} + \mathbf{r}_j \times \mathbf{F}_{ji} \right). \tag{1.14}$$

Because the Newton forces $\mathbf{F}_{ji} = -\mathbf{F}_{ij}$, then

$$\mathbf{N}_{in} = \frac{1}{2} \sum_i \sum_j \left(\mathbf{r}_i - \mathbf{r}_j \right) \times \mathbf{F}_{ij}. \tag{1.15}$$

Taking into account that central forces $\mathbf{F}_{ij}$ are parallel to $\mathbf{r}_i - \mathbf{r}_j$, then

$$\left(\mathbf{r}_i - \mathbf{r}_j \right) \times \mathbf{F}_{ij} = 0,$$

from where the torque of the inner forces is equal to zero.

$$\mathbf{N}_{in} = 0. \tag{1.16}$$

Assuming that the inner forces $\mathbf{F}_{in} = 0$, then from (1.9), (1.10) and (1.16) one finds that

$$\frac{d}{dt} \mathbf{L}_\Sigma = \mathbf{N}_{ex} \tag{1.17}$$

$$\mathbf{L}_\Sigma = \mathbf{L}_c + \mathbf{R} \times \mathbf{P}. \tag{1.18}$$

Here $\mathbf{L}_c$ is also the moment of momentum relative to the mass center, and $\mathbf{R} \times \mathbf{P}$ is the moment of momentum of the mass center relative to an arbitrarily taken origin.

For practice it is often convenient to select the geometric center of the mass as an origin. In this case the derivative from the moment of momentum relative to the mass center is the torque of the outer forces, i.e.,

$$\frac{d}{dt} \mathbf{L}_c = \mathbf{N}_{ex}. \tag{1.19}$$

It is seen from the above classical consideration that in the model of two interacting mass points reduced to a common mass center, which Newton used for solution of Kepler's problem relating to the planets' motion around the Sun, the inner forces and torques, being in the central force field, are really equal to zero. The torque, which is a derivative with respect to time from the moment of momentum of the body's material particles, is determined here by the resultant of the outer forces and the planets' orbits in the central force field that exists in the same plane. This conclusion follows from Kepler's laws of the planets' motion.

Passing to the problem of the Earth's dynamics, Newton had no choice for the formulation of new conditions. The main conditions were determined already in the two-body problem where the planet appeared in the central force field of the reduced masses. The only difference here is that the mass point has a finite dimension.

The condition of zero equality of the inner forces and torques of the rotating planet should mean that the motion could result from the forces among which the known were only the Galilean inertial forces. Such a choice followed from the inertial condition of two-body motion which he had already applied. The second part of the problem related to reduction of the two bodies to their common center of masses and to the central force that appeared accordingly as predetermining the choice of the equation of state. Being in the outer uniform central force field, it became the hydrostatic equilibrium of the body state. The physical conception and mathematical expression of hydrostatic equilibrium of an object based on Archimedes' laws (third century BP) and the Pascal law (1663) were well known in that time. This is the story of the sphere model with equatorial and polar channels filled in by a uniform liquid mass in the state of hydrostatic equilibrium at inertial rotation.

In Newton's time the dynamics of the Earth in its direct sense has been not founded as it is absent up to now. The planet, rotating as an inertial body and deprived of its own inner forces and torques, appeared to be a dead-alive creature. But up to now, the hydrostatic equilibrium condition, proposed by Newton, is the only theoretical concept of the planet's dynamics because it is based on the two-body problem solution which satisfies Kepler's laws and in practice plays the role of Hooke's law of elasticity.

In spite of the discrepancies noted above, the problem of determining the Earth's oblateness was the first step towards the formulation and solution of the very complicated task of determining the planet's shape, an effort on which theoretical and experimental study continues up to the present time. As to the value of polar oblateness of the Earth, it appears to be much higher than believed before. More recent observations and measurements show that the relative flattening has a smaller value and Newton's solution needs to have further development.

1.7 Clairaut's Model of Hydrostatic Equilibrium of a Non-uniform Earth

Aleksi Klod Clairaut (1713–1765), a French mathematician and astronomer, continued work on Newton's solution of the problem of the Earth's shape based on hydrostatics (Clairaut, 1947). The degree measurements in the equatorial and northern regions made in the eighteenth century by French astronomers proved Newton's conclusion about the Earth's oblateness, which at that time was regarded with skepticism. But the measured value of the relative flattening appeared to be different. In the equatorial zone it was equal to 1/314, and in the northern region to 1/214 (Grushinsky, 1976). Clairaut himself took part in the expeditions and found that Newton's results are not correct. It was also known to him that the Earth is not a uniform body. Because of that he focused on taking into account consideration of this effect. Clairaut's model was represented by an inertial rotating body filled with liquid of a fluctuating density. In its structure such a model was closer to the real

Earth having a shell structure. But the hydrostatic equilibrium condition and inertial
rotation was remained to be as previous the physical basis for the problem solution.
Clairaut introduced a number of assumptions in the formulation of the problem.
In particular, since the velocity of inertial rotation and the value of the oblateness
are small, then the boundary areas of the shells and their equilibrium were taken as
ellipsoidal figures with a common axis of rotation. Clairaut's solution comprised
obtaining a differential equation for the shell-structured ellipsoid of rotation relative
to geometric flattening of its main section. Such an equation was found in the form
(Melchior, 1972)

$$\frac{d^2e}{da^2} + \frac{d\rho a^2}{\int_0^a \rho a^2 da}\frac{de}{da} + \left(\frac{2\rho a}{\int_0^a \rho a^2 da} - \frac{6}{a^2}\right)e = 0, \tag{1.20}$$

where $e = (b - a)/a$ is the geometric flattening; a and b are the main axes; ρ is the
density.

The difficulty in solving the above equation was due to the absence of a density
radial distribution law for the Earth. Later on, by application of seismic data, researchers succeeded in obtaining a picture of the planet's shell structure. But quantitative interpretation of the seismic observations relative to the density appeared
to be possible again, based on the same idea of hydrostatic equilibrium of the body
masses. In spite of that, as a result of analysis of Clairaut's equation, a number of
dynamic criteria for a rotating Earth were obtained. In particular, the relationship
between the centrifugal and the gravity forces on the equator was found, the ratio
between the moments of inertia of the polar and equatorial axes (dynamical oblateness) was obtained and also the dependence of the gravity force on the latitude of
the surface area was derived. That relationship is as follows:

$$g = g_e(1 + \beta sin^2\varphi), \tag{1.21}$$

where φ is the latitude of the observation point; g_e is the acceleration of the gravity force: $\beta = 5/2q - e$; $q = \omega^2 a/g_e$ is the ratio of the centrifugal force to the gravity
force on the equator; ω is the angular velocity of the Earth's rotation; e is the geometric oblateness of the planet; a is the semi-major axis.

The solutions obtained by Clairaut and further developed by other authors became a theoretical foundation for practical application in search of the planet's
shape, for interpretation of seismic observations relative to the structure and density distribution of the Earth and also for analysis of the observed natural dynamic
processes.

Later on, the quantitative values of the geometric and dynamic oblateness of the
Earth, and the Moon, different in values, were obtained by Clairaut's equation and
with the use of satellite data. This fact underlies the conclusion that the Earth and
the Moon do not stay in hydrostatic equilibrium.

1.8 Euler's Model of the Rigid Earth Rotation

Leonhard Euler (1707–1783), a prominent Swiss mathematician, mechanic and physicist, possessed a great capacity for work, fruitful creativity and extreme accuracy and strictness in problem solution. There are about 850 titles in the list of his publications and their collection comprises 72 volumes. Half of them were prepared in Russia. He was twice invited to work in the St-Petersburg Academy of Sciences where he spent more than 30 years. The spectrum of Euler's scientific interests was very wide. In addition to mathematics and physics they included theory of elasticity, theory of machines, ballistics, optics, shipbuilding, theory of music and even the insurance business. But 3/5 of the works were devoted to mathematical problems.

In mechanics Euler developed a complete theory of motion of the rigid (non-deformable) body. His dynamic and kinematics equations became the main mathematical instrument in solution of rigid body problems. These equations, with use of the known law of a body rotation, enable determination of the acting forces and torques. And vice versa, by the applied outer forces one may find the laws of motion (rotation, precession, nutation) of a body.

On the basis of Newton's equations of motion for rotational motion of a rigid body, whose axes of coordinates x, y, z in the rotating reference system are matched with the main axes connected with the body, Euler's dynamical equations have the form:

$$I_x \dot{\omega}_x + (I_z - I_y)\omega_y \omega_z = N_x,$$
$$I_y \dot{\omega}_y + (I_x - I_z)\omega_x \omega_z = N_y, \qquad (1.22)$$
$$I_z \dot{\omega}_z + (I_y - I_x)\omega_x \omega_y = N_z,$$

where I_x, I_y, I_z are the moments of inertia of the body relative to the main axes; ω_x, ω_y, ω_z, are the components of the instantaneous angular velocities on the axes; N_x, N_y, N_z, are the main torques of the acting forces relative to the same axes: $\dot{\omega}_x, \dot{\omega}_y, \dot{\omega}_z$ are the derivatives with respect to time from the angular velocities.

The Euler kinematics equations are written as follows:

$$\omega_x = \dot{\psi} \sin\theta \sin\varphi + \dot{\theta} \cos\varphi,$$
$$\omega_y = \dot{\psi} \sin\theta \cos\varphi - \dot{\theta} \sin\varphi, \qquad (1.23)$$
$$\omega_z = \dot{\varphi} + \dot{\psi} \cos\theta.$$

The Eulerian angles φ, ψ and θ determine the position of a rigid body which has a fixed point relative to the fixed rectangular axes of coordinates. At hard linkage of the axes with the body and specification of the line of crossed planes of corresponding angles, they fix the rotation angle, the angle of precession and the angle of nutation of the rotation axis.

For a uniform sphere, such as the Earth is, according to Newton, $I_x = I_y = I_z$. Then the Eulerian equations of motion (1.22) acquire the form

$$
\begin{aligned}
I\dot{\omega}_x &= N_x, \\
I\dot{\omega}_y &= N_y, \\
I\dot{\omega}_z &= N_z.
\end{aligned}
\tag{1.24}
$$

At free (inertial by Newton) rotation of a uniform Earth, which is not effected by the torque, $N_x = N_y = N_z = 0$. In that case it follows from (1.24) that the components of the instantaneous velocities of their axes become constant and the angular velocity $\omega = const$. Thus, angular velocity of a body at non-perturbed rotation is equal to a constant value.

Newton found that the Earth is flattened relative to the polar axis by centrifugal inertial force and Clairaut has agreed with that. Then from the symmetry of the body having the form of an ellipsoid of rotation, it is found that $N_x = N_y \neq N_z$ and only $\omega_z = const$. From this in the case of absence of the outer torque, equation (1.22) is reduced to

$$
\dot{\omega}_x + \Omega\omega_y = 0,
\tag{1.25}
$$

$$
\dot{\omega}_y - \Omega\omega_x = 0,
\tag{1.26}
$$

where Ω is the angular velocity of free rotation which at $I_x = I_y$ is equal to

$$
\Omega = \frac{I_z - I_x}{I_x}\omega_z.
\tag{1.27}
$$

After transformation of equations (1.25)–(1.26) one obtains their solution in the form of ordinary equations of the harmonic oscillation

$$
\omega_x = A\cos\Omega t,
\tag{1.28}
$$

$$
\omega_z = A\sin\Omega t,
\tag{1.29}
$$

where A is the constant value representing the amplitude of oscillation.

Thus, the component ω_z of the angular velocity along the body's axis of rotation is a constant value and the component perpendicular to the axis rotates with angular velocity Ω. So the whole body, while rotating by inertia relative to the geometric axis with angular velocity ω_z, in accordance with (1.27) is wobbling with the frequency Ω. The oscillations described by equations (1.28) and (1.29) are observed in reality and are called nutation of the rotating axis or a variation of latitude. The numerical value of the ratio of inertia moments (1.28) for the Earth is known and equal to

$$
(I_x - I_z)/I_x = 0.0032732
$$

and the value of the angular velocity (free precession) is

$$
\Omega = \omega_z/305.5.
$$

For the known value $\omega_z = 7.29 \times 10^{-5}$ s^{-1} the period of Euler's free precession is equal to 305 days or about 10 months. But analysis of results of the long series of observation done by the American researcher Chandler has shown that, together with the annual component of the forced nutation, there is one more component having a period of about 420 days which was called free wobbling of the rotation axis. This component differs substantially from Euler's free precession. The nature of the latter has not been understood up to now.

Euler developed also a complete theory of motion of a perfect liquid in hydromechanics, where differential equations in his variables become the basis for solution of hydrodynamic problems. Euler's hydrodynamic equations for a perfect liquid in the rectangular Cartesian reference system x, y, z based on Newton's equations of motion have the form

$$\frac{\partial u}{\partial t} + u\frac{\partial u}{\partial x} + v\frac{\partial u}{\partial y} + \omega\frac{\partial u}{\partial z} = X - \frac{1}{\rho}\frac{\partial p}{\partial x},$$

$$\frac{\partial v}{\partial t} + u\frac{\partial v}{\partial x} + v\frac{\partial v}{\partial y} + \omega\frac{\partial v}{\partial z} = Y - \frac{1}{\rho}\frac{\partial p}{\partial y}, \qquad (1.30)$$

$$\frac{\partial \omega}{\partial t} + u\frac{\partial \omega}{\partial x} + v\frac{\partial \omega}{\partial y} + \omega\frac{\partial \omega}{\partial z} = Z - \frac{1}{\rho}\frac{\partial p}{\partial z},$$

where u, v, ω are the components of the velocity of liquid particles; p is the liquid pressure; ρ is the density; X, Y, Z are the components of the volumetric forces.

Solution of the hydrodynamic problems is reduced to determination of the components of velocities u, v, ω, the pressure and the density as a function of the coordinates with known values of X, Y, Z and the given boundary conditions. For that purpose in addition to equations (1.30) the equation of continuity is written in the form

$$\frac{\partial \rho}{\partial t} + \frac{\partial(\rho u)}{\partial x} + \frac{\partial(\rho v)}{\partial y} + \frac{\partial(\rho \omega)}{\partial z} = 0. \qquad (1.31)$$

If the density of liquid depends only on pressure, then the extra equation of state will be presented by the relation $\rho = f(p)$, and for the incompressible liquid it is $\rho = const$.

Because the Earth is a system with continuous distribution of its masses, we will use the Eulerian hydrodynamic equations repeatedly.

1.9 Jacobi's *n* Body Problem

Carl Jacobi (1804–1851) was an eminent German mathematician who was called the Euler of the nineteenth century. He is one of the authors of elliptic functions theory, the author of a number of discoveries in the theory of numbers, in the calculus of variations, in integral calculus and the theory of differential equations, in the study of a class of orthogonal polynomials.

In 1842–1843, when Jacobi was a professor at Königsberg. University, he delivered a special series of lectures on dynamics. The lectures were devoted to the dynamics of a system of n mass points, the motion of which depends only on mutual distance between them and is independent of velocities. In this connection, by deriving the law of conservation of energy, where the force function is a homogeneous function of space co-ordinates, Jacobi gave an unusual form and a new content to this law. In transforming the equations of motion, he introduced an expression for the system's center of mass. Then, following Lagrange, he separated the motion of the center of mass from the relative motion of the mass points. Making the center of mass coincident with the origin of the co-ordinate system, he obtained the following equation (Jacobi, 1884):

$$\frac{d^2}{dt^2} \left(\sum m_i r_i^2 \right) = -(2k+4)\,U + 4E,$$

where m_i is the mass point i; $r_i = \sqrt{x_i^2 + y_i^2 + z_i^2}$ is the distance between the points and the center of mass; k is the degree of homogeneity of the force function; U is the system's potential energy; and E is its total energy.

When $k = -1$, which corresponds to the interaction of mass points according to Newton's law, and writing

$$\frac{1}{2} \sum m_i r_i^2 = \Phi,$$

Jacobi obtained

$$\ddot{\Phi} = U + 2T = 2E - U,$$

where Φ is the Jacobi function (the polar moment of inertia).

This is Jacobi's generalized (non-averaged) virial equation. In the Russian scientific literature it is known as the Lagrange–Jacobi equation since Jacobi derived it by applying Lagrange's method of separation of the motion of the mass center from the relative motion of mass points.

On the right-hand side of the virial equation there is a classical expression of the virial theorem, i.e., the relation between potential and kinetic energy. In the case of constancy of its left-hand side, when motion of the system happens with a constant velocity, the equation acquires conditions of hydrostatic equilibrium of a system in the outer force field. The left-hand side of the equation, i.e., the second derivative with respect to the Jacobi function, expresses oscillation of the polar moment of inertia of the system, which, in fact, is kinetic energy of the inner volumetric torques of the interacting mass points moving in accordance with Kepler's laws.

Jacobi did not pay attention to the physics of his equation, which expresses kinetic energy of the interacting volumetric particles in the form of their oscillation. He used the equation for a quantitative analysis of stability of the Solar System and noted that the system's potential and kinetic energies should always oscillate within certain limits. In the contemporary literature of celestial mechanics and analytical dynamics, Jacobi's virial equation is used for the same purposes (Whittaker, 1937; Duboshin, 1975). Since this equation contains two independent variables, it found no other practical applications. As was earlier mentioned, it will be shown in this work that there is a functional relationship between the potential (kinetic) energy and the polar moment of inertia. On that basis the rigorous solution of the equation will be found and applied to study the Earth's dynamics.

1.10 The Clausius Virial Theorem

Rudolf Clausius (1822–1888), a German physicist, is one of the founders of thermodynamics and the molecular kinetic theory of heat. Simultaneously with W. Thomson (Lord Kelvin) he formulated the second law of thermodynamics in the following form: *"Heat can not be transferred by any continuous, self-sustaining process from a cold to a hotter body"* without some changes, which should compensate that transfer. Clausius introduced the conception of entropy to thermodynamics.

In 1870, based on studies of the process and mechanism of Carnot's thermal machine work, Clausius proved the virial theorem, according to which for a closed system the mean kinetic energy of a perfect gas's particle motion is equal to half of their potential energy. The virial relation between the potential and kinetic energy was found to be a universal condition of the hydrostatic equilibrium for describing dynamics of the natural systems in all branches of physics and mechanics.

The Clausius virial theorem is a conceptual basis for our work with physical clarification of some effects related to celestial bodies. Its derivation for different models of natural systems is presented in Sect. 2.5 of Chap. 2.

1.11 De Broglie's Wave Theory

Louis-Victor de Broglie (1892–1987), a famous French Nobel Prize physicist, is one of the founders of quantum mechanics. In his doctor's thesis, *Researches on Quantum Theory,* de Broglie extended the wave-particle duality theory of light on matter. This hypothesis was based on the works of Albert Einstein and Max Planck. Three years later on, in 1927 his idea was fully confirmed by Davisson and Germer who discovered electron diffraction by crystals. Thus, de Broglie's theory became the basis for developing present day wave mechanics for matter on an atomic scale. The particles of greater mass, which are the subject of classical mechanics, have mainly corpuscular properties. And the idea to create a unified field theory, which was a dream of Albert Einstein, has not yet been realized.

1.12   Other Approaches to Dynamics of the Planet Based on Hydrostatics

The model of the Earth proposed by Newton and developed by Clairaut was in the form of a spheroid, rotating by inertia, and filled with a non-uniform liquid, the mass of which resides in hydrostatic equilibrium in the outer force field. This model became generally accepted, commonly used and in principal has not changed up to now. Its purpose was to solve the problem of the planet's shape, i.e., the form of the planet's surface, and this goal was reached in first approximation. Moreover, the equation obtained by Clairaut of surface changes in acceleration of the gravity force as a function of the Earth latitude opened the way to experimental study of oblateness of the spheroid of rotation by means of measuring the outer gravity force field. Later on, in 1840, Stokes solved the direct and reverse task concerning the surface gravity force for a rotating body and, above its level, applying the known parameters, namely, the mass, radius and angular velocity. The above parameters uniquely determined the gravity force at surface level, which is taken as the quiet ocean's surface, and in all of outer space. By that task the relation between the Earth's shape and the gravity force was determined. In the middle of the last century M. Molodensky proposed the idea to consider the real surface of the Earth as a reduced surface and solved the corresponding boundary problem. The doctrine of the spheroidal figure of the Earth has found common understanding and researchers, having armed themselves with theoretical knowledge, started to refine the dimensions and other details of the ellipsoid of rotation and to derive the corresponding corrections.

The Earth's dynamics were always of interest not only to researchers of its shape. Fundamentals of all the Earth, planetary and the Solar System sciences are defined first of all by the laws of motion of the Earth itself, where the confidence limit of the laws can be checked by observation. Moreover, all sense of human life is connected with this planet. As far as the techniques and instruments for observation were developed, geodesists, astronomers and geophysicists have noticed that in the planet's inertial rotation some irregularities and deviations relative to the accepted standard parameters and hydrostatic conditions have appeared. Those irregularities or, as they are often called, inaccuracies, the number of which is said to be more than ten, finally were incorporated into two problems, namely, variation of the angular velocity in the daily, monthly, annually and secular time scale, and variation in the motion of poles in the same time scales. Just after the problems became evident and did not find resolution in the frame-work of the accepted physical and theoretical conceptions of celestial mechanics the latter lost interest in the problems of Earth's dynamics. In this connection the well-known German theoreticians in dynamics, Klein and Sommerfeld, stated that the Earth's mechanics appear to be more complicated than celestial mechanics and represents "*some confused labyrinths of geophysics*" (Klein and Sommerfeld, 1903). The geophysicists themselves started to solve their own problems. They had no other way except to search for the causes of the observed inaccuracies. In order to study irregular velocity of the Earth's rotation and the pole motion, numerous projects of observation and regular monitoring were organized by the planetary network. As it was always in such cases, the cause of the observed effects was sought in the effects of perturbations coming from the Moon and the Sun, and also in the influence of dynamical effects of their own shells, such as the atmosphere, the oceans and the liquid core, existence of which is justified by many researchers. In some works the absence of hydrostatic equilibrium in distribution of the masses and strength in the planet's body is named as the reason of irregular velocity of the Earth's rotation.

Many publications were devoted to analysis of the observed inaccuracies in the Earth's rotation together with explanation of their possible causes, based on experimental data and theoretical solutions. The most popular review work in the twentieth century was the book of the known English geophysicist Harold Jeffreys "*The Earth: Its Origin, History and Physical Constitution*". The first publication of the book happened in 1922 and later four more editions appeared, including the last one in 1970. Jeffreys was a great expert and direct participant in development of the most important geophysical activities. The originality of his methodological approach to describing the material lies in that, after formulation and theoretical consideration of the problem, he writes a chapter devoted to the experimental data and facts on the theme of the comparison with analytical solutions and discussion.

Remaining with Newton's and Clairaut's models, Jeffreys considers the planet as an elastic body and describes the equation of the force equilibrium from the hydrostatic pressure, which appears from the outer uniform central force field, and exhibits strength at a given point in the form

$$\rho f_i = \rho X_i + \sum_{k=1,2,3} \frac{\partial p_{ik}}{\partial x_k}, \tag{1.32}$$

where ρ is the density; f_i is the acceleration component; $p_{ik} = p_{ki}$ is the stress component from the hydrostatic pressure; X_i is the gravity force on the unit mass from the outer force field.

Additionally the equation of continuity (like the continuity equation in hydrodynamics) is written as the condition of equality of velocity of the mass inflow and outflow from an elementary volume in the form

$$\frac{\partial \rho}{\partial t} = -\sum_i \frac{\partial}{\partial x_i}(\rho v_i), \tag{1.33}$$

where v_i is the velocity component in the direction of x_i.

Further, applying the laws of elasticity theory, he expresses elastic properties of the matter by the Lamé coefficients and writes the basic equations of the strength state of the body, which links the strengths and the deformations in the point as

$$\rho \frac{\partial^2 u_i}{\partial t^2} = (\lambda + \mu) \frac{\partial \Delta}{\partial x_i} + \nabla^2 u_i, \tag{1.34}$$

where u_i is the displacement component; λ and μ are the Lamé coefficients; Δ is the component of the relative displacement; ∇ is the Laplacian operator.

One may see that Jeffreys reduced Newton's effects of gravitation to the effects of Hooke's elasticity. The author introduces a number of supplementary physical ideas related to the properties of the Earth's matter, assuming that it is not perfectly elastic. With development of stresses the matter reaches its limit of resistance and passes to the stage of plastic flow with a final effect of a break in the matter's continuity. This break leads to a sharp local change in the strength state, which, in turn, leads to appearance of elastic waves in the planet's body, causing earthquakes. For this case equation (1.34) after the same corresponding transformations is converted into the form of plane longitudinal and transversal waves, which propagate in all directions from the break place. Such is the physical basis of earthquakes which was a starting point of development of seismology as a branch of geophysics studying propagation of elastic longitudinal and transversal waves in the Earth's body. By means of seismic study, mainly by strong earthquakes and based on differences in velocity of propagation of the longitudinal and transversal waves through the shells having different elastic properties, the shell-structured body of the planet was identified.

Jeffreys has analyzed the status of study in the theory of the shapes of Earth and the Moon following Newton's basic concepts. Namely, the planet has an inner and outer gravitational force field. The gravitational pressure is formed on the planet's surface and affects both the outer space and the planet's center. The Earth's shape is presented as an ellipsoid of rotation which is perturbed from the side by inaccuracies in the density distribution, as well as from the side of the Moon's perturbations. The problem is to find the axes of the ellipsoid under action of both perturbations which occur because of a difference in the gravity field for the real Earth and the spherical body. It is accepted that the oceans' level is close to a spherical surface with deviation by a value of the first order of magnitude, and geometric oblateness of the ellipsoid is close to the value of $e \approx 1/297$. But the value squares of deviation cannot always be ignored because the value e^2 differs substantially from the value e. The observed data cannot be compared with theoretical solutions because the formulas depending on the latitudes give precise expressions neither for the radius vector from the Earth center to the sea level nor for the value of the gravity force. The problem of the planet's mass density distribution finds its resolution from the condition of the hydrostatic pressure at a known velocity of rotation. The value of oblateness of the outer spheroid can be found from the observed value of the precession constant with a higher accuracy than one can find from the theory of the outer force field. A weak side of such an approach is the condition of the hydrostatic stresses, which however are very small in comparison with the pressure at the center of the Earth. The author also notes that deviation of the outer planet's gravitational field from spherical symmetry doesn't satisfy the condition of the inner hydrostatic stresses. Analysis of that discrepancy makes it possible to assess errors in the inner strengths related to the hydrostatics. Because of the Earth's ellipticity, the attraction of the Sun and the Moon creates a force couple applied to the center, which forces the instantaneous axis of rotation to depict a cone around the pole of the ecliptic and to cause the precession phenomenon. The same effect initiates an analogous action on the Moon's orbit.

These are the main physical fundamentals that Jeffreys used for an analysis and theoretical consideration of the planet's shape problem and for determination of its oblateness and of its semi-major axis size. The author has found that the precession constant $H = 0.003\ 272\ 93 \pm 0.000\ 000\ 75$ and the oblateness $1/e = 297.299 \pm 0.071$. He assumes that the above figures could be accepted as a result which gives the hydrostatic theory. But in conclusion he says that the theory is not correct. If it is correct then the solid Earth would be a bench mark of the planet's surface covered by oceans. There are some other data confirming that conclusion. But this is the only and the most precise method for determining the spheroid flattening which needs non-hydrostatic corrections to be found. An analogous conclusion was made by the author relative to the Moon's oblateness, where the observed and calculated values have much more contrast.

The other review works on the irregularity of rotation and the pole motion of the Earth are the monographs of Munk and MacDonald (1960), Melchior (1972), Sabadini and Vermeersten (2004). The authors analyze there the state of the art and geophysical causes leading to the observed incorrectness in the planet's rotation and wobbling of the poles. They draw attention of readers to the practical significance of the two main effects and designate about ten causes of their initiation. Among them are seasonal variations of the air masses, moving of the continents, melting and growing of the glaciers, elastic properties of the planet, convective motion in the liquid core. The authors stressed that solution of any part of the above geophysical task should satisfy the dynamical equations of motion of the rotating body and the equations, which determine a relationship between the stresses and deformations inside the body. Theoretical formulation and solution of a task should be considered on the hydrostatic basis, where the forces, inducing stresses and deformations are formed by the outer uniform force field and the deformations occur in accordance with the theory of elasticity for the elastic body model, and in the frame-work of rheology laws for the elastic and viscous body model. The perturbation effects used are the wind forcing, the ocean currents and convective flows in the core and in the shells.

The causes of axis rotation wobbling and pole motion are considered in detail. The authors find that the problem of precession and nutation of the axis of rotation has been discussed for many years and does not generate any extra questions. The cause of the phenomena is explained by the Moon and the Sun perturbation of the Earth which has an equatorial swelling and obliquity of the axis to the ecliptic. The Euler equations for the rigid body form a theoretical basis for the problem's solution. In this case the free nutation of the rigid Earth according to Euler is equal to 10 months.

1.13   The Observation Results

The effects of the Earth's oblateness and the related problems of irregularity in the rotation and the planet's pole motion and also the continuous changes in the gravity and electromagnetic field have a direct relation to solution of a wide range of scientific and practical problems in Earth dynamics, geophysics, geology, geodesy, oceanography, physics of the atmosphere, hydrology, and climatology. In order to understand the physical meaning and regularities of these phenomena, regular observations are carried out. Newton's first attempts to find the quantitative value of the Earth's oblateness were based on degree measurements done by Norwood, Pikar, Kassini. As mentioned above, by his calculation of the Paris latitude the oblateness value appears to be 1/230. Very soon some analogous measurements were taken in the equatorial zone in Peru and in the northern zone in Lapland. Clairaut, Mopertui, Buge, and other known astronomers also took part in these works. They confirmed the fact of the Earth's oblateness as calculated by Newton. The degree of the arc in the northern latitudes appeared to be maximal and the oblateness was equal to 1/214. In the equatorial zone the arc length was minimal and the oblateness was equal to 1/314. So the Earth's pole axis from these measurements was found to be shorter than the equatorial approximately by 20 km.

As of the end of the first part of the twentieth century, more than twenty large degree measurements were done from which the values of the oblateness and dimension of the semi-major axis were found. The data of the measurements are presented in Table 1.1, and in Table 1.2 the parameters of the triaxial ellipsoid are shown (Grushinsky, 1976).

It is worth noting that in geodesy, a practical application of the triaxial ellipsoid has not been found, because it needs more complicated theoretical calculations and more reliable experimental data. In the theory this important fact is ignored because it is not inscribed in the hydrostatic theory of the body.

Table 1.1 Parameters of the Earth's oblateness by degree measurement data

Author	Year	a, m	e	e_e	λ
D'Alembert	1800	6 375 553	1/334.00		
Valbe	1819	376 895	1/302.78		
Everest	1830	377 276	1/300.81		
Eri	1830	376 542	1/299.33		
Bessel	1841	377 397	1/299.15		
Tenner	1844	377 096	1/302.5		
Shubert	1861	378 547	1/283.0		
Clark	1866	378 206	1/294.98		
Clark	1880	378 249	1/293.47		
Zhdanov	1893	377 717	1/299.7		
Helmert	1906	378 200	1/298.3		
Heiford	1909	378 388	1/297.0		
Heiford	1909	378 246	1/298.8	1/38 000	38°E
Krasovsky	1936	378 210	1/298.6	1/30 000	10°E
Krasovsky	1940	378 245	1/298.3		
International	1967	378 160	1/298.247		

Here e is the oblateness of the polar axis; a is the semi-major axis; e_e is the equatorial oblateness; λ is the longitude of the maximal equatorial radius

Table 1.2 Parameters of the Earth's equatorial ellipsoid

Author	Year	$a_1 - a_2$, m	λ
Helmert	1915	230 ± 51	17°W
Berrot	1916	150 ± 58	10°W
Heyskanen	1924	345 ± 38	18°E
Heyskanen	1929	165 ± 57	38°E
Hirvonen	1933	139 ± 16	19°W
Krasovsky	1936	213	10°E
Isotov	1948	213	15°E

Here a_1 and a_2 are the semi-major and semi-minor axes of the equatorial ellipsoid

In addition to the local degree measurements, which allow determination of the Earth's geometric oblateness, more precise integral data can be obtained by observation of the precession and nutation of the planet's axis of rotation. It is assumed that the oblateness depends on deflection of the body's mass density distribution from spherical symmetry and is initiated by a force couple that appears to be an interaction of the Earth with the Moon and the Sun. The precession of the Earth's axis is proportional to the ratio of the spheroid's moments of inertia relative to the body's axis of rotation in the form of the dynamical oblateness ε:

$$\varepsilon = \frac{C - A}{C}.$$

At the same time the retrograde motion of the Moon's nodes (points of the ecliptic intersection by the Moon's orbit) is proportional to the second spherical harmonics coefficient J_2 of the Earth's outer gravitational potential in the form

$$J_2 = \frac{C - A}{Ma^2}.$$

It is difficult to obtain a rigorous value of geometric oblateness from its dynamic expression because we do not know the radial density distribution. Moreover, the Moon's mass is known up to a fraction of a percent but it is inconvenient to calculate analytically the joint action of the Moon and the Sun on the precession. In spite of that, some researchers succeeded in making such calculations, assuming that the Earth's density is increasing proportionally to the depth. Their data are the following:

by Newcomb $\varepsilon = 1/305.32 = 0.0032753;$ $e = 1/297.6;$
by de Sitter $\varepsilon = 1/304.94 = 0.0032794;$ $e = 1/297.6;$
by Bullard $\varepsilon = 1/305.59 = 0.00327236;$ $e = 1/297.34;$
by Jeffreys $\varepsilon = 1/305.54 = 0.00327293;$ $e = 1/297.3.$

After appearance of the Earth's artificial satellites and some special geodetic satellites, the situation with observation procedures has in principle changed. The satellites made it possible to determine directly, by measuring of the even zonal moments, the coefficient J_n in expansion of the Earth's gravitational potential by spherical functions. In this case at hydrostatic equilibrium the odd and all the tesseral moments should be equal to zero. It was assumed before the satellite era that the correction coefficients of a higher degree of J_2 will decrease and the main expectations to improve the calculation results were focused on the coefficient J_4. But it has appeared that all the gravitational moments of higher degrees are the values proportional to the square of oblateness, i.e., $\sim(1/300)^2$ (Zharkov, 1978).

On the basis of calculated harmonics, the coefficients of the expanded gravitational potential of the Earth published by the Smithsonian Astrophysical observatory and the Goddard cosmic center of the USA, the fundamental parameters of the gravitational field and the shape of the so-called "*Standard Earth*" were determined. Among them are the coefficient of the second zonal harmonic $J_2 = 0.001\ 082\ 7$, equatorial radius of the Earth's ellipsoid $a_e = 6\ 378\ 160$ m, angular velocity of the Earth's rotation $\omega_3 = 7.292 \times 10^{-5}$ rad/s, equatorial acceleration of the gravity force $\gamma_e = 978\ 031.8$ mgl, and oblateness $1/e = 1/298.25$ (Grushinsky, 1976; Melchior, 1972). At the same time, if the Earth stays in hydrostatic equilibrium, then, applying the solutions of Clairaut and his followers, the planet's geometric oblateness should be equal to $e' = 1/299.25$. On the basis of that contradiction Melchior (1972) concluded, that the Earth does not stay in hydrostatic equilibrium. It represents either a simple equilibrium of the rigid body, or there is equilibrium of a liquid and not static but dynamic with an extra hydrostatic pressure. Coming to interpretation of the density distribution inside the Earth by means of the Wiliamson–Adams equation, Melchior (1972) adds, that in order to eliminate there the hydrostatic equilibrium, one needs a supplementary equation. Since such an equation is absent, we are obliged to accept the previous conditions of hydrostatics.

The situation with absence of hydrostatic equilibrium of the Moon is much more striking. The polar oblateness e_p of the body is (Grushinsky, 1976)

$$e_{\mathrm{p}} = \frac{b - c}{r_0} = 0.94 \times 10^{-5},$$

and the equatorial oblateness e_e is

$$e_{\mathrm{e}} = \frac{a - c}{r_0} = 0.375 \times 10^{-4},$$

where a, e and c are the equatorial and polar semi-axes; r_0 is the body's mean radius.

It was found by observation of the Moon libration, that

$$e_p = 4 \times 10^{-4} \quad \text{and} \quad e_e = 6.3 \times 10^{-4}.$$

The calculation of the ratio of theoretical values of the dynamic oblateness $e_d = e_p/e_e = 0.25$ substantially differs from observation, which is $0.5 \le e_d \le 0.75$. At the same time the difference of the semi-axes is $a_1 - a_3 = 1.03$ km and $a_2 - a_3 = 0.83$ km, where a_1 and a_2 are the Moon's equatorial semi-axes.

After the works of Clairaut, Stokes and Molodensky, on the basis of which the relationship between the gravity force change at sea level and on the real Earth surface with an angular velocity of rotation was established, one more problem arose. During measurements of the gravity force at any point of the Earth's surface, two effects are revealed. The first is an anomaly of the gravity force, and the second is a declination of the plumb line from the normal at a given point.

Analysis of the gravity force anomalies and the geoid heights (a conventional surface of a quiet ocean) based on the existent schematic maps compiled from the calculated coefficients of expansion of the Earth's gravity potential and ground level gravimetric measurements, allows derivation of some specific features related to the parameter forming the planet. As Grushinsky (1976) notes, elevation of the geoid over the ellipsoid of rotation with the observed oblateness reaches 50–70 m only in particular points of the planet, namely, in the Bay of Biscay, North Atlantic, near the Indonesian Archipelago. In the case of a triaxial ellipsoid the equatorial axis is passed near those regions with some asymmetry. The maximum of the geoid heights in the western part is shifted towards the northern latitudes, and the maximum in the eastern part remains in the equatorial zone. The western end of the major radius reaches the latitudes of 0–10° to the west of Greenwich and the western end falls on the latitudes of about 30–40° to the west of a meridian of 180°. This also indicates asymmetry in distribution of the gravity forces and the forming masses. And the main feature is that the tendency to asymmetry of the northern and the southern hemispheres as a whole is observed. The region of the geoid's northern pole rises above the ellipsoid up to 20 m, and the Antarctic region is situated lower by the same value. The asymmetry in planetary scale is traced from the north-west of Greenland to the south-east through Africa to Antarctica with positive anomalies, and from Scandinavia to Australia through the Indian Ocean with negative anomalies up to 50 mgl. Positive anomalies up to 30 mgl are fixed within the belt from Panama to Fiery Land and to the peninsula Grechem in Antarctica. The negative anomalies are located on both sides, which extend from the Aleut bank to the south-east of the Pacific Ocean and from Labrador to the south of the Atlantic. The structure of the positive and negative anomalies is such that their nature can be interpreted as an effect of spiral curling of the northern hemisphere relative to the southern one.

As to the plumb line declination, this effect is considered only in geodesy from the point of view of practical application in the corresponding geodetic problems. Physical aspects of the problem are not touched.

In Russia the problem of the Earth's oblateness is studied and the corresponding measurements were carried out by M.V. Lomonosov, F.P. Litke (1826–1829, $e = 1/288$), I.F. Parrot and V.Ya. Struve (1829, $e = 1/279.3$ and $e = 1/279.3$), A.N. Savich (1865–1868, $e = 1/296$), F.A. Sludsky (1862–1863, $e = 1/292.7$ and for the triaxial ellipsoid $e = 1/297.1$), A.A. Ivanov (1898–1903, $e = 1/297.2$). Measurements of the gravity force and declination of the plumb line in the Russian regions were done by O.V. Struve, I.I. Steblitsky and P.P. Kulberg (1876), B.Ya. Shveitser (1853–1861), P.K. Shternberg and F.A. Bredichin (1831–1904, 1916–1917), A.A. Michailov, A.I. Kasansky, F.N. Krasovsky (1930–1936), M.S. Molodensky, V.F. Eremeev, M.I. Yurkina (1953–1960).

The problem of the Earth's rotation was discussed at the NATO workshop (Cazenave, 1986). It was stated that both aspects of the problem still remain unsolved. The problems are variations in the day's duration and the observed Chandler's wobbling of the pole with a period of 14 months in comparison with 10 months, given by the Euler rigid body model. Chandler's results are based on an analysis of 200 year observational data of motion of the Earth's axis of rotation, done in the USA in the 1930s. He found that there is an effect of free wobbling of the planet's axis with a period of about 420 days. Since that time the discovered effect remains the main obstacle in explanation of the nature and theoretical justification of the pole's motion.

Summing up the above short excursion into the problem's history we found the situation as follows. The majority of researchers dealing with dynamics of the Earth and its shape came to the unanimous conclusion that the theories based on hydrostatics do not give satisfactory results in comparison with observations. For instance, Jeffreys straightforwardly says that the theories are incorrect. Munk and Macdonald more delicately note that a dozen of the observed effects can be cited which do not satisfy the hydrostatic model. It means that dynamics of the Earth as a theory is absent. The above state of the art and the conclusion motivated the authors to search for a novel physical basis for dynamics of the Earth. The first steps in this direction are presented in the next chapters.

Chapter 2
Irrelevance of the Hydrostatics Model and the Earth's Dynamic Equilibrium

It was shown in the previous short review of development of the theoretical basis for the Earth's shape and dynamics that the roots of hydrostatic fundamentals for solution of the problem date back to the distant past and are related to the names of founders of modern science. But even at that early time, these pioneers well understood that applicability of hydrostatic equilibrium to a body's dynamic problems is restricted by certain boundary conditions. Thus, Newton in his "*Principia*" (see Sect. 1.5 of Book III), while considering the conditions of attraction in the planets writes: "*The attraction being spread from the sphere surface downwards is approximately proportional to distance of the center. Be the planet's matter uniform in density, then this proportion would have exact value. It follows from here that the error is caused by non-uniformity in density*". At that time the thoughts of scientists were engaged with how to solve principle problems of a body's orbital motion. The main problems being considered as solved, the goals of farther studies changed to different aspects. In our time, however, when the details of motion of the planets are of renewed interest and new tools of investigation abound, the formulation and the methods of solution needed to be developed farther.

2.1 Hydrostatic Equilibrium Conditions

We recall briefly the conditions of the Earth's hydrostatic equilibrium. By definition, hydrostatics is a branch of hydromechanics, which studies the equilibrium of a liquid and a gas and the effects of a stationary liquid on immersed bodies relative to the chosen reference system. For a liquid equilibrated relative to a rigid body, when its velocity of motion is equal to zero and the field of densities is steady, the equation of state follows from the Eulerian and Navier-Stokes equations in the form (Sedov, 1970)

$$grad\ p = \rho F, \tag{2.1}$$

where p is the pressure; ρ is the density; F is the mass force.

V. I. Ferronsky, S. V. Ferronsky, *Dynamics of the Earth*,
DOI 10.1007/978-90-481-8723-2_2, © Springer Science+Business Media B.V. 2010

In the Cartesian system of reference eq. (2.1) is written as

$$\frac{\partial p}{\partial x} = \rho F_x,$$
$$\frac{\partial p}{\partial y} = \rho F_y, \qquad\qquad (2.2)$$
$$\frac{\partial p}{\partial z} = \rho F_z.$$

If the outer mass forces are absent, i.e., $F_x = F_y = F_z = 0$, then

$$grad\ p = 0.$$

In this case, in accordance with Pascal's law, the pressure at all liquid points will be the same.

For a uniform incompressible liquid, when $\rho = const$, its equilibrium can be only in the potential field of the outer forces. For the general case of an incompressible liquid and potential field of the outer forces from (2.1) one has

$$dp = \rho\ dU, \qquad\qquad (2.3)$$

where U is the forces' potential.

It follows from eq. (2.3), that for an equilibrated liquid in a potential force field, its density and pressure appear to be a function only of the potential U.

For a gravity force field, when only these forces act on the steady-state liquid, one has

$$F_x = F_y = 0, \quad F_z = -g, \quad U = -gz + const \quad \text{and} \quad p = p(z), \quad \rho = \rho(z).$$

Here the surfaces of constant pressure and density appear as horizontal planes. Then eq. (2.3) is written in the form

$$\frac{dp}{dz} = -\rho g < 0. \qquad\qquad (2.4)$$

It means that with elevation the pressure falls and with depth it grows. From here it follows that

$$p - p_0 = -\int_{z_0}^{z} \rho g dz = -\rho g(z - z_0) \qquad\qquad (2.5)$$

where g is the acceleration of the gravity force.

If a spherical vessel is filled with incompressible liquid and rotates around its vertical axis with constant angular velocity ω, then for determination of the equilibrated free surface of the liquid in eq. (2.2) the centrifugal inertial forces should be introduced in the form

$$\frac{\partial p}{\partial x} = \rho\omega^2 x,$$
$$\frac{\partial p}{\partial x} = \rho\omega^2 y, \qquad (2.6)$$
$$\frac{\partial p}{\partial x} = -\rho g.$$

From here, for the rotating body with radius $r^2 = x^2 + y^2$, one finds

$$p = -\rho g z + \frac{\rho\omega^2 r^2}{2} + C. \qquad (2.7)$$

For the points on the free surface $r = 0$, $z = z_0$ one has $p = p_0$. Then

$$C = p_0 + \rho g z_0, \qquad (2.8)$$

$$p = p_0 + \rho g(z_0 - z) + \frac{\rho\omega r^2}{2}. \qquad (2.9)$$

The equation of the liquid free surface, where $p = p_0$, has a paraboloidal shape

$$z - z_0 = \frac{\omega^2 r^2}{2g}. \qquad (2.10)$$

These facts determine the principal physical conditions and equations of hydrostatic equilibrium of a liquid. They remain a basis for modern dynamics and the theory of the Earth's shape. The attempt to harmonize these conditions with the planet's motion conditions has failed, which was proved by observation. It will be shown below in Sects. 2.4–2.6 of this chapter, that the main obstacle for such harmonization is rejection of the planet's inner force field without which the hydrostatics is unable to provide equilibrium between the body's interacting forces as Newton's third law requires. The Earth is a self-gravitating body. Its matter moves in its own force field which is generated by mass particle interaction. The mass density distribution, rotation and oscillation of the bodies' shells result from the inner force field. And the orbital motion of the planet is controlled by interaction of the outer force fields of the planet and the Sun in accordance with Newton's theory.

Let us look for more specific effects determining the absence of the Earth's hydrostatic equilibrium and more realistic conditions of its equilibrium based on the results of the Earth's satellite orbit motion.

2.2 Relationship Between Moment of Inertia and Gravitational Force Field According to Satellite Data

Let us come back to the fact of absence of the Earth's hydrostatic equilibrium found by the satellite data. The initial factual material for the problem study is presented by the observed orbit elements of the geodetic satellites which move on perturbed Kepler's orbits. The satellite motion is fixed by means of observational stations located within zones of a visual height range of 1 000–2 500 km, which is optimal for the planet's gravity field study. It was found, that the satellite's perturbed motion at such a close distance from the Earth's surface is connected with the non-uniform distribution of mass density, the consequences of which are the non-spherical shape in the figure and the corresponding non-uniform distribution of the outer gravity field around the planet. These non-uniformities cause corresponding changes in trajectories of the satellite's motion, which are fixed by tracking stations. Thus, distribution of the Earth's mass density determines an adequate equipotential trajectory in the planet's gravity field, which follows the satellite. The main goal of the geodetic satellites launched under different angles relative to the equatorial plane is in measurement of all deviations in the trajectory from the unperturbed Kepler's orbit.

The satellite orbits data for solving the Earth's oblateness problem are interpreted on the basis of the known (in celestial mechanics) theory of expansion of the gravity potential of a body, the structure and the shape of which do not much differ from the uniform sphere. The expression of the expansion, by spherical functions, recommended by the International Union of Astronomy, is the following equation (Grushinsky, 1976):

$$U(r, \varphi, \lambda) = \frac{GM}{r} \left[1 - \sum_{n=2}^{\infty} J_n \left(\frac{R_e}{r} \right)^n P_n (\sin \varphi) \right. $$
$$\left. + \sum_{n=2}^{\infty} \sum_{m=1}^{n} \left(\frac{R_e}{r} \right)^n P_{nm} (\sin \varphi) (C_{nm} \cos m\lambda + S_{nm} \sin m\lambda) \right], \quad (2.11)$$

where r, φ and λ are the heliocentric polar co-ordinates of an observation point; G is the gravity constant; M and R_e are the mass and the mean equatorial radius of the Earth; P_n is the Legendre polynomial of n order; $P_{nm}(\sin \varphi)$ is the associated spherical functions; J_n, C_{nm}, S_{nm} are the dimensionless constants characterizing the Earth's shape and gravity field.

The first terms of eq. (2.11) determine the zero approximation of Newton's potential for a uniform sphere. The constants J_n, C_{nm}, S_{nm} represent the dimensionless gravitational moments, which are determined through analyzing the satellite orbits. The values J_n express the zonal moments, and C_{nm} and S_{nm} are the tesseral moments. In the case of hydrostatic equilibrium of the Earth as a body of rotation, in the expression of the gravitational potential (2.11) only the even n-zonal moments J_n are rapidly decreased with growth, and the odd zonal and all tesseral moments turn into zero, i.e.

$$U = \frac{GM}{r}\left[1 - J_2\left(\frac{R_e}{r}\right)^2 P_2\left(\cos\theta\right) - \sum_{n=3}^{\infty} J_n \left(\frac{R_e}{r}\right)^{n+1} P_n\left(\cos\theta\right)\right], \quad (2.12)$$

where θ is the angle of the polar distance from the Earth's pole.

Here the constant J_2 represents the zonal gravitational moment, which characterizes the axial planet's oblateness and makes the main contribution to correction of the unperturbed potential. That constant determines the dimensionless coefficient of the moment of inertia relative to the polar axis and equal to

$$J_2 = \frac{C - A}{MR_e^2}, \quad (2.13)$$

where C and A are the Earth's moments of inertia with respect to the polar and equatorial axes accordingly, and R_e is the equatorial radius.

For expansion by spherical functions of the Earth's gravity forces potential, the rotation of which is taken to be under action of the outer inertial forces, but not of its own force field, the centrifugal force potential is introduced into Eq. (2.12). Then for the hydrostatic condition with the even zonal moments J_n one has

$$W = \frac{GM}{r}\left[1 - J_2\left(\frac{R_e}{r}\right)^2 P_2\left(\cos\theta\right) - \sum_{n=3}^{\infty} J_n \left(\frac{R_e}{r}\right)^{n+1} P_n\left(\cos\theta\right)\right]$$
$$+ \frac{\omega^2 r^2}{3}\left[1 - P_2\left(\cos\theta\right)\right], \quad (2.14)$$

where W is the potential of the body of rotation; $\omega^2 r^2$ is the centrifugal potential. The first two terms and the term of the centrifugal force in Eq. (2.14) express the normal potential of the gravity force

$$W = \frac{GM}{r}\left[1 - J_2\left(\frac{R_e}{r}\right)^2 P_2\left(\cos\theta\right)\right] + \frac{\omega^2 r^2}{3}\left[1 - P_2\left(\cos\theta\right)\right]. \quad (2.15)$$

The potential (2.15) corresponds to the spheroid's surface which within oblateness coincides with the ellipsoid of rotation. Rewriting term $P_2(\cos\theta)$ in this equation through the sinus of the heliocentric latitude and the angular velocity – through the geodynamic parameter q, one can find the relationship of the Earth's oblateness ε with the dynamic constant J_2. Then the equation of the dynamic oblateness ε is obtained in the form (Grushinsky, 1976; Melchior, 1972)

$$\varepsilon = \frac{3}{2}J_2 + \frac{q}{2}, \quad (2.16)$$

where the geodynamic parameter q is the ratio of the centrifugal force to the gravity force at the equator

$$q = \frac{\omega^2 R}{GM/R^2}. \tag{2.17}$$

Geodynamic parameter J_2, found by satellite observation in addition to the oblateness calculation, is used for determination of a mean value of the Earth's moment of inertia. For this purpose the constant of the planet's free precession is also used, which represents one more observed parameter expressing the ratio of the moments of inertia in the form:

$$H = \frac{C - A}{C}. \tag{2.18}$$

This is the theoretical base for interpretation of the satellite observations. But its practical application gave very contradictory results (Grushinsky, 1976; Melchior, 1972; Zharkov, 1978). In particular, the zonal gravitation moment calculated by means of observation was found to be $J_2 = 0.0010827$, from where the polar oblateness $\varepsilon = 1/298.25$ appeared to be short of the expected value and equal to $1/297.3$. The all zonal moments J_n, starting from J_3, which relate to the secular perturbation of the orbit, were close to constant value and equal, by an order of magnitude, to the square of the oblateness i.e., $\sim(1/300)^2$ and slowly decreasing with an increase of n. The tesseral moments C_{nm} and S_{nm} appeared to be not equal to zero, expressing the short-term nutational perturbations of the orbit. In the case of hydrostatic equilibrium of the Earth at the found value of J_2, the polar oblateness ε should be equal to $1/299.25$. On this basis the conclusion was made that the Earth does not stay in hydrostatic equilibrium. The planet's deviation from the hydrostatic equilibrium evidenced that there is a swelling in the planet's equatorial region with an amplitude of about 70 m. It means that the Earth body is forced by normal and tangential forces which develop corresponding stresses and deformations. Finally, by the measured tesseral and sectorial harmonics, it was directly confirmed that the Earth has an asymmetric shape with reference to the axis of rotation and to the equatorial plane.

Because the Earth does not stay in hydrostatic equilibrium, then the above described initial physical fundamentals for interpretation of the satellite observations should be recognized as incorrect and the related physical concepts cannot explain the real picture of the planet's dynamics.

The question is raised of how to interpret the obtained actual data and where the truth should be sought. First of all we should verify correctness of the oblateness interpretation and the conclusion about the Earth's equatorial swelling. It is known from observation that the Earth is a triaxial body (see Table 1.2). Theoretical application of the triaxial Earth model was not considered because it contradicts the hydrostatic equilibrium hypothesis. But after it was found that the hydrostatic equilibrium is absent, the alternative with the triaxial Earth should be considered first.

Let us analyze Eq. (2.16). It is known from the observation data, that the constant of the centrifugal oblateness q is equal to

$$q = \frac{\omega_3^2}{GM/R^3} = \left(\frac{1}{17.01}\right)^2 = \frac{1}{289.37}. \tag{2.19}$$

Determine a difference between the centrifugal oblateness constant q and the polar oblateness ε' found by the satellite orbits, assuming that the desired value has a relationship with the perturbation caused by the equatorial ellipsoid

$$\varepsilon' = \frac{a-c}{a} - \frac{b-c}{a} = \frac{a-b}{a} = \frac{1}{289.37} - \frac{1}{298.25}$$

$$= \frac{1}{9720} = 1.713\left(\frac{1}{289.37}\right)^2, \tag{2.20}$$

where a, b and c are the semi-axes of the triaxial Earth.

The differences between the major and minor equatorial semi-axes can be found from Eq. (2.20). If the major semi-axis is taken in accordance with recommendation of the International Union of Geodesy and Geophysics as $a = 6\ 378\ 160$ m, then the minor equatorial semi-axis b can be equal to:

$$a - b = 6378160/9720 = 656 \text{ m}; \quad b = 6377504 \text{ m}.$$

There is a reason now to assume, that the value of equatorial oblateness $\varepsilon' = 1/9\ 720$ is a component in all the zonal gravitation moments J_n, related to the secular perturbations of the satellite orbits including J_2. They are perturbed both by the polar and the equatorial oblateness of the Earth. This effect ought to be expected because it was known long ago from observation that the Earth is a triaxial body. If our conclusion is true, then there is no ground for discussion about the equatorial swelling. And also the problem of the hydrostatic equilibrium is closed automatically because in this case the Earth is not a figure of rotation; and the nature of the observing fact of rotation of the Earth should be looked for rather in the action of its own inner force field but not in the effects of the inertial forces. As to the nature of the Earth's oblateness, then for its explanation later on the effects of perturbation arising during separation of the Earth's shells by mass density differentiation and separation of the Earth itself from the Protosun will be considered. In particular, the effect of heredity in creation of the body's oblateness is evidenced by the ratio of kinetic energy of the Sun and the Moon expressed through the ratio of square frequencies of oscillation ε'' of their polar moments of inertia (see Sects. 6.5 and 6.10 in Chap. 6), which is close to the planet's equatorial oblateness:

$$\varepsilon'' = \frac{\omega_c^2}{\omega_\pi^2} = \frac{\left(10^{-4}\right)^2}{\left(0.96576 \times 10^{-2}\right)^2} = \left(\frac{1}{96.576}\right)^2 = 1,73\left(\frac{1}{289.3}\right)^2,$$

where $\omega_c = 10^{-4}$ s^{-1} and $\omega_n = 0.965\ 76 \times 10^{-2}$ s^{-1} are the frequencies of oscillation of the Sun's and the Moon's polar moment of inertia correspondingly.

By observation the Moon is also a triaxial body. In addition, the retrograde motion of the nodes of the Earth, the Moon and the artificial satellites is registered and is explained by rotation of the bodies' orbits. Later on it will be shown, that the above remarkable phenomenon is explained by rotation of the body's inner masses together with their gravity fields, the periods of which are equal to the periods of the precession of their oblique axes. The observed body rotation is valid only for the upper shells, which were separated during mass density differentiation in their own force fields and stay in that field in a suspended state of equilibrium.

The most prominent effect, which was discovered by investigation of the geodetic satellite orbits, is the fact of a physical relationship between the Earth's mean (polar) moment of inertia and the outer gravity field. That fact without exaggeration can be called a fundamental contribution to understanding the nature of the planet's self-gravity. The planet's moment of inertia is an integral characteristic of the mass density distribution. Calculation of the gravitational moments based on measurement of elements of the satellite orbits is the main content of satellite geodesy and geophysics. Short-periodic perturbations of the gravity field fixed at revolution of a satellite around the Earth, the period of which is small compared to the planet's period, provides evidence about oscillation of the moment of inertia or, to be more correct, about oscillating motion of the interacting mass particles. It will be shown, that oscillating motion of the interacting particles forms the main part of a body's kinetic energy and the moment of inertia itself is the periodically changing value.

Oscillation of the Earth's moment of inertia and also the gravitational field is fixed not only during the study by artificial satellites. Both parameters have also been registered by surface seismic investigations. Consider briefly the main points of those observations.

2.3   Oscillation of the Moment of Inertia and the Inner Gravitational Field Observed During Earthquakes

The study of the Earth's eigenoscillation started with Poisson's work on oscillation of an elastic sphere, which was considered in the framework of the theory of elasticity. In the beginning of the twentieth century Poisson's solution was generalized by Love in the framework of the problem solution of a gravitating uniform sphere of the Earth's mass and size. The calculated values of periods of oscillation were found to be within the limit of some minutes to one hour.

In the middle of the twentieth century, during the powerful earthquakes in 1952 and 1960 in Chile and Kamchatka, an American team of geophysicists headed by Beneoff, using advanced seismographs and gravimeters, reliably succeeded in recording an entire series of oscillations with periods from 8.4 m up to 57 m. Those oscillations in the form of seismograms have represented the dynamical effects of the interior of the planet as an elastic body, and the gravimetric records have shown the "tremor" of the inner gravitational field (Zharkov, 1978). In fact, the effect of the simultaneous action of the potential and kinetic energy in the Earth's interior was fixed by the above experiments.

About one thousand harmonics of different frequencies were derived by expansion of the line spectrum of the Earth's oscillation. These harmonics appear to be integral characteristics of the density, elastic properties and effects of the gravity field, i.e. of the potential and kinetic energy of separate volumetric parts of the non-uniform planet. As a result two general modes of the Earth's oscillations were found by the above spectral analysis, namely, spherical with a vector of radial direction and torsion with a vector perpendicular to the radius.

From the point of view of the existing conception about the planet's hydrostatic equilibrium, the nature of the observed oscillations was considered to be a property of the gravitating non-uniform (regarding density) body in which the pulsed load of the earthquake excites elementary integral effects in the form of elastic gravity quanta (Zharkov, 1978). Considering the observed dynamical effects of earthquakes, geophysicists came close to a conclusion about the nature of the oscillating processes in the Earth's interior. But the conclusion itself still has not been expressed because it continues to relate to the position of the planet's hydrostatic equilibrium.

Now we move to one of the main problems related to the Earth's equilibrium or, more correctly, to the absence of the Earth's equilibrium if it is considered on the basis of hydrostatics.

2.4 Imbalance Between the Earth's Potential and Kinetic Energies

We discovered the most likely serious cause, for which even formulation of the problem of the Earth's dynamics based on the hydrostatic equilibrium is incorrect. The point is that the ratio of kinetic to potential energy of the planet is equal to ~1/300, i.e. the same as its oblateness. Such a ratio does not satisfy the fundamental condition of the virial theorem, the equation of which expresses the hydrostatic equilibrium condition. According to that condition the considered energies' ratio should be equal to 1/2. Taking into account that kinetic energy of the Earth is presented by the planet's inertial rotation, then assuming it to be a rigid body rotating with the observed angular velocity $\omega_r = 7.29 \times 10^{-5}$ s^{-1}, the mass $M = 6 \times 10^{24}$ kg, and the radius $R = 6.37 \times 10^6$ m, the energy is equal to:

$$T_e = 0.6MR^2\,\omega_r^2 = 0.6 \times 6 \times 10^{24} \times (6.37 \times 10^6)^2 \times (7.29 \times 10^{-5})^2$$
$$= 7.76 \times 10^{29}J = 7.76 \times 10^{36} \text{ erg.}$$

The potential energy of the Earth at the same parameters is

$$U_e = 0.6 \times GM^2/R = 0.6 \times 6.67 \times 10^{-11} \times (6 \times 10^{24})^2/6.37 \times 10^6$$
$$= 2.26 \times 10^{32}J = 2.26 \times 10^{39} \text{erg.}$$

The ratio of the kinetic and potential energy comprises

$$T_e/U_e = 7.76 \times 10^{29}/2.26 \times 10^{32} = 1/292.$$

One can see that the ratio is close to the planet's oblateness. It does not satisfy the virial theorem and does not correspond to any condition of equilibrium of a really existing natural system because, in accordance with the third Newton's law, equality between the acting and the reacting forces should be satisfied. The other planets, the Sun and the Moon, the hydrostatic equilibrium for which is also accepted as a fundamental condition, stay in an analogous situation. Since the Earth in reality exists in equilibrium and its orbital motion strictly satisfies the ratio of the energies, then the question arises where the kinetic energy of the planet's own motion has disappeared. Otherwise the virial theorem for the Earth is not valid. Moreover, if one takes into account that the energy of inertial rotation does not belong to the body, then the Earth and other celestial bodies equilibrium problem appears to be out of discussion.

Thus, we came to the problem of the Earth equilibrium from two positions. From one side, the planet by observation does not stay in hydrostatic equilibrium, and from the other side, it does not stay in general mechanical equilibrium because there is no reaction forces to counteract to the acting potential forces. The answer to both questions is given below while deriving an equation of the dynamical equilibrium of the planet by means of generalization of the classical virial theorem.

2.5 Equation of Dynamical Equilibrium

The main methodological question arises in which the state of equilibrium of the Earth exists. The answer to the question results from the generalized virial theorem for a self-gravitating body, i.e. the body which itself generates the energy for motion by interaction of the constituent particles having innate moments. The guiding effect which we use here is the motion observed by an artificial satellite that is the functional relationship between changes in the gravity field of the Earth and its mean (polar) moment of inertia. The deep physical meaning of this relationship is as follows. The planet's mean (polar) moment of inertia is an integral (volumetric) parameter, which does not represent location of the interacting mass particles, but expresses changes in their motion under the constrained energy. The virial theorem of Clausius for a perfect gaseous cloud or a uniform body is presented in its averaged form. In order to generalize the theorem for a non-uniform body we introduce there the volumetric moments of interacting particles, taking into account their volumetric nature. Moreover, the interacting mass particles of a continuous medium generate volumetric forces (pressure or capacity of energy) and volumetric momentums, which, in fact, generate the motion in the form of oscillation and rotation of masses. The oscillating form of motion of the Earth and other celestial bodies is the dominating part of their kinetic energy which up to now has not been taken into account. We wish to fill in this gap in dynamics of celestial bodies.

The virial theorem is the analytical expression of the hydrostatic equilibrium condition and follows from Newton's and Euler's equations of motion. Let us recall its derivation in accordance with classical mechanics (Goldstein, 1980).

Consider a system of mass points, the location of which is determined by the radius vector $\mathbf{r}_i$ and the force $\mathbf{F}_i$ including the constraints. Then equations of motion of the mass points through their moments $\mathbf{p}_i$ can be written in the form

$$\dot{\mathbf{p}}_i = \mathbf{F}_i, \tag{2.21}$$

The value of the moment of momentum is

$$Q = \sum_i \mathbf{p}_i \cdot \mathbf{r}_i,$$

where the summation is done for all masses of the system. The derivative with respect to time from that value is

$$\frac{dQ}{dt} = \sum_i \dot{\mathbf{r}}_i \cdot \mathbf{p}_i + \sum_i \dot{\mathbf{p}}_i \cdot \mathbf{r}_i. \tag{2.22}$$

The first term in the right-hand side of (2.22) is reduced to the form

$$\sum_i \dot{\mathbf{r}}_i \cdot \mathbf{p}_i = \sum_i m_i \cdot \dot{\mathbf{r}}_i \cdot \dot{\mathbf{r}}_i = \sum_i m_i v_i^2 = 2T,$$

where T is the kinetic energy of particle motion under action of the forces $\mathbf{F}_i$. The second term in Eq. (2.22) is

$$\sum_i \dot{\mathbf{p}}_i \cdot \mathbf{r}_i = \sum_i \mathbf{F}_i \cdot \mathbf{r}_i.$$

Now Eq. (2.22) can be written as

$$\frac{d}{dt} \sum_i \mathbf{p}_i \cdot \mathbf{r}_i = 2T + \sum_i \mathbf{F}_i \cdot \mathbf{r}_i. \tag{2.23}$$

The mean values in (2.23) within the time interval τ are found by their integration from 0 to t and division by τ:

$$\frac{1}{\tau} \int_0^t \frac{dQ}{dt} dt = \overline{\frac{dQ}{dt}} = \overline{2T} + \overline{\sum_i \mathbf{F}_i \cdot \mathbf{r}_i}$$

or

$$\overline{2T} + \overline{\sum_i \mathbf{F}_i \cdot \mathbf{r}_i} = \frac{1}{\tau}[Q(\tau) - Q(0)]. \tag{2.24}$$

For the system in which the co-ordinates of mass point motion are repeated through the period τ, the right-hand side of Eq. (2.24) after its averaging is equal to zero. If the period is too large, then the right-hand side becomes a very small quantity. Then, the expression (2.24) in the averaged form gives the relation

$$-\overline{\sum_i \mathbf{F}_i \cdot \mathbf{r}_i} = 2T, \tag{2.25}$$

or in mechanics it is written in the form

$$2T = -U.$$

Equation (2.25) is known as the virial theorem, and its left-hand side is called the virial of Clausius (German *virial* is from the Latin *vires* which means forces). The virial theorem is a fundamental relation between the potential and kinetic energy and is valid for a wide range of natural systems, the motion of which is provided by action of different physical interactions of their constituent particles. Clausius proved the theorem in 1870 when he solved the problem of work of the Carnot thermal machine, where the final effect of the water vapor pressure (the potential energy) was connected with the kinetic energy of the piston motion. The water vapor was considered as a perfect gas. And the mechanism of the potential energy (the pressure) generation at the coal burning in the firebox was not considered and was not taken into account.

The starting point in the above-presented derivation of the virial theorem in mechanics is the moment of the mass point system, the nature of which is not considered either in mechanics or by Clausius. By Newton's definition the moment "*is the measure of that determined proportionally to the velocity and the mass*". The nature of the moment by his definition is "*the innate of the matter*". By his understanding that force is an inertial force, i.e. the motion of a mass continues with a constant velocity.

The observed (by satellites) relationship between the potential and the kinetic energy of the gravitation field and the Earth's moment of inertia provides evidence that the kinetic energy of the interacting mass particle motion, which is expressed as a volumetric effect of the planet's moment of inertia, is not taken into account. The evidence of that was given in the previous section in the quantitative calculation of a ratio between the kinetic and potential energies, equal to ~1/300.

In order to remove that contradiction, the kinetic energy of motion of the interacting particles should be taken into account in the derived virial theorem. Because any mass has volume the moment $\mathbf{p}$ should be written in volumetric form:

$$\mathbf{p}_i = \sum_i m_i \dot{\mathbf{r}}_i. \tag{2.26}$$

Now the volumetric moment of momentum acquires a wave nature and is presented as

$$Q = \sum_i \mathbf{p}_i \cdot \mathbf{r}_i = \sum_i m_i \cdot \dot{\mathbf{r}}_i \cdot \mathbf{r}_i = \frac{d}{dt}\left(\sum_i \frac{m_i r_i^2}{2}\right) = \frac{1}{2}\dot{I}_p, \qquad (2.27)$$

where I_p is the polar moment of inertia of the system (for the sphere it is equal to 3/2 of the axial moment).

The derivative from that value with respect to time is

$$\frac{dQ}{dt} = \frac{1}{2}\ddot{I}_p = \sum_i \dot{\mathbf{r}}_i \cdot \mathbf{p}_i + \sum_i \dot{\mathbf{p}}_i \cdot \mathbf{r}_i. \qquad (2.28)$$

The first term in the right-hand part of (2.28) remains without change

$$\sum_i \dot{\mathbf{r}}_i \cdot \mathbf{p}_i = \sum_i m_i \cdot \dot{\mathbf{r}}_i \cdot \dot{\mathbf{r}}_i = \sum_i m_i v_i^2 = 2T. \qquad (2.29)$$

The second term represents the potential energy of the system

$$\sum_i \dot{\mathbf{p}}_i \cdot \mathbf{r}_i = \sum_i \mathbf{F}_i \cdot \mathbf{r}_i = U. \qquad (2.30)$$

Equation (2.28) is written now in the form

$$\frac{1}{2}\ddot{I}_p = 2T + U. \qquad (2.31)$$

Expression (2.31) represents a generalized equation of the virial theorem for a mass point system interacting by Newton's law. Here in the left-hand side of (2.31) the previously ignored inner kinetic energy of interaction of the mass particles appears. Solution of Eq. (2.31) gives a variation of the polar moment of inertia within the period τ. For a conservative system averaging expression (2.28) by integration from 0 to t within time interval τ gives

$$\frac{1}{\tau}\int_0^t \frac{dQ}{dt} dt = \overline{\frac{dQ}{dt}} = \overline{2T} + \overline{U} = \ddot{I}_p. \qquad (2.32)$$

Eq. (2.32) at $\ddot{I}_p = 0$ gives $\dot{I}_p = E = const.$, where E is the total system's energy. It means that the interacting mass particles of the system move with constant velocity. In the case of a dissipative system, equation (2.32) is not equal to zero and the interacting mass particles move with acceleration. Now the ratio between the potential and kinetic energy has a value in strict accordance with equation (2.31). Kinetic energy of the interacting mass particles in the form of oscillation of the polar moment of inertia in that equation is taken into account. And now in the frame of the law of energy conservation the ratio of the potential to kinetic energy of a celestial body has a correct value.

Expression (2.31) appears to be an equation of dynamical equilibrium of the self-gravitating Earth and the Sun inside and between the bodies. The hydrostatic equilibrium presents here as a particular case and takes also into account the kinetic energy of the continuous motion of the interacting particles which generate energy due to their inner potential. The integral effect of the moving particles is fixed by the satellite orbits in the form of changing zonal, sectorial and tesseral gravitational moments. We used the resulting energy of the initial moment (2.26) for derivation of the generalized virial theorem. The initial moments form the inner, or "*innate*" by Newton's definition, energy of the body which has an inherited origin. The nature of Newton's centripetal forces and the mechanism of their energy generation will be discussed in some detail in Chap. 8.

Thus, we obtained a differential equation of the second order (2.31), which describes the Earth's dynamics and its dynamical equilibrium.

The virial equation (2.31) was obtained by Jacobi already one and a half century ago from Newton's equations of motion in the form (Jacobi, 1884)

$$\ddot{\Phi} = U + 2T, \tag{2.33}$$

where Φ is the Jacobi function (the polar moment of inertia).

At that time Jacobi was not able to consider the physical meaning of the equation. For that reason he assumed that as there are two independent variables Φ and U in the equation, then it can not be resolved. We succeeded to find an empirical relationship between the two variables and at first obtained an approximate, and later on rigorous, solution of the equation (Ferronsky et al., 1978, 1987; Ferronsky, 2005). The relationship is proved now by means of the satellite observation.

We can explain now the cause of the discrepancy between the geometric (static) and dynamic oblateness of the Earth. The reason is as follows. The planet's moments of inertia with respect to the main axes and their integral form of the polar moment of inertia do not stay in time as constant values. The polar moment of inertia of a self-gravitating body has a functional relation with the potential energy, the generation of which results by interaction of the mass particles in a regime of periodic oscillations. The hydrostatic equilibrium of a body does not express the picture of the dynamic processes from which, as it follows from the averaged virial theorem, the energy of the oscillating effects was lost from consideration. Because of that it was not possible to understand the nature of the energy. The main part of the body's kinetic energy of the body's oscillation was also lost. As to the rotational motion of the body's shells, it appears only in the case of the non-uniform radial distribution of the mass density. The contribution of rotation to the total body's kinetic energy comprises a very small part.

The cause of the accepted incorrect ratio between the Earth's potential and kinetic energy is the following. Clairaut's equation (1.20), derived for the planet's hydrostatic equilibrium state and applied to determine the geometric oblateness, because of the above reason, has no functional relationship between the force function and the moment of inertia. Therefore for the Earth's dynamical problem the equation gives only the first approximation. In formulation of the Earth oblateness problem, Clairaut accepted Newton's model of action of the centripetal forces from the surface of the planet to its geometric center. In such a physical conception the total effect of the inner force field becomes equal to zero. Below in Sect. 2.6 it will be shown, that the force field of the continuous body's interacting masses represents volumetric pressure, but not a vector force field. That is why the accepted postulate relating to the planet's inertial rotation appears to be physically incorrect.

The question was raised about how it happened, that geodynamic problems and first of all the problem of stability of the Earth's motion up to now were solved without knowing the planet's kinetic energy. The probable explanation of that seems to lie in the history of the development of science. In Kepler's problem and in Newton's two-body problem solution the transition from the averaged parameters of motion to real conditions is implemented through the mean and the eccentric anomalies, which by geometric procedures indirectly take into account the above energy of motion. In the Earth's shape problem this procedure of Kepler is not applied. Therefore, the so-called "inaccuracies" in the Earth's motion appear to be the regular dynamic effects of a self-gravitating body, and the hydrostatic model in the problem is irrelevant. The hydrostatic model was accepted by Newton for the other problem, where just this model allowed discovery and formulation of the general laws of the planets' motion around the Sun. Newton's centripetal forces in principle satisfy Kepler's condition when the distance between bodies is much more than their size accepted as mass points. Such a model gives a first approximation in the problem solution.

Kepler's laws express the real picture of the planets and satellites motion around their parent bodies in averaged parameters. All the deviations of those averaged values related to the outer perturbations are not considered as it was done in the Clausius' virial theorem for a perfect gas.

Newton solved the two-body problem, which had been already formulated by Kepler. The solution was based on the heliocentric world system of Copernicus, on the Galilean laws of inertia and free fall in the outer force field and on Kepler's laws of the planet's motion in the central force field considered as a geometric plane task. The goal of Newton's problem was to find the force in which the planet's motion resulted. His centripetal attraction and the inertial forces in the two-body problem satisfy Kepler's laws.

As it was mentioned, Newton understood the physical meaning of his centripetal or attractive forces as a pressure, which is accepted now like a force field. But by his opinion, for mathematical solutions the force is a more convenient instrument. And in the two-body problem the force-pressure is acting from the center (of a point) to outer space.

It is worth discussing briefly Newton's preference given to the force but not to the pressure. In mechanics the term "*mass point*" is understood as a geometric point of space, which has no dimension but possesses a finite mass. In physics a small amount of mass is called by the term "*particle*", which has a finite value of size and mass. But very often physicists use models of particles, which have neither size nor mass. A body model like mass point has been known since ancient times. It is simple and convenient for mathematical operations. The point is an irreplaceable geometric symbol of a reference point. The physical point, which defines inert mass of a volumetric body, is also suitable for operations. But the interacting and physically active mass point creates a problem. For instance, in the field theory the point value is taken to denote the charge, the meaning of which is no better understood than is the gravity force. But it is considered often there, that the point model for mathematical presentation of charges is not suitable because operations with it lead to zero and infinite values. Then for resolution of the situation the concept of charge density is introduced. The charge is presented as an integral of density for the designated volume and in this way the solution of the problem is resolved.

The point model in the two-body problem allowed reduction of it to the one-body problem and for a spherical body of uniform density to write the main seven integrals of motion. In the case when a body has a finite size, then not the forces but the pressure becomes an effect of the body particle interaction. The interacting body's mass particles form a volumetric gravitational field of pressure, the strength of which is proportional to the density of each elementary volume of the mass. In the case of a uniform body, the gravitational pressure should be also uniform within the whole volume. The outer gravitational pressure of the uniform body should be also uniform at the given radius. The non-uniform body has a non-uniform gravitational pressure of both inner and outer field, which has been observed in studying the real Earth field. Interaction of mass particles results appears in their collision, which leads to oscillation of the whole body system. In general if the mass density is higher, then the frequency of body oscillation is also higher.

It was known from the theory of elasticity, that in order to calculate the stress and the deformation of a beam from a continuous load, the latter can be replaced by the equivalent lumped force. In that case the found solution will be approximate because the beam's stress and deformation will be different. The question is what degree of approximation of the solution and what kind of error is expected. Volumetric forces are not summed up by means of the parallelogram rule. Volumetric forces by their nature can not be reduced for application either to a point, or to a resultant vector value. Their action is directed to the 4π space and they form inner and outer force fields. The force field by its action is proportional to action of the energy. This is because the force is the derivative of the energy.

The centrifugal and Coriolis forces are also proved to be inertial forces as a consequence of inertial rotation of the body. And the Archimedes force has not found its physical explanation, but it became an observational fact of hydrostatic equilibrium of a body mass immersed in a liquid.

Such is the short story of appearance and development of the hydrostatic equilibrium of the Earth in an outer uniform gravity field. The force of gravity of a body mass is an integral value. In this connection Newton's postulate about the gravity center as a geometric point should be considered as a model for presentation of two interacting bodies, when their mutual distance is much more of the body size. It is shown in the next section, that the reduced physical, but not geometrical, gravity center of a volumetric body is represented by an envelope of the figure, which draws an averaged value of radial density distribution of the body.

Because of the above discussion, the theorem of classical mechanics cited in Sect. 1.5 stating that if a body is found in the central force field, then the sum of their inner forces and torques are equal to zero, from the mathematical point of view is correct in the frame of the given initial conditions. As in the case of the derived virial theorem, the moment of momentum L in expression (1.5) can be presented by the first derivative from the polar moment of inertia. And then the torque equal to zero in the central field will be presented by oscillation of the polar moment of inertia not equal to zero.

The problem of dynamics of the Earth as a self-gravitating body, including its shape problem, in its formulation and solution needs a higher degree of approximation. Generalized virial theorem (2.31) satisfies the condition of the Earth's dynamical equilibrium state and creates a physical and theoretical basis for farther development of theory. It follows from the theorem that the hydrostatic equilibrium state there is the particular case of dynamics. Solution of the problem of the Earth's dynamics based on the equation of dynamical equilibrium appears to be the next natural and logistic step from the hydrostatic equilibrium model to a more perfect method without loss of the previous preference.

Below we consider the problem of "decentralization" of its own force field for a self-gravitating body.

2.6 Reduction of Inner Gravitation Field to the Resultant Envelope of Pressure

Consider the Earth as a self-gravitating sphere with uniform and one-dimensional interacting media. The motion of the Earth proceeds both in its own and in the Sun's force fields. It is known from theoretical mechanics that any motion of a body can be represented by a translation motion of its mass center, rotation around that center and motion of the body mass related to its changes in shape and structure (Duboshin, 1975). In the two-body problem the last two effects are neglected due to their smallness.

In order to study the Earth's motion in its own force field the translational (orbital) motion relative to the fixed point (the Sun) should be separated from the two other components of motion. After that one can consider the rotation around the geometric center of the Earth's masses under action of its own force field and changes in the shape and structure (oscillation). Such separation is required only for the moment of inertia, which depends on what frame of reference is selected. The force function depends on a distance between the interacting masses and does not depend on selection of a frame of reference (Duboshin, 1975). The moment of inertia of the Earth relative to the solar reference frame should be split into two parts. The first is the moment of the body mass center relative to the same frame of reference and the second is moment of inertia of the planet's mass relative to its own mass center.

So, set up the absolute Cartesian coordinates $O_c\xi\eta\zeta$ with the origin at the center of the Sun and transfer it to the system $Oxyz$ with the origin at the geometrical center of the Earth's mass (Fig. 2.1).

Fig. 2.1 Body motion in its
own force field

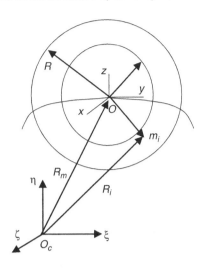

Then, the moment of inertia of the Earth in the solar frame of reference is

$$I_c = \sum m_i R_i^2, \tag{2.34}$$

where m_i is the Earth mass of particles; R_i is its distance from the origin in the same frame.

The Lagrange method is applied to separate the moment of inertia (2.34). The method is based on his algebraic identity

$$\left(\sum_{1 \leq i \leq n} a_i^2 \right) \left(\sum_{1 \leq i \leq n} b_i^2 \right) = \left(\sum_{1 \leq i \leq n} a_i b_i \right)^2 + \frac{1}{2} \sum_{1 \leq i \leq n} \sum_{1 \leq j \leq n} \left(a_i b_j - b_i a_j \right)^2, \tag{2.35}$$

where a_i and b_i are some chosen values; n is any positive number.

Jacobi in his "*Vorlesungen über Dynamik*" was the first who performed the mathematical transformation for separation of the moment of inertia of n interacting mass points into two algebraic sums (Jacobi, 1884; Duboshin, 1975; Ferronsky et al., 1987). It was shown that if we write (Fig. 2.1)

$$\xi_i = x_i + A; \quad \eta_i = y + B; \quad \zeta_i = z + C;$$

$$\sum m_i = M; \quad \sum m_i \xi_i = MA; \quad \sum m_i \eta_i = MB; \quad m_i \zeta_i = MC, \tag{2.36}$$

where A, B, C are the coordinates of the mass center in the solar frame of reference.

Then, using identity (2.35), one has

$$\sum m_i r_i^2 = \sum m_i \xi_i^2 + \sum m_i \eta_i^2 + \sum m_i \zeta_i^2 = \sum m_i x_i^2 + 2A \sum m_i x_i A^2 \sum m_i$$

$$+ \sum m_i y_i^2 + 2B \sum m_i y_i + B^2 \sum m_i$$

$$+ \sum m_i z_i^2 + 2C \sum m_i z_i + C^2 \sum m_i.$$

Since

$$MA = \sum m_i \xi_i = \sum m_i x_i + \sum m_i A = \sum m_i x_i + MA,$$

then

$$\sum m_i x_i = 0, \text{ and also } \sum m_i y_i = 0, \sum m_i z_i = 0.$$

Now, the moment of inertia (2.34) acquires the form

$$\sum m_i R_i^2 = M(A^2 + B^2 + C^2) + \sum m_i \left(x_i^2 + y_i^2 + z_i^2\right), \tag{2.37}$$

where

$$M(A^2 + B^2 + C^2) = MR_m^2, \tag{2.38}$$

$$\sum m_i \left(x_i^2 + y_i^2 + z_i^2\right) = Mr_m^2, \tag{2.39}$$

M is the Earth's mass; R_m and r_m are the radii of inertia of the Earth in the Sun's and the Earth's frame of reference.

Thus, we have separated the moment of inertia of the Earth, rotating around the Sun in the inertial frame of reference, into two algebraic terms. The first one (2.38) is the Earth's moment of inertia in the solar reference system $O_c\xi\eta\zeta$. The second term (2.39) presents the moment of inertia of the Earth in its own frame of reference $Oxyz$. The Earth mass here is distributed over the spherical surface with the reduced radius of inertia r_m. In literature the geometrical center of mass O in the Earth reference system is erroneously identified with the center of inertia and center of gravity of the planet.

For farther consideration of the problem of the Earth's dynamics we accept the polar frame of reference with its origin at center O. Then expression (2.39) for the Earth's polar moment of inertia I_p acquires the form

$$I_p = \sum m_i \left(x_i^2 + y_i^2 + z_i^2\right) = \sum m_i r_i^2 = Mr_m^2. \tag{2.40}$$

Now the reduced radius of inertia r_m, which traces out a spherical surface, is

$$r_m^2 = \frac{\sum m_i r_i^2}{M}. \tag{2.41}$$

Here $M = \sum m_i$ is the Earth's mass relative to its own frame of reference.

Taking into account the spherical symmetry of the uniform and one-dimensional Earth, we consider the sphere as a concentric spherical shell with mass $dm(r) = 4\pi r^2 \rho(r)dr$. Then the expression (2.41) in the polar reference system can be rewritten in the form

$$r_m{}^2 = \frac{1}{M} \int_0^R r^2 4\pi r^2 \rho(r) dr = \frac{4\pi R^2}{MR^2} \int_0^R r^4 \rho(r) dr, \tag{2.42}$$

or

$$\frac{4\pi r_m^2}{4\pi R^2} = \frac{4\pi \displaystyle\int_0^R r^4 \rho(r) dr}{MR^2} = \frac{\beta^2 MR^3}{MR^2} = \beta^2, \tag{2.43}$$

from where

$$r_m{}^2 = \beta^2 R^2,$$

where $\rho(r)$ is the law of radial density distribution; R is the radius of the sphere; β^2 is the dimensionless coefficient of the reduced spheroid (ellipsoid) of inertia $\beta^2 MR^2$.

The value of β^2 depends on the density distribution $\rho(r)$ and is changed within the limits of $1 \geq \beta^2 > 0$. Earlier (Ferronsky et al., 1987) it was defined as a structural form-factor of the polar moment of inertia.

Analogously, the reduced radius of gravity r_g, expressed as a ratio of the potential energy of interaction of the spherical shells with density $\rho(r)$ to the potential energy of interaction of the body mass is distributed over the shell with radius R. The potential energy of the sphere is written as

$$U = 4\pi G \int_0^R r\rho(r) m(r) dr = \alpha^2 \frac{GM^2}{R},$$

from where

$$\alpha^2 = \frac{4\pi G \displaystyle\int_0^R r\rho(r) m(r) dr}{\frac{GM^2}{R}} = \frac{4\pi\, r_g^2}{4\pi\, R^2}. \tag{2.44}$$

The form-factor α^2 of the inner force field, which controls its reduced radius, can be written as

$$\alpha^2 = \frac{4\pi\, r_g^2}{4\pi\, R^2} = \frac{\frac{4\pi G}{r} \displaystyle\int_0^R r\rho(r) m(r) dr}{\frac{GM^2}{R^2}}, \tag{2.45}$$

where in expressions (2.44) and (2.45) $m(r) = 4\pi \int_0^r r^2 \rho(r) dr$, and $r_g{}^2 = a^2 R^2$.

Table 2.1 Numerical values of form-factors α^2 and β^2 for radial distribution of mass density and for polytropic models

Distribution law Index of polytrope	α^2	$\beta_\perp^2$	β^2
Radial distribution of mass density			
$\rho(r) = \rho_0$	0.6	0.4	0.6
$\rho(r) = \rho_0(1 - r/R)$	0.74	0.27	0.4
$\rho(r) = \rho_0(1 - r^2/R^2)$	0.71	0.29	0.42
$\rho(r) = \rho_0 \exp(1 - kr/R)$	0.16k	$8/k^2$	$12/k^2$
$\rho(r) = \rho_0 \exp(1 - kr^2/R^2)$	$\sqrt{(k/2\pi)}$	$1/k$	1.5/k
$\rho(r) = \rho_0\delta(1 - r/R)$	0.5	0.67	1.0
Politrope models			
0	0.6	0.4	0.6
1	0.75	0.26	0.38
1.5	0.87	0.20	0.30
2	1.0	0.15	0.23
3	1.5	0.08	0.12
3.5	2.0	0.045	0.07

The value of β^2 depends on the density distribution $\rho(r)$ and is changed within the limits of $1 \geq \alpha^2 > 0$. Earlier (Ferronsky et al., 1987) it was defined as a structural form-factor of the force function.

Numerical values of the dimensionless form-factors α^2 and β^2 for a number of density distribution laws $\rho(r)$, obtained by integration of the numerators in Eqs. (2.43) and (2.44) for the polar moment of inertia and the force function, are presented in Table 2.1 (Ferronsky et al., 1987). Note, that the values of the polar I_p and axial I_a moments of inertia of a one-dimensional sphere are related as $I_p = 3/2I_a$.

It follows from Table 2.1 that for a uniform sphere with $\rho(r) = const$ its reduced radius of inertia coincides with the radius of gravity. Here both dimensionless structural coefficients α^2 and β^2 are equal to 3/5, and the moments of gravitational and inertial forces are equilibrated and because of that the rotation of the mass is absent (Fig. 2.2a).

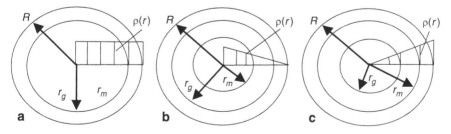

Fig. 2.2 Radius of inertia and radius of gravity for a uniform (**a**) and a non-uniform sphere with density increased to the center (**b**) and from the center (**c**)

Thus

$$\frac{r_m^2}{R^2} = \frac{r_g^2}{R^2} = \frac{3}{5},$$ (2.46)

from where

$$r_m = r_g = \sqrt{3/5R^2} = 0.7745966R.$$ (2.47)

For a non-uniform sphere at $\rho(r) \neq const$ from Eqs. (2.43)–(2.45) one has

$$0 < \frac{4\pi r_m^2}{4\pi R^2} < \frac{3}{5} < \frac{4\pi r_g^2}{4\pi R^2} < 1.$$ (2.48)

It follows from inequality (2.48) and Table 2.1 that in comparison with the uniform sphere, the reduced radius of inertia of the non-uniform body decreases and the reduced gravity radius increases (Fig. 2.2b). Because of $r_m \neq r_g$ and $r_m < 0.77R < r_g$ the torque appears as a result of an imbalance between gravitational and inertial volumetric forces of the shells. Then from Eq. (2.48) it follows that

$$r_m = r_{mo} - \delta r_{mt} \quad \text{И} \quad r_g = r_{go} + \delta r_{gt},$$ (2.49)

where subscripts o and t relate to the uniform and non-uniform sphere.

In accordance with (2.48) and (2.49) rotation of shells of a one-dimensional body should be hinged-like and asynchronous. In the case of increasing mass density towards the body surface, then the signs in (2.48) and (2.49) are reversed (Fig. 2.2c). This remark is important because the direction of rotation of a self-gravitating body is a function of its mass density distribution.

The main conclusion from the above consideration is that the inner force field of a self-gravitating body is reduced to a closed envelope (spheroid, ellipsoid or more complicated curve) of gravitational pressure, but not to a resulting force passing through the geometric center of the masses. In the case of a uniform body the envelopes have a spherical shape and both gravitational and inertial radii coincide. For a non-uniform body the radius of inertia does not coincide with the radius of gravity, the reduced envelope is closed but has non-spherical (ellipsoidal or any other) shape. Analytical solutions done below justify the above.

So, we accept the force pressure as an effect of mass interaction which is the property producing work in the form of motion. In other words, the pressure of interacting masses appears to be the force function or the flux of the potential energy.

Now we pass to derivation of the equation of dynamical equilibrium (Jacobi virial equation) for the well-known physical models of natural systems where it is correct for description of those systems. The only restriction here is the requirement of uniformity of the potential energy function of the system relative to the frame of reference. But that requirement appears to be not always obligatory. A specific physical model which is used for description of the system's dynamic in classical mechanics, hydrodynamics, statistic mechanics, quantum mechanics, theory of relativity in that case will not be an important factor.

Chapter 3
Fundamentals of the Theory of Dynamic Equilibrium

3.1 The Generalized Virial Theorem As the Equation of Dynamic Equilibrium of the Earth's Oscillating Motion

We have defined the classical virial theorem for a system moving in the outer uniform force field, which determines the relationship between mean values of the potential and kinetic energy within a certain period of time to be the averaged virial theorem. Conversely, the virial theorem for a system moving in its own force field and establishing a relationship between the potential and kinetic energy of the oscillating polar moment of inertia, is defined as the generalized (non-averaged) virial theorem or the equation of dynamical equilibrium of a system.

We come to the conclusion that a physical basis of hydrostatic equilibrium does not satisfy the demands of geophysics and geodesy in obtaining geodynamic parameters of a planet's motion. As it was shown in the previous chapter, hydrostatic equilibrium, expressed by the averaged virial theorem, does not take into account kinetic energy of the interacting mass particles of a self-gravitating Earth and does not provide fundamentals for study of its dynamics. A dynamic equilibrium state based on the generalized virial theorem is however ensured by study of the following physical and dynamical planetary problems:

- Earth's shell oscillation and rotation;
- Interpretation of satellite data with respect to the Earth's precession, nutation and pole wobbling, non-tidal variation in the planet's angular velocity, geopotential, sea level changes etc.;
- The Sun and the Moon perturbation effects based on analysis of the dynamical equilibrium of the interacting outer force fields of the Earth, the Sun and the Moon;
- Relationship between the gravitational and electromagnetic interaction of the mass particles and the nature of the gravity forces;
- Other dynamical effects arising from action of the inner force field, which earlier was not taken into account.

V. I. Ferronsky, S. V. Ferronsky, *Dynamics of the Earth,*
DOI 10.1007/978-90-481-8723-2_3, © Springer Science+Business Media B.V. 2010

The theory presented in this book is based on solution of the equation of dynamical equilibrium. The Earth by its structure presents a system that includes gaseous, liquid and solid shells. The planet's body itself is represented by a solid crust and hypothetically viscous-elastic shells of the mantle and the core, whose matter and phase state are still not fully understood. Because of that, in this chapter we present a derivation of the equation of dynamical equilibrium for possible physical models of the Earth's shells. In particular, derivation of the equation of dynamical equilibrium from the equations of Newton, Euler, Hamilton, Einstein and also from the equations of quantum mechanics is presented. In this part of the work we justify physical applicability of the above fundamental equation for study of the dynamics and structure of the Earth's shells. The main idea of derivation by introduction of volumetric forces and moments into the transformed equations, as was done in Eqs. (2.27) and (2.28), is to show that the effect of matter interaction in nature is unique, namely, the motion by energy.

3.1.1 The Averaged Virial Theorem

Derivation of the theorem and a general approach to derive the equation of dynamical equilibrium was discussed in Sect. 2.5 of the previous chapter. Here we present only some supplementary comments.

Rewrite the averaged equation of virial theorem (2.25)

$$-\overline{\sum_i F_i \cdot r_i} = 2T. \tag{3.1}$$

That equation was used first of all in the kinetic theory of gases for derivation of an equation of state for perfect gases in the outer force field of the Earth. we assume that the a specific perfect gas is found in a vessel of volume V and consists of N uniform particles (atoms or molecules). The mean kinetic energy of a particle of that gas at temperature T_0 is equal to $3kT_0/2$, where k is Boltzmann's constant. Then the virial theorem (3.1) is written in the form

$$-\frac{1}{2}\overline{\sum_i F_i \cdot r_i} = \frac{3}{2}NkT_0. \tag{3.2}$$

In this case the effect of interaction of gas atoms and molecules among themselves is negligibly small and all the gas energy is realized by its interaction with the vessel's wall. The gas pressure p inside the vessel appears only because of the walls, elastic reaction of which plays the role of inertial forces. The pressure is expressed through the energy of the motion by molecules and atoms in the vessel, and expression (3.2) is written as the Clapeyron–Mendeleev equation of state for a perfect gas in the form

$$\frac{3}{2}pV = \frac{3}{2}NkT_0,$$

or

$$pV = NkT_0. \tag{3.3}$$

Equation (3.3) is the generalized expression of the laws of Boyle and Mariotte, Gay-Lussac and Avogadro and represents the averaged virial theorem. Its left-hand side represents the potential energy of interaction of the gas particles and the right-hand side is the kinetic energy of the gas pressure on the walls. In astrophysics this equation is used as the equation of hydrostatic equilibrium state of a star, which is accepted as a gas and plasma system, where the gas pressure is equilibrated by the gravity forces of the attracted masses. In this case the gravity forces play the role of the vessel's walls or the outer force field, where the kinetic energy of motion of the interacting particles is not taken into account. Later on it will be shown, that for the natural gaseous and plasma self-gravitating systems only the generalized virial equation (2.31) can be used as the equation of state.

For celestial bodies including the Earth, other planets and satellites, whose mass particles interact by the reverse square law and the forces of interaction are characterized by the potential $U(r)$ as the uniform function of co-ordinates, the averaged virial theorem (3.1) is reduced to a relation between the potential and kinetic energy in the form (Goldstein, 1980)

$$T = \frac{1}{2}\sum_i \overline{\nabla U \cdot r_i}. \tag{3.4}$$

For a particle moving in the central force field expression (3.4) is

$$T = \frac{1}{2}\overline{\frac{\partial U}{dr}r}. \tag{3.5}$$

If U is the force function of r^n, then

$$T = \frac{n+1}{2}\overline{U}.$$

Or, taking into account Euler's theorem about uniform functions and Newton's law of interaction, when $n = -2$, one has

$$T = -\frac{1}{2}\overline{U}. \tag{3.6}$$

Relationship (3.6) is valid only for a system that is found in the outer uniform force field. As it follows from (3.1), relation (3.6) expresses only mean values of the potential and kinetic energy per the period τ without effect of the inner kinetic energy of the interacting particles.

For a uniform sphere in an outer uniform force field ρF at inner isotropic pressure p, relation (3.6) represents the condition of hydrostatic equilibrium written by means of Euler's equation in the form

$$\frac{\partial p}{\partial r} = \rho F_r.$$

Here the left-hand side of the equation is the potential energy and the right-hand side represents the kinetic energy of the sphere in the frame-work of the averaged virial theorem.

3.1.2 The Generalized Virial Theorem

As it was shown earlier, the Earth is accepted to be an inert body deprived of kinetic energy and therefore remains without physical basis to study its dynamics. The cause of this was the unsolved problem of the nature and mechanism of energy generation by interacting masses of a self-gravitating body. It was shown in Sect. 2.6 of the previous chapter, that the interacting volumetric mass particles of a body create a volumetric field of pressure, which is reduced to a resultant envelope representing the inner force field. Kinetic effects of the force field are expressed by the generalized virial theorem or equation of dynamical equilibrium (2.31), which in fact is a fundamental basis for study of the planet's dynamics under action of the inner field of pressure.

Karl Jacobi derived equation (2.31) in 1884 while considering the problem of interaction of n mass points by Newton's law and obtained it in the form of a generalized virial equation:

$$\ddot{\Phi} = U + 2T \tag{3.7}$$

where Φ is the Jacobi function of a system (polar moment of inertia); U and T are its potential and kinetic energy.

In the Russian literature this equation is called the equation of Lagrange–Jacobi, because for its derivation Jacobi used Lagrange's method and his identity (Duboshin, 1975). We will refer to it as the equation of dynamical equilibrium or Jacobi's virial equation. Full-length derivation of the equation from the classical equations of motion, which creates a theoretical basis for our dynamics, is done below. Note that the physical meaning of the derivations below consists in transformation of vector forces and moments of the classical equations into their volumetric scalar values. As a result, we unify in dynamics the existing natural system models and resolve the most complicated problem of the reference system transformation.

3.2   Derivation of Jacobi's Virial Equation from Newtonian Equations of Motion

Throughout this section the term 'system' is defined as an ensemble of material mass points $m_i(i = 1, 2, 3,..., n)$ which interact by Newton's law of universal attraction. This physical model of a natural system forms the basis for a number of branches of physics, such as classical mechanics, celestial mechanics, and stellar dynamics.

We shall not present the traditional introduction in which the main postulates are formulated; we shall simply state the problem (see, for example, Landau and Lifshitz, 1973a). We start by writing the equations of motion of the system in some absolute Cartesian co-ordinates ξ, η, ζ. In accordance with the conditions imposed, the mass point m_i is not affected by any force from the other $n-1$ points except that of gravitational attraction. The projections of this force on the axes of the selected co-ordinates ξ, η, ζ can be written (Fig. 3.1):

$$
\Xi_i = Gm_i \sum_{1 \leq j \leq n, i \neq j} \frac{m_j(\xi_j - \xi_i)}{\Delta_{ij}^3},
$$

$$
H_i = Gm_i \sum_{1 \leq j \leq n, i \neq j} \frac{m_j(\eta_j - \eta_i)}{\Delta_{ij}^3}, \tag{3.8}
$$

$$
Z_i = Gm_i \sum_{1 \leq j \leq n, i \neq j} \frac{m_j(\zeta_j - \zeta_i)}{\Delta_{ij}^3},
$$

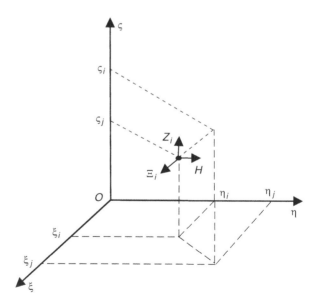

Fig. 3.1 Absolute Cartesian co-ordinate system $O\xi\,\eta\,\zeta$

where G is the gravitational constant and

$$\Delta_{ji} = \sqrt{(\xi_j - \xi_i)^2 (\eta_j - \eta_i)^2 + (\zeta_j - \zeta_i)^2}$$

is the reciprocal distance between points i and j of the system.

It is easy to check that the forces affect the i-th material point of the system and are determined by the scalar function U, which is called the potential energy function of the system, and is given by

$$U = -G \sum_{1 \le i < j \le n} \frac{m_i m_j}{\Delta_{ij}}. \tag{3.9}$$

Now Eqs. (3.8) can be rewritten in the form

$$\Xi_i = -\frac{\partial U}{\partial \xi_i},$$

$$H_i = -\frac{\partial U}{\partial \eta_i},$$

$$Z_i = -\frac{\partial U}{\partial \zeta_i}.$$

Then Newton's equations of motion for the i-th point of the system take the form

$$\begin{aligned} m_i \ddot{\xi}_i &= \Xi_i, \\ m_i \ddot{\eta}_i &= H_i, \\ m_i \ddot{\zeta}_i &= Z_i. \end{aligned} \tag{3.10}$$

or

$$\begin{aligned} m_i \ddot{\xi}_i &= -\frac{\partial U}{\partial \xi_i}, \\ m_i \ddot{\eta}_i &= -\frac{\partial U}{\partial \eta_i}, \\ m_i \ddot{\zeta}_i &= -\frac{\partial U}{\partial \zeta_i}, \end{aligned} \tag{3.11}$$

where dots over co-ordinate symbols mean derivatives with respect to time.

The motion of a system is described by Eqs. (3.10) and (3.11) and is completely determined by the initial data. In classical mechanics, the values of projections ξ_{i0}, η_{i0}, ζ_{i0} and velocities $\dot{\xi}_{i0}$, $\dot{\eta}_{i0}$, $\dot{\zeta}_{i0}$ at the initial moment of time $t = t_0$ may be known from the initial data.

The study of motion of a system of n material points affected by self-forces of attraction forms the essence of the classical many-body problem. In the general case, ten classical integrals of motion are known for such a system, and they are obtained directly from the equations of motion.

Summing all the equations (3.10) for each co-ordinate separately, it is easy to be convinced of the correctness of the expressions:

$$\sum_{1\leq i\leq n} \Xi_i = 0,$$

$$\sum_{1\leq i\leq n} H_i = 0,$$

$$\sum_{1\leq i\leq n} Z_i = 0.$$

From those equations it follows that

$$\sum_{1\leq i\leq n} m_i\ddot{\xi}_i = 0,$$

$$\sum_{1\leq i\leq n} m_i\ddot{\eta}_i = 0, \tag{3.12}$$

$$\sum_{1\leq i\leq n} m_i\ddot{\zeta}_i = 0.$$

Equations (3.12), appearing as a sequence of equations of motion, can be successively integrated twice. As a result, the first six integrals of motion are obtained:

$$\sum_{1\leq i\leq n} m_i\dot{\xi}_i = a_1,$$

$$\sum_{1\leq i\leq n} m_i\dot{\eta}_i = a_2,$$

$$\sum_{1\leq i\leq n} m_i\dot{\zeta}_i = a_3,$$

$$\sum_{1\leq i\leq n} m_i\left(\xi_i - \dot{\xi}_i t\right) = b_1, \tag{3.13}$$

$$\sum_{1\leq i\leq n} m_i(\eta_i - \dot{\eta}_i t) = b_2,$$

$$\sum_{1\leq i\leq n} m_i\left(\zeta_i - \dot{\zeta}_i t\right) = b_3,$$

where $a_1, a_2, a_3, b_1, b_2, b_3$ are integration constants.

These integrals are called integrals of motion of the center of mass. The integration constants $a_1, a_2, a_3, b_1, b_2, b_3$ can be determined from the initial data by substituting their values at the initial moment of time for the values of all the co-ordinates and velocities.

Let us obtain one more group of first integrals. To do this, the second of Eqs. (3.10) can be multiplied by $-\zeta_i$, and the third by η_i. Then all expressions obtained should be added and summed over the index i. In the same way, the first of Eqs. (3.10) should be multiplied by ζ_i, and the third by $-\xi_i$ added and summed over index I. Finally, the second of Eqs. (3.10) should be multiplied by ξ_i, and the first by $-\eta_i$ added and summed over index i. It is easy to show directly that the right-hand sides of the expressions obtained are equal to zero:

$$\sum_{1 \leq i \leq n} (Z_i \eta_i - H_i \zeta_i) = 0,$$

$$\sum_{1 \leq i \leq n} (\Xi_i \zeta_i - Z_i \xi_i) = 0,$$

$$\sum_{1 \leq i \leq n} (H_i \xi_i - \Xi_i \eta_i) = 0.$$

Consequently their left-hand sides are also equal to zero:

$$\sum_{1 \leq i \leq n} m_i (\ddot{\zeta}_i \eta_i - \ddot{\eta}_i \zeta_i) = 0,$$

$$\sum_{1 \leq i \leq n} m_i (\ddot{\xi}_i \zeta_i - \ddot{\zeta}_i \xi_i) = 0, \tag{3.14}$$

$$\sum_{1 \leq i \leq n} m_i (\ddot{\eta}_i \xi_i - \ddot{\xi}_i \eta_i) = 0.$$

Integrating Eqs. (3.14) over time, three more first integrals can be obtained:

$$\sum_{1 \leq i \leq n} m_i (\dot{\zeta}_i \eta_i - \dot{\eta}_i \zeta_i) = c_1,$$

$$\sum_{1 \leq i \leq n} m_i (\dot{\xi}_i \zeta_i - \dot{\zeta}_i \xi_i) = c_2, \tag{3.15}$$

$$\sum_{1 \leq i \leq n} m_i (\dot{\eta}_i \xi_i - \dot{\xi}_i \eta_i) = c_3.$$

The integrals (3.15) are called area integrals or integrals of moments of momentum. Three integration constants c_1, c_2, c_3 are also determined from the initial data by changing over from the values of all the co-ordinates and velocities to their values at the initial moment of time.

The last of the classical integrals can be obtained by multiplying the three Eqs. (3.11) by $\dot{\xi}_i$, $\dot{\eta}_i$ and $\dot{\zeta}_i$ respectively, and adding and summing all the expressions obtained. As a result, the following equation is obtained:

$$\sum_{1 \leq i \leq n} m_i (\ddot{\xi}_i \dot{\xi}_i + \ddot{\eta}_i \dot{\eta}_i + \ddot{\zeta}_i \dot{\zeta}_i) = -\sum_{1 \leq i \leq n} \left(\frac{\partial U}{\partial \xi_i} \dot{\xi}_i + \frac{\partial U}{\partial \eta_i} \dot{\eta}_i + \frac{\partial U}{\partial \zeta_i} \dot{\zeta}_i \right). \tag{3.16}$$

It is not difficult to see that the right-hand side of Eq. (3.16) is the complete differential over time of the potential energy function U of the system as a whole. The left-hand side of the same equation is also the complete differential of some function T called the kinetic energy function of the system, and equal to

$$T = \frac{1}{2} \sum_{1 \le i \le n} m_i \left(\dot{\xi}_i^2 + \dot{\eta}_i^2 + \dot{\zeta}_i^2 \right). \tag{3.17}$$

Equation (3.16) can then be written finally in the form

$$\frac{d}{dt}(T) = -\frac{d}{dt}(U),$$

from which, after integration, one finds that

$$E = T + U, \tag{3.18}$$

where E is the integration constant, determined from the initial conditions.

Equation (3.18) is called the integral of motion or the integral of living (kinetic) forces.

To derive the equation of dynamic equilibrium, or Jacobi's virial equation, each of the equations (3.11) should be multiplied by ξ_i, η_i and ζ_i respectively; then, after summing all the expressions, one can obtain

$$\sum_{1 \le i \le n} m_i \left(\xi_i \ddot{\xi}_i + \eta_i \ddot{\eta}_i + \zeta_i \ddot{\zeta}_i \right) = - \sum_{1 \le i \le n} \left(\xi_i \frac{\partial U}{\partial \xi_i} + \eta_i \frac{\partial U}{\partial \eta_i} + \zeta_i \frac{\partial U}{\partial \zeta_i} \right). \tag{3.19}$$

We can take farther advantage of the obvious identities:

$$m_i \xi_i \ddot{\xi}_i = \frac{1}{2} \frac{d^2}{dt^2} \left(m_i \xi_i^2 \right) - m_i \dot{\xi}_i^2,$$

$$m_i \eta_i \ddot{\eta}_i = \frac{1}{2} \frac{d^2}{dt^2} \left(m_i \eta_i^2 \right) - m_i \dot{\eta}_i^2,$$

$$m_i \zeta_i \ddot{\zeta}_i = \frac{1}{2} \frac{d^2}{dt^2} \left(m_i \zeta_i^2 \right) - m_i \dot{\zeta}_i^2$$

from the Eulerian theorem concerning the homogenous functions. For the interaction of the system points, according to Newton's law of universal attraction, the degree of homogeneity of the potential energy function of the system is equal to -1, and hence

$$- \sum_{1 \le i \le n} \left(\xi_i \frac{\partial U}{\partial \xi_i} + \eta_i \frac{\partial U}{\partial \eta_i} + \zeta_i \frac{\partial U}{\partial \zeta_i} \right) = U.$$

Substituting the above expressions into the right- and left-hand side of Eq. (3.19), one obtains

$$\frac{d^2}{dt^2} \left[\frac{1}{2} \sum_{1 \le i \le n} m_i \left(\xi_i^2 + \eta_i^2 + \zeta_i^2 \right) \right] - 2 \sum_{1 \le i \le n} \frac{1}{2} m_i \left(\dot{\xi}_i^2 + \dot{\eta}_i^2 + \dot{\zeta}_i^2 \right) = U.$$

For a system of material points we now introduce the Jacobi function expressed through the moment of inertia of the system and presented in the form

$$\Phi = \frac{1}{2} \sum_{1 \le i \le n} m_i \left(\xi_i^2 + \eta_i^2 + \zeta_i^2 \right).$$

Then taking into account (3.8), the previous equation can be rewritten in a very simple form as follows:

$$\ddot{\Phi} = 2E - U. \tag{3.20}$$

This is the equation of dynamic equilibrium or Jacobi's virial equation describing both the dynamics of a system and its dynamic equilibrium using integral (volumetric) characteristics Φ and U or T.

Let us derive now another form of Jacobi's virial equation where the translational moment of the center of mass of the system is separated and all the characteristics depend only on the relative distance between the mass points of the system. For this purpose the Lagrangian identity can be used:

$$\left(\sum_{1 \le i \le n} a_i^2 \right) \left(\sum_{1 \le i \le n} b_i^2 \right) = \left(\sum_{1 \le i \le n} a_i b_i \right)^2 + \frac{1}{2} \sum_{1 \le i \le n} \sum_{1 \le j \le n} (a_i b_j - b_i a_j)^2, \tag{3.21}$$

where a_i and b_i may acquire any values and n is any positive number.

Let us now put $a_i = \sqrt{m_i}$, and b_i equal to $\sqrt{m_i}\xi_i$, $\sqrt{m_i}\eta_i$ and $\sqrt{m_i}\zeta_i$ respectively. Then three identities can be obtained from (3.21):

$$\left(\sum_{1 \le i \le n} m_i \right) \left(\sum_{1 \le i \le n} m_i \xi_i^2 \right) = \left(\sum_{1 \le i \le n} m_i \xi_i \right)^2 + \frac{1}{2} \sum_{1 \le i \le n} \sum_{1 \le j \le n} m_i m_j (\xi_j - \xi_i)^2,$$

$$\left(\sum_{1 \le i \le n} m_i \right) \left(\sum_{1 \le i \le n} m_i \eta_i^2 \right) = \left(\sum_{1 \le i \le n} m_i \eta_i \right)^2 + \frac{1}{2} \sum_{1 \le i \le n} \sum_{1 \le j \le n} m_i m_j (\eta_j - \eta_i)^2,$$

$$\left(\sum_{1 \le i \le n} m_i \right) \left(\sum_{1 \le i \le n} m_i \zeta_i^2 \right) = \left(\sum_{1 \le i \le n} m_i \zeta_i \right)^2 + \frac{1}{2} \sum_{1 \le i \le n} \sum_{1 \le j \le n} m_i m_j (\zeta_j - \zeta_i)^2.$$

In summing up one finds

$$2m\Phi = \left(\sum_{1 \le i \le n} m_i \xi_i \right)^2 + \left(\sum_{1 \le i \le n} m_i \eta_i \right)^2 + \left(\sum_{1 \le i \le n} m_i \zeta_i \right)^2 + \frac{1}{2} \sum_{1 \le i \le n} \sum_{1 \le j \le n} m_i m_j \Delta_{ij}^2.$$

Using now Eqs. (3.13), the last equality can be rewritten in the form

$$2m\Phi = \frac{1}{2} \sum_{1 \le i \le n} \sum_{1 \le j \le n} m_i m_j \Delta_{ij}^2 + (a_1 t + b_1)^2 + (a_2 t + b_2)^2 + (a_3 t + b_3)^2, \quad (3.22)$$

where

$$m = \sum_{1 \le i \le n} m_i$$

is the total mass of the system.

Let us put

$$\Phi_0 = \frac{1}{4m} \sum_{1 \le i \le n} \sum_{1 \le j \le n} m_i m_j \Delta_{ij}^2.$$

The value Φ_0 does not depend on the choice of the co-ordinate system and coincides with the value of the Jacobi function in the barycentric co-ordinate system. Moreover, from Eq. (3.22) it follows that

$$\ddot{\Phi} = \ddot{\Phi}_0 + \frac{a_1^2 + a_2^2 + a_3^2}{m}.$$

Excluding the value Φ from Jacobi's equation (3.20) with the help of the last equality, the same equation can be obtained in the barycentric co-ordinate system:

$$\ddot{\Phi}_0 = 2E_0 - U, \quad (3.23)$$

where $E_0 = T_0 + U_0$ is the total energy of the system in the barycentric co-ordinate system equal to

$$E_0 = E - \frac{a_1^2 + a_2^2 + a_3^2}{2m}.$$

We can now show that the value of E_0 does not depend on the choice of the co-ordinate system. For this purpose we can again use the Lagrangian identity (3.21). In this case $a_i = \sqrt{m_i}$, and $b_i = \sqrt{m_i}\dot{\xi}_i$, $\sqrt{m_i}\dot{\eta}_i$ and $\sqrt{m_i}\dot{\zeta}_i$. Then the following three identities can be justified:

$$\left(\sum_{1 \le i \le n} m_i\right)\left(\sum_{1 \le i \le n} m_i \dot{\xi}_i^2\right) = \left(\sum_{1 \le i \le n} m_i \dot{\xi}_i\right)^2 + \frac{1}{2} \sum_{1 \le i \le n} \sum_{1 \le j \le n} m_i m_j (\dot{\xi}_j - \dot{\xi}_i)^2,$$

$$\left(\sum_{1 \le i \le n} m_i\right)\left(\sum_{1 \le i \le n} m_i \dot{\eta}_i^2\right) = \left(\sum_{1 \le i \le n} m_i \dot{\eta}_i\right)^2 + \frac{1}{2} \sum_{1 \le i \le n} \sum_{1 \le j \le n} m_i m_j (\dot{\eta}_j - \dot{\eta}_i)^2,$$

$$\left(\sum_{1 \leq i \leq n} m_i \right) \left(\sum_{1 \leq i \leq n} m_i \dot{\zeta}_i^2 \right) = \left(\sum_{1 \leq i \leq n} m_i \dot{\zeta}_i \right)^2 + \frac{1}{2} \sum_{1 \leq i \leq n} \sum_{1 \leq j \leq n} m_i m_j \left(\dot{\zeta}_j - \dot{\zeta}_i \right)^2.$$

After summing and using (3.13) one obtains

$$2mT = \left(a_1^2 + a_2^2 + a_3^2 \right) + \frac{1}{2} \sum_{1 \leq i \leq n} \sum_{1 \leq j \leq n} m_i m_j \left[\left(\dot{\xi}_i - \dot{\xi}_j \right)^2 + \left(\dot{\eta}_i - \dot{\eta}_j \right)^2 + \left(\dot{\zeta}_i - \dot{\zeta}_j \right)^2 \right]$$

or

$$T = \frac{\left(a_1^2 + a_2^2 + a_3^2 \right)}{2m}$$
$$+ \frac{1}{2m} \left\{ \frac{1}{2} \sum_{1 \leq i \leq n} \sum_{1 \leq j \leq n} m_i m_j \left[\left(\dot{\xi}_i - \dot{\xi}_j \right)^2 + \left(\dot{\eta}_i - \dot{\eta}_j \right)^2 + \left(\dot{\zeta}_i - \dot{\zeta}_j \right)^2 \right] \right\}. \quad (3.24)$$

Here the second term on the right-hand side of Eq. (3.24) coincides with the expression for the kinetic energy T_0 of a system.

Substituting (3.24) into an expression for E_0, one obtains

$$E_0 = T_0 + U = \frac{1}{2m} \sum_{1 \leq i < j \leq n} m_i m_j \left[\left(\dot{\xi}_i - \dot{\xi}_j \right)^2 + \left(\dot{\eta}_i - \dot{\eta}_j \right)^2 + \left(\dot{\zeta}_i - \dot{\zeta}_j \right)^2 \right]$$
$$- G \sum_{1 \leq i < j \leq n} \frac{m_i m_j}{\Delta_{ij}}. \quad (3.25)$$

Thus, the total energy of the system E_0 depends only on the distance between the points of the system and on the velocity changes of these distances. But Jacobi's equation (3.23) appears to be invariant with respect to the choice of the co-ordinate system.

We can show now that the requirement of homogeneity of the potential energy function for deriving Jacobi's virial equation is not always obligatory. For this purpose we consider two examples.

3.3   Derivation of a Generalized Jacobi's Virial Equation for Dissipative Systems

Let us derive Jacobi's virial equation for a non-conservative system. We consider a system of n material points, the motion of which is determined by the force of their mutual gravitation interaction and the friction force. It is well known that the friction force always appears in the course of evolution of any natural system. It is also known that there is no universal law describing the friction force (Bogolubov and Mitropolsky, 1974). The only general statement is that the friction force acts in the direction opposite to the vector of velocity of a considered mass point.

Consider as an example the simplest law of Newtonian friction when its force is proportional to the velocity of motion of the mass:

$$\begin{aligned}
\Xi_f &= -km_i\dot{\xi}_i, \\
H_f &= -km_i\dot{\eta}_i, \\
Z_f &= -km_i\dot{\zeta}_i,
\end{aligned} \tag{3.26}$$

where $\dot{\xi}_i$, $\dot{\eta}_i$, $\dot{\zeta}_i$ are the components of the radius-vector of the velocity of the i-th mass point in the barycentric co-ordinate system; k is a constant independent of i; $k > 0$.

Sometimes the friction force is independent of the velocity of the mass point. There are also some other laws describing the friction force.

We derive the equation of dynamical equilibrium for a system of n material points using the equations of motion (3.11) and taking into account the friction force expressed by Eqs. (3.26):

$$\begin{aligned}
m_i\ddot{\xi}_i &= -\frac{\partial U}{\partial \xi_i} - km_i\dot{\xi}_i, \\
m_i\ddot{\eta}_i &= -\frac{\partial U}{\partial \eta_i} - km_i\dot{\eta}_i, \\
m_i\ddot{\zeta}_i &= -\frac{\partial U}{\partial \zeta_i} - km_i\dot{\zeta}_i,
\end{aligned} \tag{3.27}$$

where the value of the system's potential energy is determined by Eq. (3.9).

Multiplying each of Eqs. (3.27) by ξ_i, η_i and ζ_i, respectively, and summing through all i, one obtains

$$\sum_{1 \le i \le n} m_i\left(\xi_i\ddot{\xi}_i + \eta_i\ddot{\eta}_i + \zeta_i\ddot{\zeta}_i\right) = -\sum_{1 \le i \le n}\left(\frac{\partial U}{\partial \xi_i}\xi_i + \frac{\partial U}{\partial \eta_i}\eta_i + \frac{\partial U}{\partial \zeta_i}\zeta_i\right)$$
$$-\sum_{1 \le i \le n} km_i\left(\xi_i\dot{\xi}_i + \eta_i\dot{\eta}_i + \zeta_i\dot{\zeta}_i\right). \tag{3.28}$$

Transforming the right- and left-hand sides of Eq. (3.28) in the same way as in deriving Eq. (3.20), one obtains

$$\ddot{\Phi} - 2T = U - k\dot{\Phi}$$

or

$$\ddot{\Phi} = 2E - U - k\dot{\Phi}. \tag{3.29}$$

Let us show that the total energy E of the system is a monotonically decreasing function of time. For this purpose we multiply each of the equations (3.27) by the vectors $\dot{\xi}_i,\ \dot{\eta}_i,\ \dot{\zeta}_i$ respectively, and sum over all from 1 to n, which results in

$$\sum_{1\leq i\leq n} m_i\left(\ddot{\xi}_i\dot{\xi}_i + \ddot{\eta}_i\dot{\eta}_i + \ddot{\zeta}_i\dot{\zeta}_i\right) = -\sum_{1\leq i\leq n}\left(\frac{\partial U}{\partial\xi_i}\dot{\xi}_i + \frac{\partial U}{\partial\eta_i}\dot{\eta}_i + \frac{\partial U}{\partial\zeta_i}\dot{\zeta}_i\right)$$
$$- k\sum_{1\leq i\leq n} m_i\left(\dot{\xi}_i^2 + \dot{\eta}_i^2 + \dot{\zeta}_i^2\right).$$

The last expression can be rewritten in the form

$$\frac{d}{dt}(T) = -\frac{d}{dt}(U) - 2kT$$

or

$$dE = -2kTdt. \tag{3.30}$$

Since the kinetic energy T of the system is always greater than zero, $dE \leq 0$, i.e. the total energy of a gravitating system is a monotonically decreasing function of time. Thus the expression for the total energy $E(t)$ of the system can be written as

$$E(t) = E_0 - 2k\int_{t_0}^{t} T(t)dt = E_0[1 + q(t)],$$

where $q(t)$ is a monotonically increasing function of time.

Finally, the equation of dynamical equilibrium for a non-conservative system takes the form

$$\ddot{\Phi} = 2E_0[1 + q(t)] - U - k\dot{\Phi}. \tag{3.31}$$

The second example where the requirement of homogeneity of the potential energy function for deriving Jacobi's virial equation is not obligatory is as follows. We derive Jacobi's virial equation for a system whose mass points interact mutually in accordance with Newton's law and move without friction in a spherical homogenous cloud whose density ρ_0 is constant in time. Let, also, the geometric center of the cloud coincide with the center of mass of the considered system. The equations of motion for such a system can be written in the form:

$$m_i\frac{d^2\xi_i}{dt^2} = -\frac{4}{3}\pi G\rho_0 m_i\xi_i - \frac{\partial U}{\partial\xi_i},$$
$$m_i\frac{d^2\eta_i}{dt^2} = -\frac{4}{3}\pi G\rho_0 m_i\eta_i - \frac{\partial U}{\partial\eta_i}, \tag{3.32}$$
$$m_i\frac{d^2\zeta_i}{dt^2} = -\frac{4}{3}\pi G\rho_0 m_i\zeta_i - \frac{\partial U}{\partial\zeta_i},$$

where $i = 1, 2,\ldots, n$.

It is obvious that the above system of equations possesses the ten first integrals of motion and that Jacobi's virial equation, written in the form

$$\frac{d^2\Phi}{dt^2} = 2E - U - \frac{8}{3}\pi G\rho_0\Phi. \tag{3.33}$$

is valid for it.

The equation in the form (3.33) was first obtained by Duboshin et al. (1971). Equations (3.31) and (3.33) can be written in a more general form:

$$\ddot{\Phi} = 2E - U + X(t, \Phi, \dot{\Phi}), \tag{3.34}$$

where $X(t, \Phi, \dot{\Phi})$ is a given function of time t, the Jacobi function Φ and first derivative $\dot{\Phi}$. Moreover, we can call Eq. (3.34) a generalized equation of dynamical equilibrium.

The examples considered above justify the statement that for conditions of homogeneity of the potential energy function, required for the derivation of Jacobi's virial equation, is not always necessary. This condition is required for description of dynamics of conservative systems but not for dissipative systems or for systems in which motion is restricted by some other conditions.

3.4 Derivation of Jacobi's Virial Equation from Eulerian Equations

We now derive Jacobi's virial equation by transforming of the hydrodynamic or continuum model of a physical system. As is well known, the hydrodynamic approach to solving problems of dynamics is based on the system of differential equations of motion supplement, in the simplest case, by the equations of state and continuity, and by the appropriate assumptions concerning boundary conditions and perturbations affecting the system.

In this section, we understand by the term 'system' some given mass M of ideal gas localized in space by a finite volume V and restricted by a closed surface S. Let the gas in the system move by the forces of mutual gravitational interaction and of baric gradient. In addition, we accept the pressure within the volume to be isotropic and equal to zero on the surface S bordering the volume V. Then for a system in some Cartesian inertial co-ordinate system ξ, η, ζ, the Eulerian equations can be written in the form

$$\begin{aligned}
\rho\frac{\partial u}{\partial t} + \rho u\frac{\partial}{\partial \xi}u + \rho v\frac{\partial}{\partial \eta}u + \rho w\frac{\partial}{\partial \zeta}u &= -\frac{\partial p}{\partial \xi} + \rho\frac{\partial U_G}{\partial \xi}, \\
\rho\frac{\partial v}{\partial t} + \rho u\frac{\partial}{\partial \xi}v + \rho v\frac{\partial}{\partial \eta}v + \rho w\frac{\partial}{\partial \zeta}v &= -\frac{\partial p}{\partial \eta} + \rho\frac{\partial U_G}{\partial \eta}, \\
\rho\frac{\partial w}{\partial t} + \rho u\frac{\partial}{\partial \xi}w + \rho v\frac{\partial}{\partial \eta}w + \rho w\frac{\partial}{\partial \zeta}w &= -\frac{\partial p}{\partial \zeta} + \rho\frac{\partial U_G}{\partial \zeta},
\end{aligned} \tag{3.35}$$

where $\rho(\xi, \eta, \zeta, t)$ is the gas density; u, v, w are components of the velocity vector $\bar{v}$ (ξ, η, ζ, t) in a given point of space; $p(\xi, \eta, \zeta, t)$ is the gas pressure; U_G is Newton's potential in a given point of space.

The value U_G is given by

$$U_G = G \int_{(V)} \frac{\rho(x, y, z, t)}{\Delta} dx dy dz, \tag{3.36}$$

where G is the gravity constant; $\Delta = \sqrt{(x - \xi)^2 + (y - \eta)^2 + (z - \zeta)^2}$ is the distance between system points.

The potential energy of the gravitational interaction of material points of the system is linked to the Newtonian potential (3.36) by the relation

$$U = -\frac{1}{2} \int_{(V)} U_G \rho(\xi, \eta, \zeta, t) d\xi d\eta d\zeta.$$

To supplement the system of equations of motion we write the equation of continuity:

$$\frac{\partial p}{\partial t} + \frac{\partial}{\partial \xi} (\rho u) + \frac{\partial}{\partial \eta} (\rho v) + \frac{\partial}{\partial \zeta} (\rho w) = 0 \tag{3.37}$$

and the equation of state

$$p = f(\rho) \tag{3.38}$$

assuming at the same time that the processes occurring in the system are barotropic.

Let us obtain the ten classical integrals for the system whose motion is described by Eqs. (3.35).

We derive the integrals of the motion of the center of mass by integrating each of the equations (3.35) with respect to all the volume filled by the system. Integrating the first equation, we obtain

$$\int_{(V)} \rho \frac{du}{dt} d\xi d\eta d\zeta + \int_{(V)} \rho \left(u \frac{du}{d\xi} + v \frac{du}{d\eta} + w \frac{du}{d\zeta} \right) d\xi d\eta d\zeta = - \int_{(V)} \frac{dp}{d\xi} d\xi d\eta d\zeta$$

$$+ G \int_{(V)} \rho(\xi, \eta, \zeta, t) \left[\int_{(V)} \rho(x, y, z, t) \frac{x - \xi}{\Delta^3} dx dy dz \right] d\xi d\eta d\zeta. \tag{3.39}$$

The second term in the right-hand side of Eq. (3.39) disappears because of the symmetry of the integral expression with respect to x and ξ. In accordance with the Gauss–Ostrogradsky theorem the first term in the right-hand side of Eq. (3.39) terns to zero. In fact

$$\int\limits_{(V)} \frac{dp}{d\xi} d\xi\, d\eta\, d\zeta = \int\limits_{(S)} p\, d\eta\, d\zeta = 0 \tag{3.40}$$

as pressure p on the border of the considered system is equal to zero owing to the absence of outer effects.

Bearing in mind the possibility of passing to a Lagrangian co-ordinate system, and taking into account the law of the conservation of mass $dm = \rho dV = \rho_0 dV_0 = dm_0$, we get

$$\int\limits_{(V)} \rho \frac{du}{dt} d\xi\xi\, d\eta\, d + \int\limits_{(V)} \rho \left(u\frac{du}{d\xi} + v\frac{du}{d\eta} + w\frac{du}{d\zeta} \right) d\xi\xi\, d\eta$$

$$= \int\limits_{(V)} \rho \frac{du}{dt} dV = \int\limits_{(V_0)} \rho_0 \frac{du}{dt} dV_0 = \frac{d}{dt} \int\limits_{(V_0)} u\rho_0 dV_0 = \frac{d}{dt} \int\limits_{(V)} \rho u dV,$$

where V_0 and ρ_0 are the volume and the density in the initial moment of time t_0.

Finally, Eq. (3.39) can be rewritten as

$$\frac{d}{dt} \int\limits_{(V)} \rho u dV = 0. \tag{3.41}$$

Integrating (3.41) with respect to time and writing analogous expressions for two other equations of the system (3.35), we obtain the first three integrals of motion:

$$\int\limits_{(V)} \rho u dV = a_1,$$

$$\int\limits_{(V)} \rho v dV = a_2, \tag{3.42}$$

$$\int\limits_{(V)} \rho w dV = a_3.$$

Equations (3.42) represent the law of conservation of the system moments. Integration constants a_1, a_2, a_3 can be obtained from the initial conditions.

We consider the first equation of the system (3.42) using again the law of conservation of mass. Then it is obvious that

$$\int\limits_{(V)} u\rho dV = \int\limits_{(V)} \frac{d\xi}{dt}\rho dV = \int\limits_{(V_0)} \frac{d\xi}{dt}\rho_0 dV_0 = \frac{d}{dt}\int\limits_{(V_0)} \xi\rho_0 dV_0 = \frac{d}{dt}\int\limits_{(V)} \xi\rho dV = a_1. \tag{3.43}$$

Analogous expressions can be written for the two other equations (3.42). Integrating them with respect to time, we obtain integrals of motion of the center of mass of the system in the form

$$\int\limits_{(V)} \xi \rho dV = a_1 t + b_1,$$

$$\int\limits_{(V)} \eta \rho dV = a_2 t + b_2, \qquad (3.44)$$

$$\int\limits_{(V)} \zeta \rho dV = a_3 t + b_3.$$

We now derive three integrals of the moment of momentum of motion. For this purpose we multiply the second of Eqs. (3.35) by $-\zeta$, the third by η, and then sum and integrate the resulting expressions with respect to volume V occupied by the system. We obtain

$$\int\limits_{(V)} \rho \left(\eta \frac{dw}{dt} - \zeta \frac{dv}{dt} \right) dV = -\int\limits_{(V)} \left(\eta \frac{\partial p}{\partial \zeta} - \zeta \frac{\partial p}{\partial \eta} \right) dV$$

$$+ \int\limits_{(V)} \rho \left(\eta \frac{\partial U_G}{\partial \zeta} - \zeta \frac{\partial U_G}{\partial \eta} \right) dV. \qquad (3.45)$$

Analogously, multiplying the first of Eqs. (3.35) by ζ, the third by $-\xi$, then summing and integrating with respect to volume V, we obtain

$$\int\limits_{(V)} \rho \left(\zeta \frac{du}{dt} - \xi \frac{dw}{dt} \right) dV = -\int\limits_{(V)} \left(\zeta \frac{\partial p}{\partial \xi} - \xi \frac{\partial p}{\partial \zeta} \right) dV$$

$$+ \int\limits_{(V)} \rho \left(\zeta \frac{\partial U_G}{\partial \xi} - \xi \frac{\partial U_G}{\partial \zeta} \right) dV. \qquad (3.46)$$

Multiplying the second of Eqs. (3.35) by ξ, the first by $-\eta$, and summing and integrating as above, the third equality can be written

$$\int\limits_{(V)} \rho \left(\xi \frac{dv}{dt} - \eta \frac{du}{dt} \right) dV = -\int\limits_{(V)} \left(\xi \frac{\partial p}{\partial \eta} - \eta \frac{\partial p}{\partial \xi} \right) dV$$

$$+ \int\limits_{(V)} \rho \left(\xi \frac{\partial U_G}{\partial \eta} - \eta \frac{\partial U_G}{\partial \xi} \right) dV. \qquad (3.47)$$

We write the second integral in the right-hand side of Eq. (3.45) in the form

$$\int\limits_{(V)} \rho \left(\eta \frac{dw}{dt} - \zeta \frac{dv}{dt} \right) dV = G \int\limits_{(V)} \rho(\xi, \eta, \zeta, t) \eta d\xi d\eta d\zeta \int\limits_{(V)} \rho(x, y, z, t) \frac{z - \zeta}{\Delta^3} dx, dy, dz$$

$$- G \int\limits_{(V)} \rho(\xi, \eta, \zeta, t) \zeta d\xi d\eta d\zeta \int\limits_{(V)} \rho(x, y, z, t) \frac{y - \zeta}{\Delta^3} dx, dy, dz.$$

The integral is equal to zero owing to the asymmetry expressed by the integral expressions with respect to z, ζ and y, η. Because the pressure at the border of the domain S is equal to zero, the first term in the right-hand side of Eq. (3.45) is also equal to zero. Actually,

$$\int\limits_{(V)} \left(\eta \frac{\partial p}{\partial \zeta} - \zeta \frac{\partial p}{\partial \eta} \right) dV = \int\limits_{(V)} \left[\frac{d}{d\eta} (\xi p) - \frac{d}{d\xi} (\eta p) \right] dV = \int\limits_{(V)} [\xi p d\xi d\zeta - \eta p d\eta d\zeta] = 0.$$

Taking into account the law of mass conservation, the left-hand side of Eq. (3.45) in the Lagrange co-ordinate system can be rewritten as

$$\int\limits_{(V)} \rho \left(\eta \frac{dw}{dt} - \zeta \frac{dv}{dt} \right) dV = \int\limits_{(V)} \rho \frac{d}{dt} (\eta w - \zeta v) dV$$

$$= \frac{d}{dt} \int\limits_{(V)} \rho (\eta w - \zeta v) dV = 0. \tag{3.48}$$

Integrating this equation with respect to time, the first of the three integrals is obtained:

$$\int\limits_{(V)} \rho (\eta w - \zeta v) dV = C_1.$$

The other two integrals can be obtained analogously. Thus the system of integrals of the moment of momentum has the form

$$\int\limits_{(V)} \rho (\eta w - \zeta v) dV = C_1,$$

$$\int\limits_{(V)} \rho (\zeta u - \xi w) dV = C_2, \tag{3.49}$$

$$\int\limits_{(V)} \rho (\xi v - \eta u) dV = C_3.$$

To derive the tenth integral of motion representing the law of energy conservation, we multiply each of the system of equations (3.35) by u, v, and w accordingly, and then sum and integrate the equality obtained with respect to the system volume

$$\int\limits_{(V)} \rho \left(\frac{du}{dt} u + \frac{dv}{dt} v + \frac{dw}{dt} w \right) dV = -\int\limits_{(V)} \left(\frac{\partial p}{\partial \xi} u + \frac{\partial p}{\partial \eta} v + \frac{dp}{d\zeta} w \right) dV$$

$$+ \int\limits_{(V)} \rho(\xi, \eta, \zeta, t) \left(\frac{\partial U_G}{\partial \xi} u + -\frac{\partial U_G}{\partial \eta} v + \frac{\partial U_G}{\partial \zeta} w \right) dV. \tag{3.50}$$

Applying the law of mass conservation for an elementary volume, it can easily be seen that the left-hand side of Eq. (3.50) expresses the change of the velocity of kinetic energy of the system:

$$\int\limits_{(V)} \rho \left(\frac{du}{dt} u + \frac{dv}{dt} v + \frac{dw}{dt} w \right) dV = \frac{d}{dt} \left[\frac{1}{2} \int\limits_{(V)} (u^2 + v + w^2) dV \right] = \frac{d}{dt}(T).$$

The first integral in the right-hand side of Eq. (3.50) can be transferred into

$$-\int\limits_{(V)} \left(\frac{\partial p}{\partial \xi} u + \frac{\partial p}{\partial \eta} v + \frac{dp}{d\zeta} w \right) dV = 3 \frac{d}{dt} \int\limits_{(V)} p dV$$

and gives the change of velocity of the internal energy of the system.

The second integral in the right-hand side of the same equation expresses the velocity of the potential energy change:

$$\int\limits_{(V)} \rho\,(\xi,\eta,\zeta,t)\,d\xi\xi d\eta d \left(\frac{\partial U_G}{\partial \xi} \frac{d\xi}{dt} + -\frac{\partial U_G}{\partial \eta} \frac{d\eta}{dt} + \frac{\partial U_G}{\partial \zeta} \frac{d\zeta}{dt} \right)$$

$$= \frac{d}{dt} \left[-\frac{1}{2} \int\limits_{(V)} \rho(\xi,\eta,\zeta,t)\,d\xi\xi d\eta dU_G \right] = -\frac{d}{dt}(U).$$

Finally, the law of energy conservation can be written in the form

$$T + U + W = E = const \tag{3.51}$$

where W is the internal energy of the system.

We now derive Jacobi's virial equation for a system described by Eqs. (3.35)–(3.38). For this purpose we multiply each of Eqs. (3.35) by ξ, η and ζ respectively, summing and integrating the resulting expressions with respect to the volume of the system:

$$\int\limits_{(V)} \rho \left(\frac{du}{dt} \xi + \frac{dv}{dt} \eta + \frac{dw}{dt} \zeta \right) dV = -\int\limits_{(V)} \left(\frac{\partial p}{\partial \xi} \xi + \frac{\partial p}{\partial \eta} \eta + \frac{dp}{d\zeta} \zeta \right) dV$$

$$+ \int\limits_{(V)} \rho \left(\frac{\partial U_G}{\partial \xi} \xi + -\frac{\partial U_G}{\partial \eta} \eta + \frac{\partial U_G}{\partial \zeta} \zeta \right) dV. \tag{3.52}$$

Using the obtained identities considered in the previous section, we have

$$\frac{du}{dt}\xi = \frac{1}{2}\frac{d^2}{dt^2}(\xi^2) - u^2,$$

$$\frac{dv}{dt}\eta = \frac{1}{2}\frac{d^2}{dt^2}(\eta^2) - v^2,$$

$$\frac{dw}{dt}\zeta = \frac{1}{2}\frac{d^2}{dt^2}(\zeta^2) - w^2.$$

Taking into account the law of conservation of mass for elementary volume, we transform the left-hand side of Eq. (3.52) as follows:

$$\int_{(V)}\rho\left(\frac{du}{dt}\xi + \frac{dv}{dt}\eta + \frac{dw}{dt}\zeta\right)dV = \frac{1}{2}\int_{(V)}\rho\frac{d^2}{dt^2}(\xi^2 + \eta^2 + \zeta^2)dV$$

$$- \int_{(V)}\rho(u^2 + v^2 + w^2)dV = \ddot{\Phi} - 2T,$$

$$(3.53)$$

where

$$\Phi = \frac{1}{2}\int_{(V)}\rho(\xi^2 + \eta^2 + \zeta^2)dV$$

is the Jacobi function and

$$T = \frac{1}{2}\int_{(V)}\rho(u^2 + v^2 + w^2)dV$$

is the kinetic energy of the system.

We now transform the first integral in the right-hand side of Eq. (3.52). Using the Gauss–Ostrogradsky theorem and the equality with zero pressure at the border of the system, we can write

$$-\int_{(V)}\left(\frac{\partial p}{\partial \xi}\xi + \frac{\partial p}{\partial \eta}\eta + \frac{dp}{d\zeta}\zeta\right)dV = -\int_{(V)}\left[\frac{\partial}{\partial \xi}(p\xi) + \frac{\partial}{\partial \eta}(p\eta) + \frac{d}{d\zeta}(p\zeta)\right]dV$$

$$+ 3\int_{(V)}pdV = 3\int_{(V)}pdV.$$

$$(3.54)$$

The obtained equation expresses the doubled internal energy of the system.

The second integral in the right-hand side of Eq. (3.52) is equal to the potential energy of the gravitational interaction of mass particles within the system

$$\int_{(V)}\rho\left(\frac{\partial U_G}{\partial \xi}\xi + -\frac{\partial U_G}{\partial \eta}\eta + \frac{\partial U_G}{\partial \zeta}\zeta\right)dV = U.$$

$$(3.55)$$

Substituting Eqs. (3.53), (3.54) and (3.55) into (3.52), Jacobi's virial equation is obtained in the form

$$\ddot{\Phi} - 2T = 3 \int\limits_{(V)} p dV + U. \tag{3.56}$$

Taking into account the law of conservation of energy (3.51), we rewrite Eq. (3.56) in a form which will be used farther:

$$\ddot{\Phi} = 2E - U, \tag{3.57}$$

where $E = T + U + W$ is the total energy of the system.

3.5 Derivation of Jacobi's Virial Equation from Hamiltonian Equations

Let the system of material points be described by Hamiltonian equations of motion. Let also the considered system consist of n material points with masses m_i. Its generalized co-ordinates and moments are q_i and $p_i = m_i(dq_i/dt)$. Hamiltonian equations for such a system can be written as

$$\dot{p}_i = -\frac{\partial H}{dq_i},$$
$$\dot{q}_i = \frac{\partial H}{dp_i}, \tag{3.58}$$

where $H(p, q)$ is the Hamiltonian; $i = 1, 2,..., n$.

Using values q_i and p_i, we can construct the moment of momentum

$$\sum_{i=1}^{n} p_i q_i = \sum_{i=1}^{n} m_i q_i \dot{q}_i = \frac{d}{dt}\left(\sum_{i=1}^{n} \frac{m_i q_i^2}{2}\right).$$

Now the Jacobi function may be introduced

$$\sum_{i=1}^{n} p_i q_i = \dot{\Phi}. \tag{3.59}$$

Differentiating Eq. (3.59) with respect to time, Jacobi's virial equation is obtained in the form

$$\ddot{\Phi} = \sum_{i=1}^{n} \dot{p}_i q_i + \sum_{i=1}^{n} p_i \dot{q}_i. \tag{3.60}$$

Substituting expressions for $\dot{p}_i$ and $\dot{q}_i$ taken from the Hamiltonian equations (3.58) into the right-hand side of (3.60), we obtain Jacobi's virial equation written in Hamiltonian form:

$$\ddot{\Phi} = \sum_{i=1}^{n} \left(-\frac{\partial H}{\partial q_i} q_i + \frac{\partial H}{\partial p_i} p_i \right). \tag{3.61}$$

The Hamiltonian of the system of material points interacting according to the law of the inverse squares of distance is a homogeneous function in terms of moments p_i with a degree of homogeneity of the function equal to 2, and in terms of co-ordinates q_i with a degree of homogeneity equal to -1. It follows from this

$$H(p, q) = T(p) + U(q)$$

and hence

$$\sum_{i=1}^{n} p_i \frac{\partial H}{\partial p_i} = 2T.$$

$$\sum_{i=1}^{n} q_i \frac{\partial H}{\partial q_i} = -U.$$

Taking these relationships into account, Eq. (3.61) acquires the usual form of Jacobi's virial equation (3.57) for the system of mass points interacting according to the law of inverse squares of distance.

Equation (3.61) is more general then Eq. (3.57). The use of generalized co-ordinates and moments as independent variables permits us to obtain the solution of Jacobi's virial equation, taking into account gravitational and electromagnetic perturbations as well as quantum effects, both in the framework of classical physics and in terms of the Hamiltonian written in an operator form. In the general case, Eq. (3.61) can be reduced to (3.57) as the potential energy of interaction of the system's points is a homogenous function of its co-ordinates.

3.6 Derivation of Jacobi's Virial Equation in Quantum Mechanics

It is known (Landau and Lifshitz, 1963) that in quantum mechanics some physical value L by definition takes the linear Hermitian operator $\hat{L}$. Any physical state of the system take the normalized wave function ψ. The physical value of L can take the only eigenvalues of the operator $\hat{L}$. The mathematical expectation $\hat{L}$ of the value L at state ψ is determined by the diagonal matrix element

$$\bar{L} = <\psi|\hat{L}|\psi>. \tag{3.62}$$

The matrix element of the operators of the Cartesian co-ordinates $\hat{x}_i$ and the Cartesian components of the conjugated moments $\hat{p}_k$ calculated within wave functions f and g of the system satisfy Hamilton's equations of classical mechanics:

$$\frac{d}{dt} <f|\hat{p}_i|g> = -<f\left|\frac{\partial\hat{H}}{d\hat{x}_i}\right|g>, \qquad (3.63)$$

$$\frac{d}{dt} <f|\hat{x}_i|g> = -<f\left|\frac{\partial\hat{H}}{\partial\hat{p}_i}\right|g>, \qquad (3.64)$$

where $\hat{H}$ is the operator which corresponds to the classical Hamiltonian.

Operators $\hat{p}_i$ and $\hat{x}_k$ satisfy the commutation relations

$$\begin{aligned}
\left[\hat{p}_i, \hat{x}_k\right] &= i\hbar\delta_{ik}, \\
\left[\hat{p}_i, \hat{p}_k\right] &= 0, \\
\left[\hat{x}_i, \hat{x}_k\right] &= 0,
\end{aligned} \qquad (3.65)$$

where $\hbar$ is Planck's constant; δ_{ik} is the Kronecker's symbol; $\delta_{ik} = 1$ at $i = k$ and $\delta_{ik} = 0$ at $i \neq k$.

Operator components of momentum $\hat{p}_i$ for the functions whose arguments are Cartesian co-ordinates $\hat{x}_i$ have the form

$$\hat{p}_i = i\hbar\frac{\partial}{\partial x_i} \qquad (3.66)$$

and reverse vector

$$\hat{p} = -i\hbar\nabla.$$

The derivative taken from the operator with respect to time does not depend explicitly on time; it is defined by the relation

$$\hat{\dot{L}} = -\frac{i}{\hbar}[\hat{L}, \hat{H}], \qquad (3.67)$$

where $\hat{H}$ is the Hamiltonian operator that can be obtained from the Hamiltonian of classical mechanics in accordance with the correspondence principle.

We have already noted that in the classical many-body problem the translational motion of the center of mass can be separated from the relative motion of the mass points if only the inertial forces affect the system. We can show that in quantum mechanics the same separation is possible.

The Hamiltonian operator of a system of n particles which is not affected by external forces in co-ordinates is

$$\hat{H} = -\frac{\hbar^2}{2} \sum_{i=1}^{n} \frac{\nabla_i^2}{m_i} + \frac{1}{2} \sum_{i=1}^{n} \sum_{i=1}^{n} U_{ik} \left(x_i - x_k, y_i - y_k, z_i - z_k \right). \qquad (3.68)$$

Let us replace in (3.68) the three n co-ordinates x_i, y_i, z_i by co-ordinates X, Y, Z of the center of mass and by co-ordinates ξ_λ, η_λ, ζ_λ, which determine the position of a particle λ ($\lambda = 1, 2,..., n-1$) relative to particle n. We obtain

$$X = \frac{1}{M} \sum_{i=1}^{n} m_i x_i,$$

$$M = \sum_{i=1}^{n} m_i, \qquad (3.69)$$

$$\xi_\lambda = x_\lambda - x_n,$$

where $\lambda = 1, 2,..., n-1$.

Analogously the corresponding relations for Y, Z, η_λ, ζ_λ are obtained.

It is easy to obtain from (3.69) the following operator relations:

$$\frac{d}{dx_p} = \frac{m_p}{M} \frac{\partial}{\partial X} + \frac{\partial}{\partial \xi_p}, \quad p = 1, 2, \ldots, n-1,$$

$$\frac{\partial}{\partial X_n} = \frac{m_n}{M} \frac{\partial}{\partial X} - \sum_{\lambda=1}^{n-1} \frac{\partial}{\partial \xi_\lambda},$$

$$\sum_{\lambda=1}^{n-1} \frac{1}{\partial x_i} \frac{\partial^2}{\partial x_i^2} = \sum_{\lambda=1}^{n-1} \frac{1}{m_\lambda} \left(\frac{m_\lambda^2}{M^2} \frac{\partial^2}{\partial X^2} + 2 \frac{m_\lambda}{M} \frac{\partial^2}{\partial X \partial \xi_\lambda} + \frac{\partial^2}{\partial \xi_\lambda^2} \right)$$

$$+ \frac{1}{m_n} \left(\frac{m_n^2}{M^2} \frac{\partial^2}{\partial X^2} - 2 \frac{m_n}{M} \sum_{\lambda=1}^{n-1} \frac{\partial^2}{\partial X \partial \xi_\lambda} + \sum_{\mu=1}^{n-1} \sum_{\lambda=1}^{n-1} \frac{\partial^2}{\partial \xi_\mu \partial \xi_\lambda} \right)$$

$$= \frac{1}{m_n} \frac{\partial^2}{\partial X^2} + \left(\sum_{\lambda=1}^{n-1} \frac{1}{m_\lambda} \frac{\partial^2}{\partial \xi_\lambda^2} + \frac{1}{m_n} \sum_{\mu=1}^{n-1} \sum_{\lambda=1}^{n-1} \frac{\partial^2}{\partial \xi_\mu \partial \xi_\lambda} \right),$$

where summing on the Greek index is provided from 1 до $n-1$. It is seen that all the combined derivatives $\partial^2/\partial X \partial \xi_\lambda$ were mutually reduced and do not enter into the final expression. This allows the Hamiltonian to be separated into two parts:

$$H = H_0 + H_r$$

where, in the right-hand side, the first term

$$H_0 = \frac{\hbar^2}{2M} \left(\frac{\partial^2}{\partial X^2} + \frac{\partial^2}{\partial y^2} + \frac{\partial^2}{\partial Z^2} \right)$$

describes the motion of the center of mass, and the second term

$$H_r = -\frac{\hbar^2}{2} \left(\sum_{\lambda=1}^{n-1} \frac{1}{m_\lambda} \nabla_\lambda^2 + \frac{1}{m_n} \sum_{\mu=1}^{n-1} \sum_{\lambda=1}^{n-1} \nabla_\lambda \nabla_\mu \right) + U \qquad (3.70)$$

describes the relative motion of the particles.

The potential energy in (3.70), which is

$$U = \frac{1}{2} \sum_{\mu=1}^{n-1} \sum_{\lambda=1}^{n-1} U_{\lambda\mu} \left(\xi_\lambda - \xi_\mu, \eta_\lambda - \eta_\mu, \zeta_\lambda - \zeta_\mu \right) + \sum_{\lambda-1}^{n-1} U_{\lambda\mu} \left(\xi_\lambda, \eta_\lambda, \zeta_\lambda \right), \qquad (3.71)$$

also certainly does not depend on the co-ordinates of the center of mass.

Now the Schrödinger's equation

$$(H_0 + H_r)\psi = E\psi \qquad (3.72)$$

permits the separation of variables.

Assuming $\psi = \Phi(X, Y, Z)$ and $(\xi_\lambda, \eta_\lambda, \zeta_\lambda,)$, we obtain

$$-\frac{\hbar^2}{2V} \nabla^2 \Phi = E_0 \Phi, \qquad (3.73)$$

$$H_r u = E_r u, \qquad (3.74)$$

$$E_0 + E_r = E. \qquad (3.75)$$

The solution of Eq. (3.73) has the form of a plane wave:

$$\Phi = e^{i\overline{k}\overline{R}},$$
$$E_0 = \frac{\hbar^2 k^2}{2M}, \qquad (3.76)$$

where R is a vector with co-ordinates X, Y, Z.

The result obtained is in full accordance with the classical law of the conservation of motion of the center of mass. This means that the center of mass of the system moves like a material point with mass m and momentum $\hbar\overline{k}$. The mode of relative motion of the particles is determined by Eq. (3.74), which does not depend on the motion of the center of mass.

The existence in the right-hand side of Eq. (3.70) of the third term restricts further factorization of the function $u(\xi_\lambda, \eta_\lambda, \zeta_\lambda)$. Only in the two-body problem, where $n = 2$ and at $\lambda = \mu = 1$, a part of the Hamiltonian connected with the relative motion simplified and takes the form

$$H_r = -\frac{\hbar^2}{2} \left(\frac{1}{m_1} \nabla_1^2 + \frac{1}{m_2} \nabla_2^2 \right) + U_{12} (\xi_1, \eta_1, \zeta_1). \tag{3.77}$$

It seems that choosing the corresponding system of co-ordinates can lead us to an approach for separating the motion of the center of mass to the many-body problem.

Introducing into Eq. (3.77) the reduced mass m^*, which is determined as in classical mechanics by the relation

$$\frac{1}{m_1} + \frac{1}{m_2} = \frac{1}{m^*}, \tag{3.78}$$

and omitting indices in the notation for relative co-ordinates and potential energy U_{12}, we come to

$$-\frac{\hbar^2}{2m^*} \nabla^2 u + U (\xi, \eta, \zeta) u = E_r u. \tag{3.79}$$

This is Schrödinger's equation for the equivalent one-particle problem.

Considering the hydrogen atom in the framework of the one-particle problem, it is assumed that the nucleus is in ground state. In accordance with Eq. (3.79), the normalized mass of the nucleus and electron m^* should be introduced. No changes which account for the effect of the nucleus on the relative motion should be introduced. Because of the nucleus, mass m is much heavier than electron mass m_e^*; instead of Eq. (3.78) we can use its approximation

$$m^* = m \left(1 - \frac{m}{M} \right).$$

Comparing, for example, the frequency of the red line $H_\alpha (n = 3 - n' = 2)$ in the spectrum of a hydrogen atom:

$$\omega(H_\alpha) = \frac{5}{36} \frac{m_H^* e^4}{2\hbar^2 h}$$

with the frequency of the corresponding line in the spectrum of a deuterium atom:

$$\omega(D_\alpha) = \frac{5}{36} \frac{m_D^* e^4}{2\hbar^2 h},$$

and taking into account that $m_D \approx 2m_H$, for the difference of frequencies, we obtain

$$\omega(D_\alpha) - \omega(H_\alpha) = \frac{m_D^* - m_H^*}{m_H^*} \omega(H_\alpha) \approx \frac{m}{2M_H} \omega(H_\alpha).$$

This difference is not difficult to observe experimentally. At wavelength 6563 Å it is equal to 4.12 cm^{-1}. Heavy hydrogen was discovered in 1932 by Urey, Brickwedde and Murphy, who observed a weak satellite D_α in the line H_α of the spectrum of natural hydrogen. This proves the practical significance of even the first integrals of motion.

We now show that the virial theorem is valid for any quantum mechanical system of particles retained by Coulomb (outer) forces:

$$2\overline{T} + U = 0.$$

We prove this by means of scale transformation of the co-ordinates keeping unchanged normalization of wave functions of a system.

The wave function of a many-particle system with masses m_i and electron charge e_i satisfies the Schrödinger's equation:

$$-\frac{\hbar^2}{2}\sum_{i=1}^{n-1}\frac{1}{m_i}\nabla_i^2\psi + \frac{1}{2}\sum_{i=1}^{n-1}\sum_{k=1}^{n-1}\frac{e_ie_k}{r_{ik}}\psi = E\psi \tag{3.80}$$

and the normalization condition

$$\int d\tau_1 \ldots \int \psi^*\psi d\tau_n = 1. \tag{3.81}$$

The mean values of the kinetic and potential energies of a system at stage ψ are determined by the expressions

$$T = -\frac{\hbar^2}{2}\sum_{i=1}^{n-1}\frac{1}{m_i}\int d\tau_1 \ldots \int \psi^*\nabla_i^2\psi d\tau_n, \tag{3.82}$$

$$U = \frac{1}{2}\sum_{i=1}^{n-1}\sum_{i=1}^{n-1}e_ie_k\int d\tau_1 \ldots \int d\tau_n\frac{\psi^*\psi}{r_{ik}}d\tau_n. \tag{3.83}$$

The scale transformation

$$\bar{r}_i' = \lambda\bar{r}_i, \tag{3.84}$$

keeps in force the condition (3.81) and means that the wave function

$$\psi(\bar{r}_i, \ldots, \bar{r}_n) \tag{3.85}$$

is replaced by the function

$$\psi_\lambda = \lambda^{3n/2}\psi(\lambda\bar{r}_i, \ldots, \lambda\bar{r}_n). \tag{3.86}$$

Substituting (3.86) into Eqs. (3.83) and (3.82) and passing to new variables of integration (3.84), and taking into account that

$$\nabla_i^2 = \lambda^2 \nabla_i'^2, \quad \frac{1}{r_{ik}} = \lambda \frac{1}{r_{ik}'},$$

instead of the true value of the energy, $\overline{E} = \overline{T} + \overline{U}$, we obtain

$$\overline{E}(\lambda) = \lambda^2 \overline{T} + \lambda \overline{U}. \tag{3.87}$$

Equation (3.87) should have a minimum value in the case when the function which is the solution of the Schrödinger's equation is taken from the family of functions (3.86), i.e. when $\lambda = 1$. So, at $\lambda = 1$ the expression

$$\frac{\partial \overline{E}(\lambda)}{\partial \lambda} = 2\lambda^2 \overline{T} + \overline{U}$$

should turn into zero, and thus

$$2\overline{T} + \overline{U} = 0,$$

which is what we want to prove.

We now derive Jacobi's virial equation for a particle in the inner force field with the potential $U(q)$ and fulfilling the condition

$$-q\nabla U(q) = U \tag{3.88}$$

using the quantum mechanical principle of correspondence. We shall also show that in quantum mechanics Jacobi's virial equation has the same form and contents as in classical mechanics, the only difference being that its terms are corresponding operators.

In the simplest case the Hamiltonian of a particle is written

$$\hat{H} = -\frac{\hbar^2}{2m}\nabla^2 + \hat{U}, \tag{3.89}$$

and its Jacobi function is

$$\hat{\Phi} = \frac{1}{2}m\hat{q}^2. \tag{3.90}$$

It is clear that the following relations are valid:

$$\nabla\hat{\Phi} = m\hat{q},$$

$$\nabla^2\hat{\Phi} = m.$$

Following the definition of the derivative with respect to time from the operator of the Jacobi function of a particle (3.67), we can write

$$\dot{\hat{\Phi}} = -\frac{1}{\hbar}[\hat{\Phi}, \hat{H}],$$

where, after corresponding simplification, quantum mechanical Poisson brackets can be reduced to the form

$$[\hat{\Phi}, \hat{H}] = \frac{\hbar^2}{2m}\left\{\nabla^2\hat{\Phi} + 2(\nabla\hat{\Phi})\nabla\right\} = \frac{\hbar^2}{2m}(m + 2m\hat{q}\nabla). \tag{3.91}$$

The second derivative with respect to time from the operator of the Jacobi function is:

$$\ddot{\hat{\Phi}} = -\frac{1}{\hbar^2}\left\{[\hat{\Phi}, \hat{H}], \hat{H}\right\}. \tag{3.92}$$

Substituting the corresponding value of $[\hat{\Phi}, \hat{H}]$ and $\hat{H}$ from (3.91) and (3.89) into the right-hand side of (3.92), we obtain

$$\ddot{\hat{\Phi}} = -\frac{\hbar^2}{2m}\frac{1}{\hbar^2}\left[(m + 2m\hat{q}\nabla), \left(-\frac{\hbar^2}{2m}\nabla^2 + \hat{U}\right)\right]. \tag{3.93}$$

After simple transformation, the right-hand side of (3.93) will be

$$\ddot{\hat{\Phi}} - \frac{1}{2m}\left\{2\hbar^2\nabla^2 + 2m\hat{q}(\nabla\hat{U})\right\} = -\frac{2\hbar^2}{2m}\nabla^2 + \hat{U}, \tag{3.94}$$

where, in writing this expression in the right-hand side, we used condition (3.88).

Add and subtract the operator $\hat{U}$ from the right-hand side of Eq. (3.94) and, following the definition of the Hamiltonian of the system (3.89), we obtain the quantum mechanical Jacobi virial equation (equation of dynamical equilibrium of the system), which has the form

$$\ddot{\hat{\Phi}} = 2\hat{H} - \hat{U}. \tag{3.95}$$

From Eq. (3.95), by averaging with respect to time, we obtain the quantum mechanical analogue of the classical virial theorem (equation of hydrostatic equilibrium of the system). In accordance with this theorem the following relation is kept for a particle performing finite motion in space

$$2\hat{H} = \overline{\hat{U}}. \tag{3.96}$$

Analogously, one can derive Jacobi's virial equation and the classical virial theorem for a many-particle system, the interaction potential for which depends on distance between any particle pair and is a homogeneous function of the co-ordinates. In particular, Jacobi's virial equation for Coulomb interactions will have the form of Eq. (3.95).

3.7 General Covariant Form of Jacobi's Virial Equation

Jacobi's initial equation

$$\ddot{\Phi} = 2E - U,$$

which was derived in the framework of Newtonian mechanics and is correct for the system of material points interacting according to Newton and Coulomb laws, includes two scalar functions Φ and U relates to each other by a differential relation. We draw attention to the fact that neither function, in its structure, depends explicitly on the motion of the particles constituting the body. The Jacobi function Φ is defined by integrating the integrand $\rho(r)r^2$ over the volume (where $\rho(r)$ is the mass density and r is the radius vector of the material point) and is independent in explicit form of the particle velocities. The potential energy U also represents the integral of $m(r)dm(r)/r$ over the volume (where $m(r)$ is the mass of the sphere's shell of radius r; $dm(r)$ is the shell's mass) independent of the motion of the particles for the same reason.

Let us derive Jacobi's equation from Einstein's equation written in the form

$$\Delta G = 2\pi T, \tag{3.97}$$

where ΔG and T are the Einstein tensor and energy-momentum tensor accordingly.

In fact, since the covariant divergence of Einstein's tensor is equal to zero, we consider the covariant divergence of the energy-momentum tensor T only of substance and fields (not gravitational). Moreover, the ordinary divergence of the sum of the tensor T and pseudotensor t of the energy-momentum of the gravitational field can be substituted for the covariant divergence of the tensor T. This ordinary divergence leads to the existence of the considered quantities.

Let us define the sum of the tensor T and pseudotensor t through T_{ij} and derive Jacobi's equation in this notation.

The equation for ordinary divergence of the sum $T_{ij} = (T+t)_{ij}$ can be written

$$T_{0k,k} - T_{00,0} = 0, \tag{3.98}$$

$$T_{jk,k} - T_{j0,0} = 0. \tag{3.99}$$

We multiply Eq. (3.99) by x^j and integrate over the whole space (assuming the existence of a synchronous co-ordinate system). Integrating by parts, neglecting the surface integrals (they vanish at infinity), and transforming to symmetrical form with respect to indices, we obtain

$$\int T_{ij}dV = \frac{1}{2}\left[\int \left(T_{i0}x^j + T_{j0}x^i\right)dV\right] = 0, \qquad (3.100)$$

where i, j are spatial indices.

Similarly, multiplying (3.100) by $x^i x^j$ and integrating over the whole space, it follows that

$$\left[\int T_{00}x^i x^j dV\right]_{,0} = -\int \left(T_{i0}x^j + T_{j0}x^i\right)dV. \qquad (3.101)$$

From (3.100) and (3.101) we finally get

$$\int T_{ij}dV = \frac{1}{2}\left[\int T_{00}x^i x^j\right]_{,0,0}. \qquad (3.102)$$

It is worth recalling that T_{00} also includes the gravitational defect of the mass due to the pseudotensor t by definition.

The integral $\int T_{00}x^i x^j dV$ represents the generalization of the Jacobi function $\Phi = \frac{1}{2}\int \rho r^2 dV$ introduced earlier, if we take the spur (also commonly known as the trace) of Eq. (3.102). Let us clarify this operation.

In Eq. (3.102) the spur is taken by the spatial co-ordinates. It is therefore necessary either to represent the total zero spur by four indices, as happens in the case of a transverse electromagnetic field, or to represent the relationship between the reduced spur with three indices and the total spur, as happens in the case of the energy-momentum tensor of matter.

Special care should be taken while representing the spur of the pseudotensor of the energy-momentum t. Consider the post-Newtonian approximation. In this approximation, assuming the value of $2u$ to be $-g_{00}-1$, the components of the pseudotensor t are written in the form

$$t^{00} = -\frac{7}{8\pi}u_{,j,i},$$

$$t^{ij} = -\frac{1}{4\pi}\left(u_{,j,i} - \frac{1}{2}\delta_{ij}u_{,k}u_{,k}\right),$$

so that

$$S_p t = t^{00} + S_p\left(t^{ij}\right) = -\frac{1}{\pi}u_{,i}u_{,j} = \frac{1}{7}t^{00},$$

$$S_p\left(t^{ij}\right) = \frac{6}{7}t^{00}.$$

The spur in the left-hand side of Eq. (3.102) can therefore be reduced to the energy of the Coulomb field, the total energy of the transverse electromagnetic field and

the gravitational energy (when it can be separated, i.e. post-Newtonian approximation).

Finally, it follows in this case that the scalar form of Jacobi's equation holds:

$$\Phi_{,0,0} = mc^2, \tag{3.103}$$

where m is the mass, accounting for the baryon defect of the mass and the total
energy of the electromagnetic radiation. We do not take into account the radiation
of the gravitational waves.

The result obtained by Tolman for the spherical mass distribution (Tolman, 1969)
is of interest:

$$m = 4\pi \int \hat{\varepsilon} r^2 dr, \tag{3.104}$$

where r is the radius and $\hat{\varepsilon}$ is the energy density.

The integral (3.104) acquires a form which is also valid in the case of flat space–
time. This result can be explained as follows. The curvature of space–time is exactly
compensated by the mass defect. This probably explains the fact that Jacobi's virial
equation, derived from Newton's equations of motion which are valid in the case of
non-relativistic approximation for a weak gravitational field, becomes more universal than the equations from which it was derived.

We shall not study the general tensor of Jacobi's virial equation, since in the
framework of the assumed symmetry for the considered problems we are interested
only in the scalar form of the equation as applied to electromagnetic interactions. As
follows from the above remarks, in this case Jacobi's equation remains unchanged
and the energy of the free electromagnetic field is accounted for in the term defining
the total energy of the system. Total energy enters into Jacobi's equation without the
electromagnetic energy irradiated up to the considered moment of time. Moreover,
for the initial moment of time we take the moment of system formation. This irradiated energy appears also to be responsible for the growth of the gravitational mass
defect in the system, as was mentioned above.

3.8 Relativistic Analogue of Jacobi's Virial Equation

Let us derive Jacobi's virial equation for asymptotically flat space–time. We write
the expression of a 4-moment of momentum of a particle:

$$p^i x_i, \tag{3.105}$$

where $p^i = mcu^i$ is the 4-momentum of the particle; c is the velocity of light;
$u^i = dx^i/ds$ is the 4-velocity; x^i is the 4-co-ordinate of the particle; s is the interval of
events, and i is the running index with values 0, 1, 2, 3.

In asymptotically flat space–time we write

$$\frac{d}{ds}\left(p^i x_i\right) = mc\frac{d}{ds}\left(u^i x_i\right) = mc\frac{d^2}{ds^2}\left(\frac{x^i x_i}{2}\right). \tag{3.106}$$

Since

$$x^i x_i = c^2 t^2 - r^2 \quad \text{and} \quad \frac{d}{ds} = \frac{\gamma}{c}\frac{d}{dt},$$

where $\gamma = 1/\sqrt{1 - (v^2/c^2)}$, and r is the radius of mass particle.

Then we continue transformation of the Eq. (3.106):

$$mc\frac{d^2}{ds^2}\left(\frac{x^i x_i}{2}\right) = mc\frac{\gamma^2}{c^2}\frac{d^2}{dt^2}\left(\frac{c^2 t^2 - r^2}{2}\right) = mc\gamma^2 - \frac{\gamma^2}{c^2}\frac{d^2}{dt^2}\left(\frac{mr^2}{2}\right),$$

and finally

$$\frac{d}{ds}\left(p^i x_i\right) = mc\gamma^2 - \frac{\gamma^2}{c}\ddot{\Phi}, \tag{3.107}$$

where $\ddot{\Phi} = \frac{d^2}{dt^2}\left(\frac{mr^2}{2}\right)$ is the Jacobi function.

On the other hand, we have

$$\frac{d}{ds}\left(p^i x_i\right) = mc\frac{d}{ds}\left(u^i x_i\right) = mcu^i u_i + mc\frac{du^i}{ds}x_i. \tag{3.108}$$

Using the identity $u_i u^i \equiv 1$ and the geodetic equation

$$\frac{du^i}{ds} = -\Gamma^i_{k\ell}u^k u^\ell,$$

where

$$\Gamma^i_{k\ell} = \frac{1}{2}g^{im}\left(\frac{\partial g_{km}}{\partial x^\ell} + \frac{\partial g_{\ell m}}{\partial x^k} + \frac{\partial g_{k\ell}}{\partial x^m}\right)$$

are the Christoffel's symbols, and the equation (3.108) will be rewritten as

$$\frac{d}{ds}\left(p^i x_i\right) = mc - mcx_i\Gamma^i_{k\ell}u^k u^\ell. \tag{3.109}$$

The metric tensor g_{ik} for a weak stationary gravitational field is

$$g_{ik} = \eta_{ik} + \xi_{ik}, \tag{3.110}$$

where in our notation η_{ik} is the Lorentz tensor with signature $(+, -, -, -)$.

For the Schwarzschild metric tensor ξ_{ik} we write

$$\xi_{00} = -\frac{r_g}{r}; \quad \xi_{11} = -\frac{1}{1 - r_g/r} + 1 \approx -\frac{r_g}{r};$$

$$\xi_{ik} = 0 \quad \text{if } i \neq k \text{ and } i \neq 0, 1. \tag{3.111}$$

Here $r_g = 2GV/c^2$ is the Schwarzschild gravitational radius of the mass m'.

Now we can rewrite the second term in the right-hand side of Eq. (3.109), using (3.110) and (3.111)

$$mcx_i \Gamma^i_{k\ell} u^k u^\ell = mcx^m u^k u^\ell \left(\frac{\partial \xi_{km}}{\partial x^m} - \frac{1}{2} \frac{\partial \xi_{k\ell}}{\partial x^m} \right)$$

$$= mc \left(x^0 u^0 u^1 \frac{\partial \xi_{00}}{\partial x^1} + x^1 u^1 u^1 \frac{\partial \xi_{11}}{\partial x^1} - \frac{1}{2} x^1 u^0 u^0 \frac{\partial \xi_{00}}{\partial x^1} - x^1 u^1 u^1 \frac{\partial \xi_{11}}{\partial x^1} \right). \tag{3.112}$$

But $u^1 \ll u^0 = \gamma$ and $x^1 = r$.

We therefore obtain for Eq. (3.112)

$$mcx_i \Gamma^i_{k\ell} u^k u^\ell = -\frac{mc}{2} x^1 u^0 u^0 \frac{\partial \xi_{00}}{\partial x^1} = -\frac{mc}{2} r\gamma^2 \frac{r_g}{r^2}$$

$$= \frac{mc}{2} \gamma^2 \frac{2Gm'}{c^2 r} = -\frac{\gamma^2}{c} \frac{Gm'm}{r}. \tag{3.113}$$

Finally, we see that

$$\frac{d}{ds} (p^i x_i) = mc - \frac{\gamma^2}{c} U, \tag{3.114}$$

where U is the potential energy of the mass in the gravitational field of the mass m'.

Identification of the expression $(d/ds)(p^i x_i)$ obtained from Eqs. (3.107) and (3.114) gives

$$mc\gamma^2 - \frac{\gamma^2}{c} \ddot{\Phi} = mc - \frac{\gamma^2}{c} U. \tag{3.115}$$

It is easy to see that

$$mc (\gamma^2 - 1) = mc \left(\frac{1}{1 - v^2/c^2} - 1 \right) = mc \frac{v^2}{c^2} \frac{1}{1 - v^2/c^2} = \frac{\gamma^2}{c} mv^2 = \frac{\gamma^2}{c} 2T.$$

We then obtain

$$\frac{\gamma^2}{c} \ddot{\Phi} = \frac{\gamma^2}{c} U + \frac{\gamma^2}{c} 2T,$$

which gives

$$\ddot{\Phi} = U + 2T,$$

or

$$\ddot{\Phi} = 2E - U, \tag{3.116}$$

where T is the kinetic energy of the particle m and $E = U + T$ is its total energy.

Equations (3.116) are known as classical Jacobi's virial equations, and the expression (3.109) represents its relativistic analogue for asymptotically flat space–time.

3.9 Universality of Jacobi's Virial Equation for Description of the Dynamics of Natural Systems

It follows from this derivation of Jacobi's virial equation that it appears to be a universal mathematical expression for consideration of the dynamics of celestial bodies described by equations of motion for a wide range of existing physical models. The derived equation represents not only formal mathematical transformation of the initial equations of motion. Physical quintessence of mathematical transformation of the equations of motion involves change of the vector forces and moment of momentums by the volumetric forces or pressure and the oscillation of the interacted mass particles (inner energy) expressed through the energy of oscillation of the polar moment of inertia of a body. Here the potential (kinetic) energy and the polar moment of inertia of a body have a functional relationship and within the period of oscillation are inversely changed by the same law. Moreover, as it was demonstrated in Sect. 2 of Chap. 2. and will be shown in Chap. 6, the virial oscillations of a body represent the main part of the body's kinetic energy, which is lost in the hydrostatic equilibrium model. The change of the vector forces and moment of momentums by the force pressure and the oscillation of the interacting mass particles disclose the physical meaning of the gravitation and mechanism of generation of the gravitational and electromagnetic energy and their common nature, which is considered in Chap. 6. The most important advantage given by Jacobi's virial equation, is its independence from the choice of the co-ordinate system, transformation of which, as a rule, creates many mathematical difficulties.

By averaging for a uniform system the generalized virial equation $\ddot{\Phi} = 2E - U$, when the first derivative over the polar moment of inertia Φ acquires constant value, it becomes the classical virial theorem $2E = U$, or $-U = 2T$, which expresses the condition of the hydrostatic equilibrium being in the outer force field and without kinetic energy of oscillations of the interacting particles.

The starting point for derivation of the virial theorem is the particle momentum. By Newton's definition this value "*is a certain measure determined proportionally to the velocity and the mass*". This value is defined or it is found experimentally. All the other force parameters are obtained by transformation of the initial momentum and those actions are explained by physical interaction of the mass particles, which are the carrier of the momentum. In fact, we recognize the momentum to be "*innate*", according to Newton's terminology, value, i.e. the hereditary value. Under the "*innate*" value Newton understood "both the resistance and the pressure of the mass" and finally the effect acquires its status of the inertial force. But the essence does not change, because the momentum appears together with the mass. Thus, the circle of the philosophical speculations is locked by the momentum, i.e. by the mass and its oscillation. All other attributes of the motion are formed by mathematical transformations.

One more mainly physical problem that was solved in derivation in this chapter of Jacobi's virial equation by mathematical transformations is an understanding of the nature and dynamical effects of the gravitational interaction of mass particles for a continuous body. Contrary to the interaction of two bodies presented by mass points, when a dynamical effect is developed in the orbital motion by vector force and angular momentum, the dynamical effect of the interacting mass particles of the continuous body is developed in the form of volumetric pressure and volumetric oscillation. The integral effect of the mass interaction is expressed by oscillation of the polar moment of inertia. In the next chapter we consider solution of Jacobi's virial equation.

Chapter 4
Solution of Jacobi's Virial Equation
for Conservative Systems

In Chap. 3 we derived Jacobi's virial equation of dynamical equilibrium in the framework of various physical models which are used for describing the dynamics of natural systems. We showed that, instead of the traditional description of a system in co-ordinates and velocities, the problem of dynamics can be studied from the position of an external observer. In this case the system as a whole is described by a compact and elegant equation and is characterized by integral (volumetric) parameters. Such a description of the integral equation does not depend on the choice of the frame of reference. The external observer can estimate by observations only some moments of distribution of mass density, i.e. total mass and energy of a system, which are its integral characteristics. Moreover, in order to solve the problem of a body's motion in the framework of its dynamical equilibrium, we invoked the relationship between its force function and the polar moment of inertia, which is the source motion. This relationship reveals the nature of the gravitational energy. We also succeeded in reanimating the lost kinetic energy and obtaining both an equation of dynamics and an equation of dynamical equilibrium in the form of the oscillating motion during each period of time and within the whole duration of the system's evolution.

The problem is now to find the general solution of Jacobi's virial equation relative to oscillation and rotation of a body and to apply the solution to study of the Earth's dynamics. This application is not restricted only by the Earth's dynamics. The results are also valid for studying the Sun, the Moon and other celestial bodies.

In this chapter we show that Jacobi's virial equation provides, first of all, a solution for the models of natural systems, which have explicit solutions in the framework of the classical many-body problem. We shall give parallel solutions for both the classical and dynamical approaches, and in doing so we shall show that, with the dynamical approach, the solution acquires a new physical meaning. We shall also consider a general case of the solution of Jacobi's virial equation for conservative and dissipative systems.

V. I. Ferronsky, S. V. Ferronsky, *Dynamics of the Earth,*
DOI 10.1007/978-90-481-8723-2_4, © Springer Science+Business Media B.V. 2010

4.1 Solution of Jacobi's Virial Equation
in Classical Mechanics

The many-body problem is known to be the key problem in classical mechanics and especially in celestial mechanics. A particular example of this is the unperturbed problem of Keplerian motion, when the system consists of only two material points interacting by Newtonian law. The explicit solution of the problem of unperturbed Keplerian motion permits the many-body problem to be solved with some approximation by varying arbitrary constants. In this case the problem of dynamics, for example that of the Solar System, is transferred into the solution of the problem of dynamics of nine pairs of bodies in each of which one body is always the Sun and the second is each of the nine planets forming the system. Considering each planet-sun sub-system, the influence of the other eight planets of the system is taken into account by introducing the perturbation function. By the virial approach we can obtain for the Sun one characteristic period of circulation with respect to the center of mass of the system which will not coincide with any period of the planets. The dynamical approach evidences that the planet's orbital motion is performed by the central body, i.e. by the Sun, by the energy of its outer force field or by the field of the pressure. Each planet interacts with the solar force field by the energy of its own outer force field. The planet's orbit is the certain curve of its equilibrium motion which results from the two interacting fields of pressure. The planet's own oscillation and rotation perform by action of the inner fields of pressure.

Following these brief physical comments on the dynamical equilibrium motion of a planet, we now present two methods of solving the Keplerian problem: the classical and the integral.

4.1.1 The Classical Approach

The traditional way of solving the unperturbed Keplerian problem is excellently described in the university courses for celestial mechanics found in (Duboshin, 1978). Here we present only the principle ideas. The method consists in transforming the two-body problem described by the system of equations (3.10) into the one-body problem using six integrals of motion of the center of mass (3.13). The system of equations obtained is sixth order and expresses the change of barycentric co-ordinates of one point with respect to the center of mass of the system as a whole. Let us write it in the form

$$\ddot{x} = -\frac{\mu x}{r^3},$$

$$\ddot{y} = -\frac{\mu y}{r^3}, \qquad\qquad (4.1)$$

$$\ddot{z} = -\frac{\mu z}{r^3},$$

where μ is the constant depending on the number of the point and for which the second point is equal to

$$\mu = \frac{Gm_1^3}{(m_1 + m_2)^2}.$$

We then pass on from that Cartesian system of co-ordinates $OXYZ$ to orbital $\xi\,\eta\,\zeta$, using first integrals of the system of equations (4.1). Those are three integrals of the area,

$$\begin{aligned}
y\dot{z} - z\dot{y} &= c_1, \\
z\dot{x} - x\dot{z} &= c_2, \\
x\dot{y} - y\dot{x} &= c_3,
\end{aligned} \qquad (4.2)$$

the energy integral,

$$\dot{x}^2 + \dot{y}^2 + \dot{z}^2 = \frac{2\mu}{r} + h, \qquad (4.3)$$

and the Laplacian integrals,

$$\begin{aligned}
-\frac{\mu x}{r} + c_3\dot{y} - c_2\dot{z} &= f_1, \\
-\frac{\mu y}{r} + c_1\dot{z} - c_3\dot{x} &= f_2, \\
-\frac{\mu z}{r} + c_2\dot{x} - c_1\dot{y} &= f_3.
\end{aligned} \qquad (4.4)$$

As these seven integrals are not independent, we conclude that they cannot form a general solution of the system (4.1). In fact there are two relations for these integrals:

$$\begin{aligned}
c_1 f_1 + c_2 f_2 + c_3 f_3 &= 0, \\
f_1^2 + f_2^2 + f_3^2 &= \mu^2 + h\left(c_1^2 + c_2^2 + c_3^2\right),
\end{aligned}$$

showing that only five of them are independent. But the last integral needed can be found by simple quadrature. Using these integrals we can pass on to the system of orbital co-ordinates $O\xi\eta\zeta$ using the transformation relations (see Fig. 4.1):

$$\begin{aligned}
\xi &= \frac{f_1}{f}x + \frac{f_2}{f}y + \frac{f_3}{f}z, \\
\eta &= \frac{C_2 f_3 - C_3 f_2}{Cf}x + \frac{C_3 f_1 - C_1 f_2}{Cf}y + \frac{C_1 f_2 - C_3 f_1}{Cf}z, \\
\zeta &= \frac{C_1}{C}x + \frac{C_2}{C}y + \frac{C_3}{C}z.
\end{aligned} \qquad (4.5)$$

Fig. 4.1 Transition from
Cartesian co-ordinate system
OXYZ to orbital $O\xi\eta\zeta$

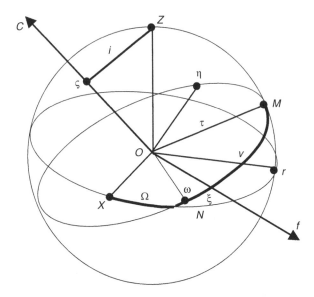

The equation of the curve along which the point moves in accordance with (4.1) has
the simplest form in the system of initial co-ordinates. The equation is

$$\zeta = 0,$$
$$\mu r = C^2 - f\xi. \tag{4.6}$$

Finally, introducing the polar orbital co-ordinates r and v, which are related to the
rectangular orbital co-ordinates ξ and η by the expressions (see Fig. 4.2)

$$\xi = r\cos v \quad \text{and} \quad \eta = r\sin v,$$

and using the integral of areas

$$r^2 v = C,$$

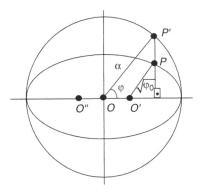

Fig. 4.2 Relationship
between the polar and the
rectangular co-ordinates

we come to the equation

$$C(t - r) = \left(\frac{C^2}{\mu}\right)^2 \int_0^v \frac{dv}{\left(1 + \frac{f}{\mu}\cos v\right)^2}. \tag{4.7}$$

The solution of Eq. (4.7) gives the change of function v with respect to time. Repetition of the transformation in the reverse order leads to solution of the problem. In doing this, we obtain the expression for the change of co-ordinates of the material point with respect to the initial data $\xi_{10}, \eta_{10}, \zeta_{10}, \xi_{20}, \eta_{20}, \zeta_{20}, \dot{\xi}_{10}, \dot{\eta}_{10}, \dot{\zeta}_{10}, \dot{\xi}_{20}, \dot{\eta}_{20}, \dot{\zeta}_{20}$. It is remarkable that if the total energy (4.3) has negative value, then the solution of Eq. (4.7) leads to the Keplerian equation

$$E' - e \sin E' = n(t - \tau), \tag{4.8}$$

where the function v is related to the variable E' by the expression

$$tg\frac{v}{2} = \sqrt{\frac{1+e}{1-e}} tg\frac{E}{2},$$

and

$$e = \frac{f}{\mu}, \quad n = \frac{\sqrt{\mu}}{a^{3/2}}, \quad p = \frac{C^2}{\mu} = a(1 - e^2).$$

Because energy by definition is the property to do work (motion) and can be only a positive value, then the physical meaning of negative total energy which defines the elliptic orbit of a body moving in the central field of the two-body problem should be revealed. In the presented solution of the two-body problem, the left-hand side of the energy integral (4.3) expresses the kinetic energy and the right-hand side means the potential energy of the mass interaction. The integral of energy (4.3) as a whole, in the co-ordinates and in the velocities, represents the averaged virial theorem, where the potential energy has formally a negative value. Here the physical meaning of the total energy determination consists in comparison of magnitude of the potential and kinetic energy. A negative value of the total energy means that the potential energy exceeds the kinetic one by that value. As it follows from analysis of the inner force field of a self-gravitating body presented in Chap. 2, the potential energy exceeds the kinetic one only in the case of non-uniform distribution of the mass density and cannot be less than it. In the case of equality of both energies the total potential energy is realized into oscillating motion. The excited part of the potential energy is used for rotation of the masses and for the dissipation. The last case is discussed in Chap. 6.

4.1.2   The Dynamic (Virial) Approach

Let us consider the solution of the problem of unperturbed motion of two material points on the basis of Jacobi's virial equation which in accordance with Eq. (3.23) is written in the form

$$\ddot{\Phi}_0 = 2E_0 - U,$$

where $E_0 = T_0 + U = const$ is the total energy of the system in a barycentric co-ordinate system;

The Jacobi function Φ_0 is expressed by (3.22):

$$\Phi_0 = \frac{m_1 m_2}{2(m_1 + m_2)}[(\xi_1 - \xi_2)^2 + (\eta_1 - \eta_2)^2 + (\zeta_1 - \zeta_2)^2],$$

and the potential energy U in accordance with (3.9) is

$$U = \frac{G m_1 m_2}{\sqrt{(\xi_1 - \xi_2)^2 + (\eta_1 - \eta_2)^2 + (\zeta_1 - \zeta_2)^2}}.$$

It is easy to see that between the Jacobi function Φ_0 and the potential energy U exists the relationship

$$|U|\sqrt{\Phi_0} = \frac{G(m_1 m_2)^{3/2}}{\sqrt{2(m_1 + m_2)}} = G^{1/2} m \mu^{3/2} = B = const, \qquad (4.9)$$

where μ is the generalized mass of the two bodies; m is the total mass of the system; B is a constant value.

The relationship (4.9) is remarkable because it is independent of the initial data, i.e. of its co-ordinates and velocities. Being an integral characteristic of the system and dependent only on the total mass and the generalized mass of the two points, the relationship permits Jacobi's virial equation to be transformed to an equation with one variable as follows:

$$\ddot{\Phi}_0 = 2E_0 + \frac{B}{\sqrt{\Phi_0}}. \qquad (4.10)$$

We consider the solution of Eq. (4.10) for the case when total energy E_0 has negative value. Introducing $A = -2E_0 > 0$, equation (4.10) can be rewritten:

$$\ddot{\Phi}_0 = -A + \frac{B}{\sqrt{\Phi_0}}. \qquad (4.11)$$

We apply the method of change of variable for solution of Eq. (4.11) and show that partial solution of two linear equations (Ferronsky et al., 1984b):

$$\left(\sqrt{\Phi_0}\right)'' + \sqrt{\Phi_0} = \frac{B}{A}, \tag{4.12}$$

$$t'' + t = \frac{4B\lambda}{(2A)^{3/2}}, \tag{4.13}$$

which include only two integration constants, is also the solution of Eq. (4.11).

We now introduce the independent variable λ into Eqs. (4.12) and (4.13), where primes denote differentiation with respect to λ. Note that time here is not an independent variable. This allows us to search for the solution of two linear equations instead of solving one non-linear equation. The solution of Eqs. (4.12) and (4.13) can be written in the form

$$\sqrt{\Phi_0} = \frac{B}{A}[1 - \varepsilon\cos(\lambda - \psi)], \tag{4.14}$$

$$t = \frac{4B}{(2A)^{3/2}}[\lambda - \varepsilon\sin(\lambda - \psi)]. \tag{4.15}$$

Let us prove that the partial solution (4.14) and (4.15) differential equations (4.12) and (4.13) is the solution of Eq. (4.10) that is sought. For this purpose we express the first and second derivatives of the function $\sqrt{\Phi_0}$ with respect to λ through corresponding derivatives with respect to time using Eq. (4.15). It follows from (4.15) that

$$\frac{dt}{d\lambda} = \frac{4B}{(2A)^{3/2}}[1 - \varepsilon\cos(\lambda - \psi)]. \tag{4.16}$$

We can replace the right-hand side of the obtained relationship by $\sqrt{\Phi_0}$ from (4.14)

$$\frac{dt}{d\lambda} = \sqrt{\Phi_0}\sqrt{\frac{2}{A}}. \tag{4.17}$$

Transforming the derivative from $\sqrt{\Phi_0}$ with respect to λ into the form

$$\frac{d\sqrt{\Phi_0}}{d\lambda} = \frac{d\sqrt{\Phi_0}}{dt}\frac{dt}{d\lambda} = \frac{\dot{\Phi}_0}{2\sqrt{\Phi_0}}\frac{dt}{d\lambda}$$

and taking into account (4.17), we can write

$$\left(\sqrt{\Phi_0}\right)' = \frac{\dot{\Phi}_0}{\sqrt{2A}}.$$

The second derivative can be written analogously:

$$\left(\sqrt{\Phi_0}\right)'' = \frac{dt}{d\lambda}\frac{d}{dt}\left(\frac{\dot{\Phi}_0}{\sqrt{2A}}\right) = \frac{\ddot{\Phi}_0}{\sqrt{2A}}\sqrt{\Phi_0}\sqrt{\frac{2}{A}} = \frac{\ddot{\Phi}_0\sqrt{\Phi_0}}{A}. \qquad (4.18)$$

Putting Eq. (4.18) into (4.12), we obtain

$$\frac{\ddot{\Phi}_0\sqrt{\Phi_0}}{A} + \sqrt{\Phi_0} = \frac{B}{A}.$$

Dividing the above expression by $\sqrt{\Phi_0}/A$, we can finally write

$$\ddot{\Phi}_0 = -A + \frac{B}{\sqrt{\Phi_0}}.$$

This shows that the partial solution of the two linear differential equations (4.12) and (4.13) appears to be the solution of the non-linear equation (4.11).

4.2 Solution of the *n*-Body Problem in the Framework of a Conservative System

After solving Jacobi's virial equation for the unperturbed two-body problem, we come to dynamics of a system of n material particles where $n \to \infty$.

Let us assume that an external observer studying the dynamics of a system of n particles in the framework of classical mechanics has the following information. He has the mass of the system, its total and potential energy and can determine the Jacobi function and its first derivative with respect to time in any arbitrary moment. Then he can use Jacobi's virial equation (4.9) and, making only the assumption needed for its solution that $|U|\sqrt{\Phi_0} = B = const$, may predict the dynamics of the system, i.e. the dynamics of its integral characteristics at any moment of time.

If the total energy E_0 of the system has negative value, the external observer can immediately write the solution of the problem of the Jacobi function change with respect to time in the form of (4.14) and (4.15):

$$\sqrt{\Phi_0} = \frac{B}{A}[1 - \varepsilon \cos(\lambda - \psi)],$$

$$t = \frac{4B}{(2A)^{3/2}}[\lambda - \varepsilon \sin(\lambda - \psi)],$$

where $A = -2E_0$; ε and ψ are constants depending on the initial values of the Jacobi function Φ_0 and its first derivative $\dot{\Phi}_0$ at the moment of time t_0.

Let us obtain the values of constants ε and ψ, in explicit form expressed through the values Φ_0 and $\dot{\Phi}_0$ at the initial moment of time t_0. For convenience we introduce a new independent variable φ, connected to λ by the relationship $\lambda - \psi = \varphi$. Then Eqs. (4.14) and (4.15) can be rewritten:

$$\sqrt{\Phi_0} = \frac{B}{A}[1 - \varepsilon \cos \varphi], \tag{4.19}$$

$$t - \frac{4B}{(2A)^{3/2}} \psi = \frac{4B}{(2A)^{3/2}}[\varphi - \varepsilon \sin \varphi]. \tag{4.20}$$

Using Eq. (4.19) we write the expression for φ:

$$\varphi = arccos \frac{1 - \frac{A}{B}\sqrt{\Phi_0}}{\varepsilon} \tag{4.21}$$

and taking into account the equality

$$\frac{d\sqrt{\Phi_0}}{d\lambda} = \frac{d\sqrt{\Phi_0}}{d\varphi},$$

substituting Eq. (4.21) into the expression

$$\frac{\dot{\Phi}_0}{\sqrt{2A}} = \frac{B}{A}\varepsilon \sin \varphi.$$

The last equation can be rewritten finally in the form

$$\frac{\dot{\Phi}_0}{\sqrt{2A}} = \frac{B}{A}\varepsilon \sqrt{1 - \left(\frac{1 - \frac{A}{B}\sqrt{\Phi_0}}{\varepsilon}\right)^2}. \tag{4.22}$$

Equation (4.22) allows us to determine the first constant of integration ε as a function of the initial data Φ_0 and $\dot{\Phi}_0$ at $t = t_0$. Solving equation (4.22) with respect to ε after simple algebraic transformation, we obtain

$$\varepsilon = \sqrt{1 - \frac{A}{2B^2}\left(-\dot{\Phi}_0 + 4B\sqrt{\Phi_0} - 2A\Phi_0\right)}\Big|_{t=t_0} = const. \tag{4.23}$$

The second constant of integration ψ can be expressed through the initial data after solving Eq. (4.20) with respect to ψ and change of value φ by its expression from

Eq. (4.21). Defining

$$t - \frac{4B}{(2A)^{3/2}}\psi = \tau,$$

we obtain

$$-\tau = \left\{ \frac{4B}{(2A)^{3/2}} \left[arccos\frac{1 - \frac{A}{B}\sqrt{\Phi_0}}{\varepsilon} \right.\right.$$

$$\left.\left. -\varepsilon\sqrt{1 - \left(\frac{1 - \frac{A}{B}\sqrt{\Phi_0}}{\varepsilon}\right)^2} \right] - t \right\} |_{t=t_0} = const. \qquad (4.24)$$

The physical meaning of the integration constants ε, τ, and the parameter $T_v = 8\pi B/(2A)^{3/2}$ can be understood after the definitions

$$T_v = \frac{8\pi B}{(2A)^{3/2}},$$

$$n = \frac{2\pi}{T_v} = \frac{(2A)^{3/2}}{4B},$$

$$a = \frac{B}{A}$$

and rewriting Eqs. (4.19) and (4.20) in the form

$$\sqrt{\Phi_0} = a\,(1 - \varepsilon\cos\varphi), \qquad (4.25)$$

$$M = \varphi - \varepsilon\sin\varphi, \qquad (4.26)$$

where $M = n(t - \tau)$.

The value $\sqrt{\Phi_0}$ draws an ellipse during the period of time $T_v = 8\pi B/(2A)^{3/2}$ (see Fig. 4.3). The ellipse is characterized by a semi-major axis a equal to B/A and by the eccentricity ε which is defined by expression (4.23). In the case considered, where $E_0 < 0$, the value ε is changed in time from 0 to 1. The value τ characterizes the moment of time when the ellipse passes the pericenter.

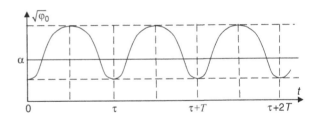

Fig. 4.3 Changes of the
Jacobi function over time

Let us obtain explicit expressions with respect to time for the functions $\sqrt{\Phi_0}$, Φ_0 and $\dot{\Phi}_0$. For this purpose we write Eq. (4.24) in the form of a Lagrangian:

$$F(\varphi) = \varphi - \varepsilon \sin \varphi - M = 0. \tag{4.27}$$

It is known (Duboshin, 1978) that by application of Lagrangian formulas we can write in the form of a series the expressions for the root of the Lagrange equation (4.27) and for the arbitrary function f which is dependent φ:

$$\varphi = \sum_{k=0}^{\infty} \frac{\varepsilon^{k-1}}{k!} \frac{d^{k-1}}{dM^{k-1}} \left[\sin^k M \right] = M + \varepsilon \sin M + \frac{\varepsilon^2}{1 \cdot 2} \frac{d}{dM} \left[\sin^2 M \right] + \cdots , \tag{4.28}$$

$$f(\varphi) = \sum_{k=0}^{\infty} \frac{\varepsilon^{k-1}}{k!} \frac{d^{k-1}}{dM^{k-1}} \left[f'(M) \sin^k M \right] = f(M) + \varepsilon f'(M) \sin M$$

$$+ \frac{\varepsilon^2}{1 \cdot 2} \frac{d}{dM} \left[f(M) \sin^2 M \right] + \cdots . \tag{4.29}$$

Using Eq. (4.29), we write expressions for $\cos \varphi$, $\cos^2 \varphi$ and $\sin \varphi$ in the form of a Lagrangian series of parameter ε power:

$$\cos \varphi = \sum_{k=0}^{\infty} \frac{\varepsilon^{k-1}}{k} \frac{d^{k-1}}{dM^{k-1}} [(-1) \sin M \sin^k M] = \cos M + \varepsilon(-1) \sin M \sin(M)$$

$$+ \frac{\varepsilon^2}{1 \cdot 2} \frac{d}{dM} [(-1) \sin(M) \sin^2 M] + \cdots = \cos M - \frac{\varepsilon}{2} + \frac{\varepsilon}{2} \cos 2M$$

$$- \frac{3}{4} \varepsilon^3 \cos M + \frac{3}{8} \varepsilon^2 \cos 3M + \cdots . \tag{4.30}$$

$$\cos^2 \varphi = \sum_{k=0}^{\infty} \frac{\varepsilon^{k-1}}{k!} \frac{d^{k-1}}{dM^{k-1}} [(-2) \sin M \cos M \sin^k M] = \cos^2 M$$

$$+ \varepsilon(-2) \sin M \cos M \sin M + \frac{\varepsilon^2}{1 \cdot 2} \frac{d}{dM} [(-2) \sin M \cos M \sin^2 M] + \cdots$$

$$= \cos^2 M - 2\varepsilon \sin^2 M \cos M + \frac{\varepsilon^2}{2} (-2)(3 \sin^2 M \cos^2 M - \sin^4 M) + \cdots \tag{4.31}$$

$$\sin \varphi = \sum_{k=0}^{\infty} \frac{\varepsilon^{k-1}}{k!} \frac{d^{k-1}}{dM^{k-1}} [\cos M \sin M] = \sin M + \varepsilon \cos M \sin M$$

$$+ \frac{\varepsilon^2}{1 \cdot 2} \frac{d}{dM} [\cos M \sin^2 M] + \cdots = \sin M + \varepsilon \cos M \sin M$$

$$+ \frac{\varepsilon^2}{1 \cdot 2} [2 \sin M \cos^2 M - \sin^3 M] + \cdots . \tag{4.32}$$

We write the expressions for $\sqrt{\Phi_0}, \Phi_0, \dot{\Phi}_0$ using Eqs. (4.25) and (4.26) in the form

$$\sqrt{\Phi_0} = a(1 - \varepsilon\cos\varphi), \tag{4.33}$$

$$\Phi_0 = a^2\left(1 - 2\varepsilon\cos\varphi + \varepsilon^2\cos^2\varphi\right), \tag{4.34}$$

$$\dot{\Phi}_0 = \sqrt{\frac{2}{A}}\varepsilon B\sin\varphi. \tag{4.35}$$

Substituting into (4.33) to (4.35) the expressions for $\cos\varphi$, $\cos^2\varphi$ and $\sin\varphi$ in the form of the Lagrangian series (4.30) to (4.32) we obtain

$$\sqrt{\Phi_0} = \frac{B}{A}\left[1 + \frac{\varepsilon^2}{2} + \left(-\varepsilon + \frac{3}{8}\varepsilon^3\right)\cos M - \frac{\varepsilon^2}{2}\cos 2M - \frac{3}{8}\varepsilon^3\cos 3M + \cdots\right], \tag{4.36}$$

$$\Phi_0 = \frac{B^2}{A^2}\left[1 + \frac{3}{2}\varepsilon^2 + \left(-2\varepsilon + \frac{\varepsilon^3}{4}\right)\cos M - \frac{\varepsilon^2}{2}\cos 2M - \frac{\varepsilon^3}{4}\cos 3M + \cdots\right], \tag{4.37}$$

$$\dot{\Phi}_0 = \sqrt{\frac{2}{A}}\varepsilon B\left[\sin M + \frac{1}{2}\varepsilon\sin 2M + \frac{\varepsilon^2}{2}\sin M\left(2\cos^2 M - \sin^2 M\right) + \cdots\right]. \tag{4.38}$$

The series of equations (4.36) to (4.38) obtained are put in order of increased power of parameter ε and are absolutely convergent at any value of M in the case when the parameter ε satisfies the condition

$$\varepsilon < \bar{\varepsilon} = 0.6627\ldots, \tag{4.39}$$

where $\bar{\varepsilon}$ is the Laplace limit.

In some cases it is convenient to expand the values $\sqrt{\Phi_0}, \Phi_0, \dot{\Phi}_0$ in the form of a Fourier series, using conventional methods (see, for example, Duboshin, 1978). Figure 4.4 demonstrates the changes of $\sqrt{\Phi_0}$ in time at $\varepsilon = 1$.

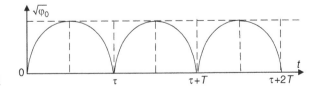

Fig. 4.4 Changes of the value $\sqrt{\Phi_0}$ in time at $\varepsilon = 1$

It is also possible to consider the case solution of Jacobi's virial equation for $E_0 = 0$ and $E_0 > 0$. The reader can find here without difficulty a full analogy of these results as well as the solution of the two-body problem.

4.3 Solution of Jacobi's Virial Equation in Hydrodynamics

Let us consider the solution of the problem of the dynamics of a homogeneous isotropic gravitating sphere in the framework of traditional hydrodynamics and the virial approach we have developed.

4.3.1 The Hydrodynamic Approach

The sphere is assumed to have radius R_0 and be filled by an ideal gas with ρ_0. We assume that at the initial time the field of velocities which has the only component is described by equation

$$u = H_0 r, \tag{4.40}$$

where u is the radial component of the velocity of the sphere's matter at the distance r from the center of mass; H is independent of the quantity r and equal to H_0 at time t_0.

We also assume that the motion of the matter of the sphere goes on only under action of the forces of mutual gravitational interaction between the sphere particles. In this case the influence of the pressure gradient is not taken into account, assuming that the matter of the sphere is sufficiently diffused. Then the symmetric spherical shells will move only under forces of gravitational attraction and will not coincide. In this case the mass of the matter of any sphere shell will keep its constant value and the condition (4.40) will be satisfied at any moment of time, and constant H should be dependent on time.

Under those conditions the Eulerian system of equations (3.35) can be written in the form

$$\rho \frac{\partial u}{\partial t} + \rho(u\nabla)u = \rho \frac{\partial U_G}{\partial r}, \tag{4.41}$$

where $\rho(t)$ is the density of the matter of the sphere at the moment of time t; u is the radial component of the velocity of matter at distance r from the sphere's center; U_G is the Newtonian potential for the considered point of the sphere.

The expression for the Newtonian potential U_G (3.36) can be written as follows:

$$U_G = G \frac{4}{3}\pi \rho r^2, \tag{4.42}$$

and the continuity equation will be

$$\frac{\partial \rho}{\partial t} + \rho \frac{\partial U}{\partial r} = 0. \tag{4.43}$$

Within the framework of the traditional approach, the problem is to define the sphere radius R and the value of the constant H at any moment of time, if the radius R_0, density ρ_0 and the value of the constant H_0 at the initial moment of time t_0 are given. If we know the values $H(t)$ and $R(t)$, we can then obtain the field of velocities of the matter within the sphere which is defined by Eq. (4.40), and also the density ρ of matter at any moment of time, using the relationship

$$\frac{4}{3} \pi R_0^3 \rho_0 = \frac{4}{3} \pi R^3 \rho = m.$$

Hence the formulated problem is reduced to identification of the law of motion of a particle which is on the surface of the sphere and within the field of attraction of the entire sphere mass $m = 4/3\pi\rho_0 R_0^3$.

The equation of motion for a particle on the surface of the sphere, which follows from Eq. (4.41) after transforming the Eulerian co-ordinates into a Lagrangian, has the form

$$\frac{d^2 R}{dt^2} = -G \frac{m}{R^2}. \tag{4.44}$$

It is necessary to determine the law of change of $R(t)$, resolving Eq. (4.44) at the initial data:

$$R(t_0) = R_0,$$
$$\left. \frac{dR}{dt} \right|_{t=t_0} = H_0 R_0. \tag{4.45}$$

We reduce the order of Eq. (4.44). To do so we multiply it by dR/dt:

$$\frac{dR}{dt} \frac{d^2 R}{dt^2} = -\frac{dR}{dt} \frac{Gm}{R^2}$$

and integrate with respect to time:

$$\int_{t_0}^{t} \frac{1}{2} \frac{d}{dt} (\dot{R})^2 = \int_{t_0}^{t} \frac{d}{dt} \left(\frac{Gm}{R} \right) dt.$$

After integration we obtain

$$\frac{1}{2} \dot{R}^2 - \frac{1}{2} \dot{R}_0^2 = \frac{Gm}{R} - \frac{Gm}{R_0}$$

or

$$\frac{1}{2}\dot{R}^2 = \frac{Gm}{R} + k,$$ (4.46)

where the constant k is determined as

$$k = \frac{1}{2}\dot{R}_0^2 - \frac{Gm}{R_0} = \frac{1}{2}H_0^2 R_0^2 - G\frac{4\pi}{3}\rho_0\frac{R_0^3}{R_0}$$

$$= \frac{1}{2}H_0^2 R_0^2\left[1 - \frac{8\pi}{3}\frac{G\rho_0}{H_0^2}\right] = \frac{1}{2}H_0^2 R_0^2\left[1 - \Omega\right] = const$$ (4.47)

Here the quantity $\Omega = \rho_0/\rho_{cr}$, where $\rho_{cr} = 3H_0^2/8\pi G$.

Note that Eq. (4.46) obtained after reduction of the order of the initial equation (4.44) is in its substance the energy conservation law. Equation (4.46) permits the variables to be divided and can be rewritten in the form

$$\int_{R_0}^{R} \frac{dR}{\sqrt{\frac{2Gm}{R} + 2k}} = \int_{t_0}^{t} dt.$$ (4.48)

The plus sign before the root is chosen assuming that the sphere at the initial time is expanding, i.e., $H_0 > 0$.

The differential equation (4.46) has three different solutions at $k = 0$, $k > 0$ and $k < 0$ depending on the sign of the constant k, which is in its turn defined by the value of the parameter Ω at the initial moment of time. First we consider the case when $k = 0$ which relates, by analogy with the Keplerian problem, to the parabolic model at $k = 0$. Equation (4.46) is easily integrated and for the expression case, i.e., $\dot{R} > 0$, we obtain

$$\dot{R}^2 = \frac{2Gm}{R},$$

$$\dot{R} = \frac{(2Gm)^{1/2}}{R^{1/2}},$$

from which it follows that

$$R^{1/2}dR = (2Gm)^{1/2}dt$$

or

$$\frac{2}{3}R^{3/2} = (2Gm)^{1/2}\,t + const.$$ (4.49)

We choose as initial counting time $t = 0$, the moment when $R = 0$. In this case the integration constant disappears:

$$R = \left(\frac{9}{2}Gm\right)^{1/3} t^{2/3}. \tag{4.50}$$

The density of the matter changes in accordance with the law

$$\rho(t) = \frac{m}{\frac{4}{3}\pi R^3} = \frac{1}{6\pi Gt^2}, \tag{4.51}$$

and the quantity $H(t)$, as a consequence of (4.50), has the form

$$H(t) = \frac{\dot{R}}{R} = \frac{2}{3}\frac{1}{t}. \tag{4.52}$$

For the case when $k > 0$, which corresponds to so-called hyperbolic motion, the solution of Eq. (4.46) can be written in parametric form (Zeldovich and Novikov, 1967)

$$R = \frac{Gm}{2k}(ch\eta - 1),$$
$$t = \frac{Gm}{(2k)^{3/2}}(sh\eta - \eta), \tag{4.53}$$

where the constants of integration in (4.53) have been chosen so that $t = 0$, $\eta = 0$ at $R = 0$.

Finally we consider the case when $k < 0$, which corresponds to elliptic motion. At $k < 0$ the expansion of the sphere cannot continue for unlimited time and the expansion phase should be changed by attraction of the sphere.

The explicit solution of Eq. (4.46) at $k < 0$ can be written in parametric form (Zeldovich and Novikov, 1967)

$$R = \frac{Gm}{2|k|}(1 - ch\eta),$$
$$t = \frac{Gm}{(2|k|)^{3/2}}(\eta - sh\eta). \tag{4.54}$$

The maximum radius of the sphere is determined from Eq. (4.46) on the condition $dR/dt = 0$ and equals

$$R_{max} = \frac{Gm}{|E|}. \tag{4.55}$$

The time needed for expansion of the sphere from $R_0 = 0$ at $t_0 = 0$ to R_{max} is

$$t_{max} = \frac{\pi Gm}{(2|k|)^{3/2}}.$$ (4.56)

So the sphere should make periodic pulsations with period T_p equal to

$$T_p = \frac{2\pi Gm}{(2|k|)^{3/2}}.$$ (4.57)

The considered solution has important cosmologic applications.

4.3.2 The Virial Approach

We shall limit ourselves by formal consideration of the same problem in the framework of the condition of the dynamical equilibrium of a self-gravitating body based on the solution of Jacobi's virial equation, which we discussed earlier.

As shown in Chap. 3, Jacobi's virial equation (3.57), derived from Eulerian equations (3.35), is valid for the considered gravitating sphere. It was written in the form

$$\ddot{\Phi} = 2E - U,$$ (4.58)

where Φ is the Jacobi function for a homogeneous isotropic sphere and is defined by

$$\Phi = \frac{1}{2} \int_0^R 4\pi r^2 \rho r^2 dr = \frac{2\pi \rho R^5}{5} = \frac{3}{10} mR^2.$$ (4.59)

The potential gravitational energy of the matter of the sphere is expressed as

$$U = -4\pi G \int_0^R r\rho(r)m(r)dr = -\frac{16\pi^2}{15} G\rho^2 R^2 = -\frac{3}{5} G\frac{m^2}{R}.$$ (4.60)

The total energy of the sphere E will be equal to the sum of the potential U and kinetic T energies.

The kinetic energy T is expressed as

$$T = \frac{1}{2} \int_0^R 4\pi u^2 \rho r^2 dr = \frac{1}{2} \int_0^R 4\pi H^2 r^2 \rho r^2 dr = \frac{4\pi \rho H^2 R^5}{10} = \frac{3}{10} mH^2 R^2.$$ (4.61)

For a homogeneous isotropic gravitating sphere, the constancy of the relationship between the Jacobi function (4.59) and the potential energy (4.60) can be written:

$$|U|\sqrt{\Phi} = B = \frac{3}{5}G\frac{m^2}{R}\sqrt{\frac{3}{10}mR^2} = \frac{1}{\sqrt{2}}\left(\frac{3}{5}\right)^{3/2}Gm^{3/2}, \qquad (4.62)$$

where B has constant value because of the conservation law of mass m of the considered sphere.

The total energy E of the sphere also has a constant value:

$$E = T + U = \frac{A}{2}. \qquad (4.63)$$

Then, if the total energy of the sphere has a negative value, Jacobi's virial equation can be written in the form:

$$\ddot{\Phi} = -A + \frac{B}{\sqrt{\Phi}}. \qquad (4.64)$$

Let us consider the conditions under which the total energy of the system will have a negative value. For this purpose we write it explicitly:

$$E = T + U = -\frac{16}{15}\pi^2 G\rho^2 R^5 + \frac{2\pi\rho H^2 R^5}{5} = \frac{2}{5}\pi\rho H^2 R^5\left[1 - \frac{8\pi G\rho}{3H^2}\right]. \qquad (4.65)$$

It is clear from Eq. (4.65) that the total energy E has a negative value, when $\rho > \rho_c$, where $\rho_c = 3H^2/8\pi G$.

The general solution of Eq. (4.64) has the form of Eqs. (4.14) and (4.15):

$$\sqrt{\Phi_0} = \frac{B}{A}[1 - \varepsilon\cos(\lambda - \psi)], \qquad (4.66)$$

$$t = \frac{4B}{(2A)^{3/2}}[\lambda - \varepsilon\sin(\lambda - \psi)], \qquad (4.67)$$

where ε and ψ are constants dependent on the initial values of the Jacobi function Φ_0 and its first derivative $\dot{\Phi}_0$ at the moment of time t_0. The constants ε and ψ are determined by Eqs. (4.23) and (4.24) accordingly.

If we express all the constants in Eq. (4.23):

$$\varepsilon = \sqrt{1 - \frac{A}{2B^2}\left(-\dot{\Phi}_0 + 4B\sqrt{\Phi_0} - 2A\Phi_0\right)}\Big|_{t=t_0} = const. \qquad (4.68)$$

through mass m of the system, it is not difficult to see that

$$-\dot{\Phi}_0^2 + 4B\sqrt{\Phi_0} - 2A\Phi_0 = 0.$$

Then the constant ε will be equal to zero. Hence the solutions (4.28) and (4.29) coincide with the solution (4.54), which was obtained in the framework of the traditional hydrodynamic approach. In this case the period of eigenpulsations of the Jacobi function (the polar moment of inertia) of the sphere $T = 8\pi R/(2A)^{3/2}$ will be equal to the period of change of its radius $T_p = 2\pi Gm/(2|k|)^{3/2}$ obtained from Eq. (4.54).

4.4 The Hydrogen Atom as a Quantum Mechanical Analogue of the Two-Body Problem

Let us consider the problem concerning the energy spectrum of the hydrogen atom, which is a unique example of the complete conformity of the analytical solution with experimental results. The problem consists of a study of all the forms of motion using the postulates of quantum mechanics and based on the solution of Jacobi's virial equation.

The classical Hamiltonian in the two-body problem is written as

$$H = \frac{\bar{p}_1^2}{2m_1} + \frac{\bar{p}_2^2}{2m_2} + U(|\bar{r}_1 - \bar{r}_2|), \qquad (4.69)$$

where

$$\bar{p}_1 = \frac{\partial H}{\partial \dot{\bar{r}}_1} = m_1 \dot{\bar{r}}_1,$$

$$\bar{p}_2 = \frac{\partial H}{\partial \dot{\bar{r}}_2} = m_2 \dot{\bar{r}}_2,$$

which after separation of the center of mass can be transformed into the form

$$H = \frac{\bar{P}^2}{2M} + \frac{\bar{p}^2}{2m} + U(r), \qquad (4.70)$$

where $r = |\bar{r}_1 - \bar{r}_2|$ is the distance between two particles and

$$\bar{P} = M\dot{\bar{R}}; \quad \bar{p} = m\dot{\bar{r}}; \quad M = m_1 + m_2;$$

$$\bar{R} = \frac{m_1\bar{r}_1 + m_2\bar{r}_2}{m_1 + m_2}; \quad m = \frac{m_1 m_2}{m_1 + m_2}.$$

We obtain the Hamiltonian operator for the quantum mechanical two-body problem through changing the pulses and radii by the corresponding operators with the communication relations

$$[\hat{p}_i, \hat{p}_k] = -i\hbar\delta_{ik},$$

$$[\hat{p}_i, \hat{r}_k] = -i\hbar\delta_{ik}.$$

Then

$$\hat{H} = -\frac{\hbar^2}{2M}\Delta_R - \frac{\hbar^2}{2m}\Delta_r + \hat{U}(r).$$

The wave function $u(\bar{r}_1, \bar{r}_2) = \varphi(\bar{R})\Psi(\bar{r})$, which satisfies the Schrödinger equation

$$\hat{H}u = \varepsilon u,$$

describes the motion of the inertia center (the free motion of the particle of mass m_c is described by the function and the motion of the particle of mass m in the $U(r)$ is described by the wave function $\Psi(\bar{r})$). Subsequently we consider only the wave function of the motion of particle m.

The Schrödinger equation

$$\Delta\Psi + \frac{2m}{\hbar^2}\left[E - U(r)\right]\Psi = 0$$

written here for the stationary state in a central symmetrical field in spherical coordinates, has the form

$$\frac{1}{r^2}\frac{\partial}{\partial r}\left(r^2\frac{\partial\Psi}{\partial r}\right) + \frac{1}{r^2}\left[\frac{1}{\sin\Theta}\frac{\partial}{\partial\Theta}\left(\sin\Theta\frac{\partial\Psi}{\partial\Theta}\right) + \frac{1}{\sin^2\Theta}\frac{\partial^2\Psi}{\partial\varphi^2}\right]$$
$$+ \frac{2m}{\hbar^2}\left[E - U(r)\right]\Psi = 0. \tag{4.71}$$

Using the Laplacian operator $\hat{\ell}^2$:

$$\hat{\ell}^2 = \left[\frac{1}{\sin\Theta}\frac{\partial}{\partial\Theta}\left(\sin\Theta\frac{\partial}{\partial\Theta}\right) + \frac{1}{\sin^2\Theta}\frac{\partial^2}{\partial\varphi^2}\right],$$

we obtain

$$\frac{\hbar^2}{2m}\left[-\frac{1}{r^2}\frac{\partial}{\partial r}\left(r^2\frac{\partial\Psi}{\partial r}\right) + \frac{\hat{\ell}^2}{r^2}\Psi\right] + U(r)\Psi = E\Psi.$$

The operators $\hat{\ell}^2$ and $\hat{\ell}_z(\hat{\ell}_z = -i\partial/\partial\varphi)$ commute with the Hamiltonian $\hat{H}(r)$ and therefore there are common eigenfunctions of the operators $\hat{H}$, $\hat{\ell}^2$ и $\hat{\ell}_z$. We consider only such solutions of Schrödinger equations. This condition determines the dependence of the function Ψ on the angles

$$\Psi(r, \Theta, \varphi) = R(r)Y_{\ell k}(\Theta, \varphi),$$

where the quantity $Y_{\ell k}(\Theta, \varphi)$ is determined by the expression

$$Y_{\ell k}(\Theta, \varphi) = \frac{1}{\sqrt{2\pi}} e^{ik\varphi} (-1)^k i^\ell \sqrt{\frac{(2\ell + 1)(\ell - k)!}{2(\ell + k)!}} P_\ell^k(\cos\Theta),$$

and $P_\ell^k(\cos\Theta)$ is the associated Legendre polynomial, which is

$$P_\ell^k(\cos\Theta) = \frac{1}{2^\ell \ell!} \sin^k\Theta \frac{d^{r+\ell}}{d\cos\Theta^{r+\ell}} (\cos^2\Theta - 1)^\ell.$$

Since

$$\hat{\ell}^2 Y_{\ell k} = \ell(\ell + 1) Y_{\ell k},$$

we obtain for the radial part of the wave function $R(r)$

$$\frac{1}{r^2} \frac{\partial}{\partial r} \left(r^2 \frac{dR}{dr} \right) - \frac{\ell(\ell + 1)}{r^2} R + \frac{2m}{\hbar^2} [E - U(r)] R = 0. \qquad (4.72)$$

Equation (4.72) does not contain the value $\ell_z = m$, i.e. at the given ℓ the energy level E corresponds to $2\ell + 1$ states differing by the value ℓ_z.

The operator

$$\frac{1}{r^2} \frac{\partial}{\partial r} \left(r^2 \frac{d\Psi}{dr} \right)$$

is equivalent to the expression

$$\frac{1}{r} \frac{d^2}{dr^2} (rR)$$

and thus it is convenient to make the change of variables, assuming that

$$X(r) = rR(r).$$

So that Eq. (4.43) can be rewritten in the form

$$\frac{d^2 X}{dr^2} - \frac{\ell(\ell + 1)}{r^2} X + \frac{2m}{\hbar^2} [E - U(r)] X = 0. \qquad (4.73)$$

We now consider the demand following from the boundary conditions and related to the behavior of the wave function $X(r)$. At $r \to 0$ and the potentials satisfying the condition

$$\lim_{r \to 0;} U(r) r^2 = 0, \qquad (4.74)$$

only the first two terms play an important role in Eq. (4.73). $X(r) \sim r^\nu$ and we obtain

$$v(\nu - 1) = \ell(\ell + 1).$$

This equation has roots $\nu_1 = \ell + 1$ and $\nu_2 = -\ell$.

The requirement of normalization of the wave function is incompatible with the values $\nu = -\ell$ at $\ell \neq 0$ because the normalization integral

$$\int\limits_0^\infty |X_r^2(r)dr|$$

will be divergent for the discrete spectrum, and the condition

$$\int \Psi(\lambda, \xi)\Psi(\lambda, X)d\lambda = \delta(X - \xi)$$

does not hold for the continuous spectrum.

At $\ell = 0$ the boundary conditions are determined by the demand for the finiteness of the mean value of the kinetic energy which is satisfied only at $\nu = 1$. So, when the condition (4.74) is satisfied, then the wave function of a particle is everywhere finite and at any ℓ

$$X(0) = 0.$$

Let us consider the energy spectrum and the wave function of the bounded states of a system of two charges. The bounded states exist only in the case of the attracted particles. Such a system defines the properties of the hydrogen atom and hydrogen-like ions.

The equation for the radial wave function is

$$\frac{d^2 R}{dt^2} + \frac{2}{r}\frac{dR}{dr} - \frac{\ell(\ell + 1)}{r^2}R + \frac{2m}{\hbar^2}\left(E + \frac{\alpha}{r}\right)R = 0, \qquad (4.75)$$

where $\alpha = Ze^2$ is constant, characterizing the potential; e is the electron charge; Z is the whole number equal to the nucleus charge in the charge units.

The constants e^2, m and $\hbar$ allow us to construct the value with the dimension of length

$$a_0 = \frac{\hbar^2}{me^2} = 0.529 \times 10^{-8} cm$$

known as the Bohr radius, and the time

$$t_0 = \frac{\hbar^3}{me^4} = 0.242 \times 10^{-11} sec.$$

These quantities define the typical space and time scale for describing a system, and it is therefore convenient to use these units as the basic system of atomic units.

Equation (4.75) in atomic units (at $Z = 1$) takes the form

$$\frac{d^2R}{dt^2} + \frac{2}{r}\frac{dR}{dr} - \frac{\ell(\ell+1)}{r^2}R + 2\left(E + \frac{1}{r}\right)R = 0. \tag{4.76}$$

At $E < 0$ the motion is finite and the energy spectrum is discrete. We need the solutions (4.76) quadratically integrable with r^2. Let us introduce the specification

$$n = \frac{1}{\sqrt{-2E}}, \quad \rho = \frac{2r}{n}.$$

Equation (4.76) can be written as

$$\frac{d^2R}{dt^2} + \frac{2}{\rho}\frac{dR}{d\rho} + \left[\frac{n}{\rho} - \frac{1}{4} - \frac{\ell(\ell+1)}{\rho^2}\right]R = 0. \tag{4.77}$$

We find the asymptotic forms of the radial function $R(r)$. At $\rho \to \infty$ and omitting the terms $\sim\rho^{-1}$ and $\sim\rho^{-2}$ in (4.77), we obtain

$$\frac{d^2R}{d\rho^2} = \frac{R}{4}.$$

Therefore at high values of ρ, $R \propto e^{\pm\rho/2}$. The normalization demand is satisfied only by $R(\rho) \propto e^{-\rho/2}$. The asymptotic forms at $r \to 0$ have already been determined.

Substituting

$$R(\rho) = \rho^\ell e^{-\rho/2} w(\rho),$$

Eq. (4.77) is reduced to the form

$$\rho\frac{d^2w}{d\rho^2} + (2\ell + 2 - \rho)\frac{dw}{d\rho} + (n - \ell - 1)\,w = 0. \tag{4.78}$$

To solve this equation in the limit of $\rho = 0$, we substitute $w(\rho)$ in the form of a power series

$$w(\rho) = 1 + \frac{(0-v)}{(0+\lambda)}\rho + \frac{(0-v)(1-v)}{(0+\lambda)(1+\lambda)}\frac{\rho^2}{2!}$$
$$+ \frac{(0-v)(1-v)(2-v)}{(0+\lambda)(1+\lambda)(2+\lambda)}\frac{\rho^3}{3!} + \cdots, \tag{4.79}$$

where $\lambda = 2\ell + 2$ and $-v = -n + \ell + 1$.

At $\rho \rightarrow \infty$, the function $\omega(\rho)$ should increase, but not faster than the limiting power ρ. Then $\omega(\rho)$ has to be a polynomial of v power. So, $-n + \ell + 1 = -k$, and $n = \ell + 1 + k$ ($k = 0, 1, 2, \ldots$) at a given value of ℓ. Hence, using the definition for n, we can find the expression for the energy spectrum

$$E_n = -\frac{1}{2n^2}. \tag{4.80}$$

The number n is called the principal quantum number. In general units it has the form

$$E = -Z^2 \frac{me^4}{2\hbar^2 n^2}. \tag{4.81}$$

This formula was obtained by Bohr in 1913 on the basis of the old quantum theory, by Pauli in 1926 from matrix mechanics, and by Schrödinger in 1926 by solving the differential equations.

Let us solve the problem of the spectrum of the hydrogen atom using the equation of dynamical equilibrium of the system. In Chap. 3 we obtained Jacobi's virial equation for a quantum mechanical system of particles whose interaction is defined by the potential being a homogeneous function of the co-ordinates. This equation in the operator form is

$$\ddot{\hat{\Phi}} = 2\hat{H} - \hat{U}, \tag{4.82}$$

where $\hat{\Phi}$ is the operator of the Jacobi function, which, for the hydrogen atom, is written

$$\hat{\Phi} = \frac{1}{2} m \hat{r}^2, \tag{4.83}$$

The Hamiltonian operator is

$$\hat{H} = -\frac{\hbar^2}{2m} \Delta_r + \hat{U} \tag{4.84}$$

and the operator of the function of the potential energy for the hydrogen atom is

$$\hat{U} = -\frac{e^2}{r}. \tag{4.85}$$

We solve the problem with respect to the eigenvalues of Eq. (4.82), using the main idea of quantum mechanics. For this we use the Schrödinger equation

$$\hat{H}\Psi = E\Psi$$

and rewrite equation (4.82) in the form

$$\ddot{\hat{\Phi}} = 2E - \hat{U}. \tag{4.86}$$

This equation includes two (unknown in the general case) operator functions $\hat{\Phi}$ and $\hat{U}$. In the case of the interaction, the potential is determined by the relation (4.85), and we can use a combination of the operators $\hat{\Phi}$ and $\hat{U}$ in the form

$$|\hat{U}|\sqrt{\hat{\Phi}} = \frac{e^2 m^{1/2}}{\sqrt{2}} = B. \tag{4.87}$$

We now transform (4.86) into the form which was considered in classical mechanics:

$$\ddot{\hat{\Phi}} = 2E + \frac{B}{\sqrt{\hat{\Phi}}}. \tag{4.88}$$

Equation (4.88) is a consequence of Eq. (4.86) when the Schrödinger equation and the relationship (4.87) are satisfied. Its solution for the bounded state, i.e. when total energy E is determined in parametric form, can be written

$$\sqrt{\hat{\Phi}} = \frac{B}{2|E|} (1 - \varepsilon \cos \varphi), \tag{4.89}$$

$$\varphi - \varepsilon \sin \varphi = M, \tag{4.90}$$

where the parameter M is defined by the relation

$$M = \frac{(4|E|)^{3/2}}{4B} (t - \tau), \tag{4.91}$$

where ε and τ are integration constants and where

$$\varepsilon = \sqrt{1 - \frac{AC}{2B^2}},$$

$$C = -\dot{\hat{\Phi}}_0^2 + 4B\sqrt{\hat{\Phi}_0} - 2A\hat{\Phi}_0.$$

Moreover, the solution can be written in the form of Fourier and Lagrange series. Thus, the expression (4.37) describes the expansion of the operator $\hat{\Phi}$ into a Lagrange series including the accuracy of ε^3, and has the form

$$\Phi_0 = \frac{B^2}{A^2} \left[1 + \frac{3}{2}\varepsilon^2 + \left(-2\varepsilon + \frac{\varepsilon^3}{4} \right) \cos M - \frac{\varepsilon^2}{2} \cos 2M - \frac{\varepsilon^3}{4} \cos 3M + \cdots \right]. \tag{4.92}$$

Using the general expression for the mean values of the observed quantities in quantum mechanics

$$< \Psi|\hat{\Phi}|\Psi > = \bar{\Phi},$$

and taking into account that the mean value of the Jacobi function of the hydrogen atom should be different from zero, we find that our system has multiple eigenfrequencies $v_n = nv_0$ with respect to the basic v_0 which corresponds to the period

$$T_v = \frac{8\pi B}{(4|E|)^{3/2}}. \tag{4.93}$$

In accordance with the expression

$$E_n = \hbar\omega_n = \frac{\hbar 2\pi n}{T_0} \tag{4.94}$$

each of these frequencies corresponds to the energy level E_n of the hydrogen atom. We substitute the expression (4.93) for T_v into Eq. (4.94) and resolve it in relation to E_n:

$$|E_n| = \frac{\hbar 2\pi n (4|E_n|)^{3/2}}{8\pi B} = \frac{\hbar n (4|E_n|)^{3/2}}{\frac{4e^2 m^{1/2}}{\sqrt{2}}} = \frac{\hbar n 2\sqrt{2|E_n|}}{e^2 m^{1/2}}. \tag{4.95}$$

The expression obtained by Bohr follows from (4.95):

$$E_n = \frac{e^4 m}{2\hbar^2 n^2}. \tag{4.96}$$

This equation solves the problem.

4.5 Solution of a Virial Equation in the Theory of Relativity (Static Approach)

We consider now the solution of Jacobi's virial equation in the framework of the theory of relativity, showing its equivalence to Schwarzschild's solution.

Let us write down the known expression for the radius of curvature of space-time as a function of mass density:

$$\frac{1}{R^2} = \frac{8\pi}{3}\frac{G\rho}{c^2}, \tag{4.97}$$

where R is the curvature radius; ρ is the mass density; G is the gravitational constant and c is the velocity of light.

Equation (4.97) can also be rewritten in the form

$$\rho R^2 = \frac{3}{8\pi}\frac{c^2}{G}. \tag{4.98}$$

If the product ρR^2 in Eq. (4.98) is the Jacobi function ($\Phi = \rho R^2$ is the density of the Jacobi function) then, from T (4.98):

$$\Phi = \frac{3}{8\pi} \frac{c^2}{G}.$$ (4.99)

and it follows that the Jacobi function is a fundamental constant for the Universe. (In general relativity, the spatial distance does not remain invariant. Therefore, instead of this the Gaussian curvature is used, which has the dimension of the universe distance and is the invariant or, more precisely, the covariant.)

The constancy of the Jacobi function in this case reflects the smoothness of the description of motion in general relativity. The oscillations relative to this smooth motion described by Jacobi's equation are the gravitational waves and horizons, in particular the collapse and all types of singularity up to the process of condensation of matter in galaxies, stars etc.

Now we can show that Schwarzschild's solution in general relativity is equivalent to the solution of Jacobi's equation when $\ddot{\Phi} = 0$. Let us write the expression for the energy-momentum tensor

$$T_i^k = (\rho + p)u_i u^k + p\delta_i^k.$$ (4.100)

In the corresponding co-ordinate system, we obtain

$$u^i = \left(0, 0, 0, \frac{1}{\sqrt{-g_{00}}}\right),$$ (4.101)

where $\rho = \rho(r)$ and $p = p(r)$.

The independent field equations are written

$$G_1^1 = T_1^1, \quad G_0^0 = T_0^0,$$
$$R^{-2} = \frac{1}{3}G\rho c^2.$$ (4.102)

The expression for the metric is written in the form

$$ds^2 = \frac{dr^2}{1 - \frac{r^2}{R^2}} + r^2 (d\Omega)^2 - \left\{A - B\sqrt{1 - \frac{r^2}{R^2}}\right\}^2 c^2 r^2,$$ (4.103)

where

$$\frac{dr^2}{1 - \frac{r^2}{R^2}} + r^2 (d\Omega)^2.$$

is the spatial element.

In this case the expression for the volume occupied by the system is written

$$V = \int_0^r \int_0^\pi \int_0^{2\pi} \frac{r^2 \sin\Theta}{\sqrt{1 - \frac{r^2}{R^2}}} dr d\Theta d\Psi = \frac{4\pi R^3}{3} \left[arcsin\frac{r}{R} - \frac{r}{R}\sqrt{1 - \frac{r^2}{R^2}} \right]. \quad (4.104)$$

It can be easily verified that the right-hand side of Eq. (4.104) coincides with solution (4.14) and (4.15) of the equation of virial oscillations (4.11) at $\ddot{\Phi} = 0$, i.e.,

$$arcsinx - x\sqrt{1 - x^2} = arccos \left(\frac{\frac{A}{B}\sqrt{\Phi} - 1}{\sqrt{1 - \frac{AC}{2B^2}}} \right)$$

$$- \sqrt{1 - \frac{AC}{2B^2}} \sqrt{1 - \left(\frac{\frac{A}{B}\sqrt{\Phi} - 1}{\sqrt{1 - \frac{AC}{2B^2}}} \right)^2}. \quad (4.105)$$

In fact, Eq. (4.105) is satisfied for

$$x = \frac{\frac{A}{B}\sqrt{\Phi} - 1}{\sqrt{1 - \frac{AC}{2B^2}}} \quad \text{and} \quad x = \sqrt{1 - \frac{AC}{2B^2}},$$

i.e.,

$$\frac{A}{B}\sqrt{\Phi} - 1 = 1 - \frac{AC}{2B^2}, \quad \text{or} \quad \frac{AC}{2B^2} + \frac{A\sqrt{\Phi}}{B} = 2.$$

At $\ddot{\Phi} = 0$, the parameter of virial oscillations,

$$e = \sqrt{1 - \frac{AC}{2B^2}} \quad \text{и} \quad \sqrt{\Phi} = \frac{B}{A},$$

so the last condition is satisfied.

Schwarzschild's solution is rigorous and unique for Einstein's equation for a static model of a system with spherical symmetry.

Since this solution coincides with the solution of virial oscillations at the same conditions, the solutions (4.14) and (4.15) of Eq. (4.11), obtained in this chapter, should be considered rigorous. Thus we can conclude that the constancy of the product $U\sqrt{\Phi}$ in the framework of the static system model is proven. In Chap. 6 we will come back to this condition and will obtain another proof of the same very important relationship which is applied for study of the Earth's dynamics.

Chapter 5
Perturbed Virial Oscillations of a System

In the previous chapter we have considered a number of cases of explicitly solved problems in mechanics and physics for the dynamics of an *n*-body system and have shown that all those classical problems have also explicit solution in the framework of the virial approach. But in the latter case, the solutions acquire a new physical meaning because the dynamics of a system is considered with respect to new parameters, i.e. its Jacobi function (polar moment of inertia) and potential (kinetic) energy. In fact, the solution of the problem in terms of co-ordinates and velocities specifies the changes in location of a system or its constituents in space. The solution, with respect to the Jacobi function and the potential energy, identifies the evolutionary processes of the structure or redistribution of the mass density of the system. Moreover, the main difference of the two approaches is that the classical problem considers motion of a body in the outer central force field. The virial approach considers motion of a body both in the outer and in its own force field applying, instead of linear forces and moments, the volumetric forces (pressure) and moments (oscillations).

It appears from the cases considered that the existence of the relationship between the potential energy of a system and its Jacobi function written in the form

$$U\sqrt{\Phi} = B = const. \qquad (5.1)$$

is the necessary condition for the resolution of Jacobi's equation.

This is the only case when the scalar equation

$$\ddot{\Phi} = 2E - U$$

is transferred into a non-linear differential equation with one variable in the form

$$\ddot{\Phi} = 2E + \frac{B}{\sqrt{\Phi}}. \qquad (5.2)$$

V. I. Ferronsky, S. V. Ferronsky, *Dynamics of the Earth*,
DOI 10.1007/978-90-481-8723-2_5, © Springer Science+Business Media B.V. 2010

It was shown in Chap. 4 that if the total energy of a system $E_0 = -A/2 < 0$, then the general solution for Eq. (5.2) can be written as

$$
\sqrt{\phi_0} = \frac{B}{A} \left[1 - \varepsilon cos(\lambda - \psi) \right],
$$

$$
t = \frac{4B}{(2A)^{3/2}} \left[\lambda - \varepsilon sin(\lambda - \psi) \right],
$$

(5.3)

where ε and ψ are integration constants, the values of which are determined from initial data using Eqs. (4.23) and (4.24).

Equation (5.2) was called the equation of virial oscillation because its solution discovers a new physical effect – periodical non-linear change of the Jacobi function and hence the potential energy of a system around their mean values determined by the virial theorem. Thus, in addition to the static effects determined by the hydrostatic equilibrium, in the study of dynamics of a system the effects, determined by a condition of dynamical equilibrium expressed by the Jacobi function, are introduced.

The equation of virial oscillations (5.2) reflects physics of motion of the interacting mass particles of a body or masses of bodies themselves by the inverse square law. Its application opens the way to study the nature and the mechanism of generation of the body's energy, which performs its motion, and to search the law of change for the system's configuration, i.e. a mutual change location of particles or the law of redistribution of the mass density for the system's matter during its oscillations. This problem was considered earlier in our work (Ferronsky et al., 1987). We continue its study in the next chapter.

As described in Chap. 4, cases of solution of Eq. (5.2) relate to unperturbed conservative systems. But in reality, in nature all systems are affected by internal and external perturbations which, from a physical point of view, are developed in the form of dissipation or absorption of energy. In this connection, as shown in Chap. 3 in the right-hand side of the equation of virial oscillations (5.2), an additional term appears which is proportional to the Jacobi function Φ (indicating the presence of gravitational background or the existence of interaction between the system particles in accordance with Hook's law) and its first derivative $\dot{\Phi}$ depending on time t (indicating the existence of energy dissipation). All these and other possible cases can be formally described by a generalized equation of virial oscillations (3.34):

$$
\ddot{\Phi} = 2E + \frac{B}{\sqrt{\Phi}} + X(t, \Phi, \dot{\Phi}),
$$

(5.4)

where $X(t, \Phi, \dot{\Phi})$ is the perturbation function, the value of which is small in comparison with the term $B/\sqrt{\Phi} \neq const$.

In this chapter we consider general as well as some specific approaches to the solution of Eq. (5.4) in the framework of different physical models of a system.

5.1 Analytical Solution of a Generalized Equation of Virial Oscillations

The equation of perturbed virial oscillations is generalized in the form

$$\ddot{\Phi} = -A + \frac{B}{\sqrt{\Phi}} + X(t, \Phi, \dot{\Phi}), \qquad (5.5)$$

where $A = -2E$; B is constant; $X(t, \Phi, \dot{\Phi})$ is the perturbation function which we assume is given and dependent in general cases on time t, the Jacobi function Φ and its first derivative $\dot{\Phi}$.

We consider two ways for analytical construction of the solution of Eq. (5.5). In addition, let the function $X(t, \Phi, \dot{\Phi})$ in Eq. (5.5) depend on some small parameter e in relation to which the function can be expanded into absolutely convergent power series of the form

$$X(t, \Phi, \dot{\Phi}) = \sum_{r=1}^{\infty} e^k X^k(t, \Phi, \dot{\Phi}). \qquad (5.6)$$

Let the series be convergent in some time interval t absolute for all values of e which are satisfied to condition $|e| < \bar{e}$. Then Eq. (5.5) can be rewritten in the form

$$X(t, \Phi, \dot{\Phi}) = -A + \frac{B}{\sqrt{\Phi}} \sum_{r=1}^{\infty} e^k X^k(t, \Phi, \dot{\Phi}). \qquad (5.7)$$

We look for the solution of Eq. (5.7) also in the form of the power series of parameter e. For this purpose we write the function $\Phi(t)$ in the form of a power series, the coefficients of which are unknown:

$$\Phi(t) = \sum_{k=0}^{\infty} e^k \Phi^{(k)}(t). \qquad (5.8)$$

Putting (5.8) into (5.7), the task can be reduced to the determination of such functions $\Phi^{(k)}(t)$ which identically satisfy Eq. (5.7). In this case, the coefficient $\Phi^{(0)}(t)$ becomes the solution of the unperturbed oscillation equation (5.2), which can be obtained from (5.7) by putting $e = 0$.

One can consider the series (5.8) as a Taylor series expansion in order to determine all the other coefficients $\Phi^{(k)}(t)$, i.e.,

$$\Phi^{(k)} = \frac{1}{k!} \left(\frac{d^k \Phi}{de^k} \right) |_{e=0},$$

$$\dot{\Phi}^{(k)} = \frac{1}{k!} \left(\frac{d^k \dot{\Phi}}{de^k} \right) |_{e=0}.$$

$$(5.9)$$

Accepting the series (5.8) for introduction into Eq. (5.7), it becomes identical with respect to the parameter e. Thus we have justified the differentiation of the identity with respect to the parameter e several times assuming that the identity remains after repeated differentiation.

We next obtain

$$\frac{d^2}{dt^2}\left(\frac{d\Phi}{de}\right) = -\frac{1}{2}\frac{B}{\Phi^{3/2}}\left(\frac{d\Phi}{de}\right) + \sum_{k=1}^{\infty} k e^{k-1} X^{(k)} + \sum_{k=1}^{\infty} e^k \left(\frac{dX^{(k)}}{de}\right), \quad (5.10)$$

where $dX^{(k)}/de$ is the total derivative of the function $X^{(k)}$ with respect to parameter e, expressed by

$$\frac{dX^{(k)}}{de} = \frac{\partial X^{(k)}}{\partial \Phi}\left(\frac{d\Phi}{de}\right) + \frac{\partial X^{(k)}}{\partial \dot{\Phi}}\left(\frac{d\dot{\Phi}}{de}\right).$$

Now let $e = 0$ in (5.10). Then by taking into account (5.8) and (5.9), we obtain

$$\frac{d^2\Phi^{(1)}}{dt^2} + p_1\Phi^{(1)} = X_1, \quad (5.11)$$

where

$$p_1 = \frac{1}{2}\frac{B}{\Phi^{3/2}}\Big|_{e=0} = \frac{1}{2}\frac{B}{\Phi^{(0)3/2}}, \quad X_1 = X^1\left(t, \Phi^{(0)}, \dot{\Phi}^{(0)}\right)$$

are known functions of time, since the solution of the equation in the zero approximation (unperturbed oscillation equation (5.3)) is known.

Carrying out differentiation of Eq. (5.7) with respect to parameter e for the second, third and so on $(k-1)$ times, and assuming after each differentiation that $e = 0$, we will step by step obtain equations determining second, third and so on approximations. It is possible to show that in each succeeding approximation the equation will have the same form and the same coefficient p_1 as in Eq. (5.11). If so, the equation determining the functions $\Phi^{(k)}$ and $\dot{\Phi}^{(k)}$ has the form

$$\frac{d^2\Phi^{(k)}}{dt^2} + p_1\Phi^{(k)} = X_k\left(t, \Phi^{(0)}, \dot{\Phi}^{(0)}, \cdots, \Phi^{(k-1)}, \dot{\Phi}^{(k-1)}\right), \quad (5.12)$$

where the function X_k depends on $\Phi^{(0)}$, $\dot{\Phi}^{(0)}$,..., $\Phi^{(k-1)}$, $\dot{\Phi}^{(k-1)}$, which were determined earlier and are the functions of t and unknown functions $\Phi^{(0)}$ and $\dot{\Phi}^{(0)}$.

It is known that there is no general way of obtaining a solution for any linear differential equation with variable coefficients, but in our case we can use the following theorem of Poincaré (Duboshin, 1975). Let the general solution of the unperturbed virial oscillation equation be determined by the function $\Phi^{(0)} = f(t, C_1, C_2)$, where C_1 and C_2 are, for instance, arbitrary constants ε and Ψ in the solution (5.3) of equation (5.2). Then, Poincaré's theorem confirms that the function determined by the equalities

$$\phi_1 = \frac{\partial f}{\partial C_1},$$

$$\phi_2 = \frac{\partial f}{\partial C_2}$$

satisfies the linear homogeneous differential equation reduced by omission of the right-hand side of Eq. (5.12).

Thus, the general solution of the linear homogeneous equation

$$\frac{d^2 \Phi^{(k)}}{dt^2} + p_1 \Phi^{(k)} = 0$$

has the form

$$\phi_1 C_1^{(k)} + \phi_2 C_2^{(k)} = \phi^{(k)} \tag{5.13}$$

and the general solution of Eq. (5.12) can be obtained by the method of variation of arbitrary constants, i.e. assuming that $C_2^{(k)}$ are functions of time. Then, using the key idea of the method of variation of arbitrary constants, we obtain a system of two equations:

$$\dot{C}_1^{(k)} \phi_1 + \dot{C}_2^{(k)} \phi_2 = 0,$$
$$\dot{C}_1^{(k)} \dot{\phi}_1 + \dot{C}_2^{(k)} \dot{\phi}_2 = X_k. \tag{5.14}$$

Solving this system with respect to $\dot{C}_1^{(k)}$ and $\dot{C}_2^{(k)}$ and integrating the expression obtained, we write the general solution of Eq. (5.12) as follows:

$$\phi^{(k)}(t) = \phi_2 \int_{t_0}^{t} \frac{\phi_1 X_k dt}{\phi_1 \dot{\phi}_2 - \phi_2 \dot{\phi}_1} - \phi_1 \int_{t_0}^{t} \frac{\phi_2 X_k dt}{\phi_1 \dot{\phi}_2 - \phi_2 \dot{\phi}_1},$$

where

$$\phi_1 = \frac{\partial f(t, C_1, C_2)}{\partial C_1} \quad \text{and} \quad \phi_2 = \frac{\partial f(t, C_1, C_2)}{\partial C_2}.$$

Thus, we can determine any coefficient of the series (5.8), reducing Eq. (5.7) into an identity, and therefore write the general solution of Eq. (5.5) in the form

$$\phi = \sum_{k=0}^{\infty} e^k \phi^{(k)} = \sum_{k=0}^{\infty} e^k \left[\phi_2 \int_{t_0}^{t} \frac{\phi_1 X_k dt}{\phi_1 \dot{\phi}_2 - \phi_2 \dot{\phi}_1} - \phi_1 \int_{t_0}^{t} \frac{\phi_2 X_k dt}{\phi_1 \dot{\phi}_2 - \phi_2 \dot{\phi}_1} \right]. \tag{5.15}$$

Let us consider the second way of approximate integration of the perturbed virial equation (5.5), based on Picard's method (Duboshin, 1975). It is convenient to apply this method of integrating the equations which was obtained using the Lagrange method of variation of arbitrary constants.

We assume that the first integrals (4.23) and (4.24)

$$\varepsilon = \sqrt{1 - \frac{A}{2B^2}\left(-\dot{\phi}_0 + 4B\sqrt{\phi_0} - 2A\phi_0\right)},$$ (5.16)

$$-\tau = \left\{\frac{4B}{(2A)^{3/2}}\left[arccos\frac{1 - \frac{A}{B}\sqrt{\phi_0}}{\varepsilon} - \varepsilon\sqrt{1 - \left(\frac{1 - \frac{A}{B}\sqrt{\phi_0}}{\varepsilon}\right)^2}\right] - t\right\}$$ (5.17)

of the unperturbed virial oscillation equation (5.2) are also the first integrals of the perturbed oscillation equation (5.5). But constants ε and τ are now unknown functions of time. Let us derive differential equations which are satisfied by these functions, using the first integrals (5.16) and (5.17). For convenience, we replace the integration constant ε by C, using the expression

$$\varepsilon = \sqrt{1 - \frac{AC}{2B^2}}.$$

Now we rewrite Eq. (5.16) in the form

$$C = -\dot{\Phi}_0^2 + 4B\sqrt{\Phi_0} - 2A\Phi_0.$$ (5.18)

Then using the main idea of the Lagrange method, after variation of the first integrals (5.17) and (5.18) and replacement of $\ddot{\Phi}$ by

$$\left(-A + \frac{B}{\sqrt{\Phi}} + X\left(t, \Phi, \dot{\Phi}\right)\right)$$

we write

$$\dot{C} = -2\dot{\Phi}X\left(t, \Phi, \dot{\Phi}\right),$$ (5.19)

$$\dot{\tau} = \Psi(\Phi, C)\dot{C} = -2\dot{\Phi}X\left(t, \Phi, \dot{\Phi}\right)\Psi(\Phi, C),$$ (5.20)

where

$$\Psi(\Phi, C) = -\frac{4B}{(2A)^{3/2}} \frac{d}{dC} \left[arccos \frac{1 - \frac{A}{B}\sqrt{\Phi}}{\sqrt{1 - \frac{AC}{2B^2}}} \right.$$

$$\left. -\sqrt{1 - \frac{AC}{2B^2}} \sqrt{1 - \left(\frac{1 - \frac{A}{B}\sqrt{\Phi}}{\sqrt{1 - \frac{AC}{2B^2}}} \right)^2} \right].$$

We now express Φ and $\dot\Phi$ in explicit form through C, τ and t, using, for example, the Lagrangian series (4.37) and (4.38)

$$\Phi(t) = \frac{B^2}{A^2} \left[1 + \frac{3}{2}\varepsilon^2 + \left(-2\varepsilon + \frac{\varepsilon^3}{4} \right) \cos M - \frac{\varepsilon^2}{2} \cos 2M - \frac{\varepsilon^3}{4} \cos 3M + \cdots \right],$$

$$(5.21)$$

$$\dot\Phi(t) = \sqrt{\frac{2}{A}}\varepsilon B \left[\sin M + \frac{1}{2}\varepsilon \sin 2M + \frac{\varepsilon^2}{2}\sin M \left(2\cos^2 M - \sin^2 M \right) + \cdots \right].$$

$$(5.22)$$

Thus, taking into account Eqs. (5.21) and (5.22) for the functions Φ and $\dot\Phi$, Eqs. (5.19) and (5.20) can be rewritten as

$$\frac{dC}{dt} = F_1(t, C, \tau),$$

$$\frac{d\tau}{dt} = F_2(t, C, \tau).$$

$$(5.23)$$

To solve the system of differential equations (5.23), we use Picard's successive approximation method, obtained in the k-th approximation expressions for $C^{(k)}$ and $\tau^{(k)}$ in the form

$$C^{(k)} = C^{(0)} + \int_{t_0}^{t} F_1\left(t, C^{(k-1)}, \tau^{(k-1)}\right) dt,$$

$$(5.24)$$

$$\tau^{(k)} = \tau^{(0)} + \int_{t_0}^{t} F_2\left(t, C^{(k-1)}, \tau^{(k-1)}\right) dt,$$

$$(5.25)$$

where $C^{(0)}$ and $\tau^{(0)}$ are the values of arbitrary constants C and τ at initial time t_0, and $k = 1, 2,\ldots$.

Then, in the limit of $k \to \infty$, we obtain the solution of the system (5.23):

$$C = \lim_{k \to \infty} C^{(k)},$$
$$\tau = \lim_{k \to \infty} \tau^{(k)}.$$

(5.26)

Consider now two possible cases of the perturbation function behavior. First, assume that the perturbation function X does not depend explicitly on time. Then, since it is possible to expand functions Φ and $\dot{\Phi}$ into a Fourier series in terms of sine and cosine of argument M, the right-hand sides of the system (5.23) can also be expanded into a Fourier series in terms of sine and cosine of M.

Finally we obtain

$$\frac{dC}{dt} = \left[A_0 + \sum_{k=1}^{\infty} (A_k \cos kM + B_k \sin kM) \right],$$

(5.27)

$$\frac{d\tau}{dt} = \left[a_0 + \sum_{k=1}^{\infty} (a_k \cos kM + b_k \sin kM) \right],$$

(5.28)

where A_0, A_k, B_k, a_0, a_k, b_k are the corresponding coefficients of the Fourier series which are

$$A_0 = \frac{2}{\pi} \int_0^{2\pi} F_1(M, C)\, dM,$$

$$A_k = \frac{2}{\pi} \int_0^{2\pi} F_1(M, C)\cos kM dM,$$

$$B_k = \frac{2}{\pi} \int_0^{2\pi} F_1(M, C)\sin kM dM,$$

$$a_0 = \frac{2}{\pi} \int_0^{2\pi} F_2(M, C)\, dM,$$

$$a_k = \frac{2}{\pi} \int_0^{2\pi} F_2(M, C)\cos kM dM,$$

$$b_k = \frac{2}{\pi} \int_0^{2\pi} F_2(M, C)\sin kM dM.$$

Following Picard's method, in order to solve Eqs. (5.27) and (5.28) in the first approximation, we introduce into the right-hand side of the equations the values of arbitrary constants C and τ corresponding to the initial time t_0. Then we obtain

$$C^{(1)}(t) = C^{(0)} + A_0^{(0)}(t - t_0) + \sum_{k=1}^{\infty} \frac{1}{kn}$$

$$\left\{ A_k^{(0)} [\sin kM - \sin kM_0] + B_k^{(0)} [\cos kM - \cos kM_0] \right\} \qquad (5.29)$$

$$\tau^{(1)}(t) = \tau^{(0)} + a_0^{(0)}(t - t_0) + \sum_{k=1}^{\infty} \frac{1}{kn}$$

$$\left\{ a_k^{(0)} [\sin kM - \sin kM_0] + b_k^{(0)} [\cos kM - \cos kM_0] \right\} \qquad (5.30)$$

Thus, when the function X does not depend explicitly on time t, solutions (5.29) and (5.30) of Eq. (5.5) have three analytically different parts. The first is a constant term, depending on the initial values of the arbitrary constants. It is usually called the constant term of perturbation of the first order. The second part is a function monotonically increasing in time. It is called the secular term of a perturbation of the first order. The third part consists of an infinite set of trigonometric terms. All of them are periodic functions of M and consequently of time t. This is called periodic perturbation.

Similarly, we can obtain solutions in the second, third etc., orders. Here we limit our consideration only within the first order of perturbation theory. In practice, few terms of the periodic perturbation can be taken into account and the solution obtained becomes effective only for a short period of time.

When the perturbation function X is a periodic function of some argument M',

$$M' = n'(t - \tau'),$$

the right-hand side of the system of Eqs. (5.23) are periodic functions of the two independent arguments M and M'. Therefore, they can be expanded into a double Fourier series in terms of sine and cosine of the linear combination of arguments M and M'. Then in the first approximation of perturbation theory we obtain the following system of equations:

$$\frac{dC^{(1)}}{dt} = A_{00}^{(0)} + \sum_{k',k=-\infty}^{\infty} \left[A_{k,k'}^{(0)} \cos(kM + k'M') + B_{k,k'}^{(0)} \sin(kM + k'M') \right],$$

$$\qquad (5.31)$$

$$\frac{d\tau^{(1)}}{dt} = a_{00}^{(0)} + \sum_{k',k=-\infty}^{\infty} \left[a_{k,k'}^{(0)} \cos(kM + k'M') + b_{k,k'}^{(0)} \sin(kM + k'M') \right].$$

$$\qquad (5.32)$$

Integrating equations (5.31) and (5.32) with respect to time, we obtain a solution of the system:

$$
\begin{aligned}
C^{(1)}(t) = C^{(0)} + A_{00}^{(0)}(t - t_0) + \sum_{k',k=-\infty}^{\infty} \frac{1}{kn + k'n'} &\left\{ A_{k,k'}^{(0)} \left[cos\left(kM + k'M'\right) \right.\right. \\
- cos\left(kM_0 + k'M_0'\right) \Big] &+ B_{k,k'}^{(0)} \left[sin\left(kM + k'M'\right) - sin\left(kM_0 + k'M_0'\right) \right] \Big\}
\end{aligned}
$$
$$(5.33)$$

$$
\begin{aligned}
\tau^{(1)}(t) = \tau^{(0)} + a_{00}^{(0)}(t - t_0) + \sum_{k',k=-\infty}^{\infty} \frac{1}{kn + k'n'} &\left\{ a_{k,k'}^{(0)} \left[cos\left(kM + k'M'\right) \right.\right. \\
- cos\left(kM_0 + k'M_0'\right) \Big] &+ b_{k,k'}^{(0)} \left[sin\left(kM + k'M'\right) - sin\left(kM_0 + k'M_0'\right) \right] \Big\}
\end{aligned}
$$
$$(5.34)$$

Equations (5.33) and (5.34) have the same analytical structure as (5.29) and (5.30). At the same time, in this case, the periodic part of the perturbation can be divided into two groups, depending on the value of the divisor $kn + k'n'$. If the values of k and k' are such that the divisor is sufficiently large, then period $T_{k,k'} = 2\pi/(kn + k'n')$ of the corresponding inequality will be rather small. Such inequalities are called short-periodic. Their amplitudes are also rather small, and they can play a role only within short periods of time.

   If the values of k and k' are such that the divisor $kn + k'n'$ is sufficiently small but unequal to zero, then the period of the corresponding inequality will become large. The amplitude of such terms could also be large and play a role within large periods of time. Such terms form series of long-periodic inequalities. In the case of such k and k', when $kn + k'n' = 0$, the corresponding terms are independent of t and change the value of the secular term in the solutions (5.33) and (5.34).

5.2   Solution of the Virial Equation for a Dissipative System

In Chap. 3 we derived Jacobi's virial equation for a non-conservative system in the form

$$
\ddot{\Phi} = 2E_0 \left[1 + q(t) \right] - U - k\dot{\Phi}. \tag{5.35}
$$

At $k \ll 1$, $t \gg t_0$, $|U| \sqrt{\Phi} = B = const$, $2E_0 = -A_0$, and when the magnitude of the term $k\dot{\Phi}$ is sufficiently small, Eq. (5.35) can be rewritten in a parametric form

$$
\ddot{\Phi} = -A_0 \left[1 + q(t) \right] + \frac{B}{\sqrt{\Phi}}, \tag{5.36}
$$

where q(t) is a monotonically increasing function of time due to dissipation of energy during 'smooth' evolution of a system within a time interval $t \in [0, \tau]$.

Using the theorem of continuous solution depending on the parameter, we write
the solution of Eq. (5.36) as follows:

$$
- arccos W + arccos W_0 - \sqrt{1 - \frac{A_0 \left[1 + q(t)\right] C}{2B^2}} \sqrt{1 - W^2}
$$

$$
+ \sqrt{1 - \frac{A_0 C}{2B^2}} \sqrt{1 - W_0^2} = \sqrt{\frac{\left(2A_0 \left[1 + q(t)\right]\right)^{3/2}}{4B}} (t - t_0), \qquad (5.37)
$$

$$
arccos W - arccos W_0 + \sqrt{1 - \frac{A_0 \left[1 + q(t)\right] C}{2B^2}} \sqrt{1 - W^2}
$$

$$
- \sqrt{1 - \frac{A_0 C}{2B^2}} \sqrt{1 - W_0^2} = \sqrt{\frac{\left(2A_0 \left[1 + q(t)\right]\right)^{3/2}}{4B}} (t - t_0), \quad (5.38)
$$

where

$$
W = \frac{\frac{A_0 \left[1 + q(t)\right]}{B} \sqrt{\Phi} - 1}{\sqrt{1 - \frac{A_0 \left[1 + q(t)\right] C}{2B^2}}}; \quad W_0 = \frac{\frac{A_0}{B} \sqrt{\Phi} - 1}{\sqrt{1 - \frac{A_0 C}{2B^2}}};
$$

$$
A_0 \left[1 + q(t)\right] > 0; \quad C < \frac{2B^2}{A_0 \left[1 + q(t)\right]};
$$

$$
\left|-A_0 \left[1 + q(t)\right] \sqrt{\Phi} + B\right| < B \sqrt{1 - \frac{A_0 \left[1 + q(t)\right] C}{2B^2}};
$$

$$
C = -2A_0 \Phi_0 + 4B \sqrt{\Phi_0} - \dot{\Phi}_0^2.
$$

Equations of discriminant curves which bound oscillations of the Jacobi function Φ
by analogy with the case of the conservative system can be written as

$$
\sqrt{\Phi_1} = \frac{B}{A_0 \left[1 + q(t)\right]} \left[1 + \sqrt{1 - \frac{A_0 \left[1 + q(t)\right] C}{2B^2}}\right], \quad t \in [0, \tau], \quad (5.39)
$$

$$
\sqrt{\Phi_2} = \frac{B}{A_0 \left[1 + q(t)\right]} \left[1 - \sqrt{1 - \frac{A_0 \left[1 + q(t)\right] C}{2B^2}}\right], \quad t \in [0, \tau]. \quad (5.40)
$$

It is obvious that the solution of Jacobi's virial equation for a non-conservative system is quasi-periodic with period

$$
T_v(q) = \frac{8\pi B}{\left(2A_0 \left[1 + q(t)\right]\right)^{3/2}}, \qquad (5.41)
$$

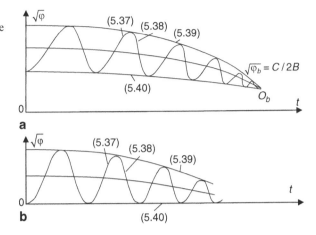

Fig. 5.1 Virial oscillations of the Jacobi function in time for a non-conservative system (a) and for the general (Wintner's) case (b)

and an amplitude of Jacobi function oscillations

$$\Delta\sqrt{\Phi} = \frac{B}{A_0\left[1+q(t)\right]}\left(1 - \frac{A_0\left[1+q(t)\right]C}{2B^2}\right)^{1/2}. \tag{5.42}$$

As $q(t)$ is a monotonically and continuously increasing parameter confined in time, the period and the amplitude of the oscillations will gradually decrease and tend to zero in the time limit.

In Fig. 5.1a the integral curves (5.37) and (5.38) and the discriminant curves (5.39) and (5.40) are shown in a general case when $0 < C < 2B^2/A_0$. At the point O_b, the integral and discriminant curves tend to coincide and the value of the amplitude of the Jacobi function (polar moment of inertia) oscillations of the system goes to zero.

When $C = 0$ (Fig. 5.1b) the discriminant line (5.39) coincides with the axis of abscissae, $\Phi_2 = 0$. In the accepted case of constancy of the system mass, the point O_b, where the integral and discriminant curves coincide, will be reached in the time limit $t \to \infty$.

Where $2B^2/A_0 \to C$ and $C < 0$, the solutions (5.37), (5.38) and (5.39), (4.40) could be complex so the processes considered are not physical.

We note that, by analogy with the case for a conservative system, considered in Chap. 4, we can show here that the asymptotic relations (4.30)–(4.32) for the solutions (5.37) and (5.38) of Jacobi's equation (5.36) in the points of contact of the discriminant line $\Phi_2 = 0$, are justified. In the points of contact for the integral curves (5.37) and (5.38) and the discriminant curves (5.39) and (5.40) for which Φ_1 and Φ_2 are not equal to zero, the following asymptotic relations are also justified:

$$\left(\sqrt{\Phi_1} - \sqrt{\Phi}\right) \propto \left(t' - t\right)^2, \tag{5.43}$$

$$\left(\sqrt{\Phi} - \sqrt{\Phi_2}\right) \propto \left(t - t'\right)^2, \tag{5.44}$$

where t' is time of a tangency point for the corresponding integral curve of the discriminant lines $\Phi_{1,2}$ when $\Phi_{1,2} \neq 0$.

5.3 Solution of the Virial Equation for a System with Friction

Let us consider the solution of Jacobi's virial equation for conservative systems, but let the relationship between its potential energy and the Jacobi function be as follows:

$$U\sqrt{\Phi} = B + k\dot{\Phi}. \tag{5.45}$$

In this case, the equation of virial oscillations (5.2) can be written

$$\ddot{\Phi} = -A + \frac{B}{\sqrt{\Phi}} - k\frac{\dot{\Phi}}{\sqrt{\Phi}}. \tag{5.46}$$

The term $-k\dot{\Phi}/\sqrt{\Phi}$ in (5.46) plays the role of perturbation function, reflecting the effect of internal friction of the matter while the system is oscillating.

In principle, Eq. (5.46) can be solved using the above perturbation theory methods. However, we can show that a particular solution exists for the system of two differential equations of the second order, which satisfies Eq. (5.46). These differential equations are as follows:

$$\left(\sqrt{\Phi}\right)'' + \sqrt{\frac{2}{A}}k\left(\sqrt{\Phi}\right)' + \sqrt{\Phi} = \frac{B}{A}, \tag{5.47}$$

$$t'' + \sqrt{\frac{2}{A}}kt' + t = \frac{4B}{(2A)^{3/2}}\lambda. \tag{5.48}$$

In Eqs. (5.47) and (5.48) we introduced a new variable λ, so the primes at Φ and t mean differentiation with respect to λ. Note also that time t here is not an independent variable. This allows us to transfer the non-linear equation into two linear equations. The partial solution of Eqs. (5.47) and (5.48) containing two integration constants is

$$\sqrt{\Phi} = \frac{B}{A}\left[1 - \varepsilon e^{-r/2\sqrt{2/A}\lambda}\cos\left(\sqrt{\frac{4A - 2k^2}{4A}}\lambda + \psi + \tau\right)\right], \tag{5.49}$$

$$t = \frac{4B}{(2A)}\left[\lambda - \varepsilon e^{-r/2\sqrt{2/A}\lambda}\sin\left(\sqrt{\frac{4A - 2k^2}{4A}}\lambda + \psi\right)\right] - \frac{4B}{(2A^{3/2})}\sqrt{\frac{2}{A}}k, \tag{5.50}$$

where ε and ψ are arbitrary constants and

$$\tau = arctg\sqrt{\frac{2}{A}}k\left(\frac{4A - 2k^2}{4A}\right)^{-1/2}.$$

To show that Eqs. (5.49) and (5.50) of the two linear differential equations (5.47) and (5.48) are also general solutions of (5.46), let us do as follows.

Differentiating (5.50) with respect to λ, we obtain

$$t' = \sqrt{\frac{2}{A}}\sqrt{\Phi}. \tag{5.51}$$

We write the derivative from function $\sqrt{\Phi}$ with respect to λ using Eq. (5.51) in the form

$$\left(\sqrt{\Phi}\right)' = \frac{\dot{\Phi}}{\sqrt{2A}}. \tag{5.52}$$

Then the second derivative from $\sqrt{\Phi}$ with respect to λ can be obtained analogously

$$\left(\sqrt{\Phi}\right)'' = \frac{\ddot{\Phi}}{\sqrt{2A}}t' = \frac{\ddot{\Phi}\sqrt{\Phi}}{A}. \tag{5.53}$$

Substituting Eqs. (5.52) and (5.53) for $\left(\sqrt{\Phi}\right)'$ and $\left(\sqrt{\Phi}\right)''$ into Eq. (5.47), we obtain

$$\frac{\ddot{\Phi}\sqrt{\Phi}}{A} + \sqrt{\frac{2}{A}}k\frac{\dot{\Phi}}{\sqrt{\Phi}} + \sqrt{\Phi} = \frac{B}{A}. \tag{5.54}$$

Dividing Eq. (5.54) by $\sqrt{\Phi}/A$ we have

$$\ddot{\Phi} + k\frac{\dot{\Phi}}{\sqrt{\Phi}} + A = \frac{B}{\sqrt{\Phi}},$$

which is in fact our Eq. (5.46). This means that Eqs. (5.49) and (5.50) are the general solution of Eq. (5.46).

Note that Eq. (5.50) differs in general from Kepler's equation both by the exponential factor before the sine function and by the constant term in the right-hand side of Eq. (5.50). In addition, it follows from Eq. (5.49) that the period of virial oscillations of the Jacobi function depends on the parameter k. Therefore, when λ changes its value by $2\pi / \left[\sqrt{(4A - 2k^2)/4A}\right]$ the value of $\sqrt{\Phi}$ remains unchanged (we neglect the changes of the amplitude of virial oscillations due to existence of the exponential factor) assuming that

$$\frac{k}{2}\sqrt{\frac{2}{A}}2\pi / \sqrt{\frac{4A - 2k^2}{4A}} \ll 1.$$

It follows from Eq. (5.50) that time t changes by the relationship of $T = 8\pi B/(2A)^{3/2}\sqrt{(4A - 2k^2)/4A}$ defining the period of the damping virial oscillations. Therefore, from solutions (5.49) and (5.50) of Eq. (5.46) it follows that if during the evolution of the system the value $U\sqrt{\Phi}$ varies only slightly around the constant, this leads to damping of the virial oscillations of the integral characteristics of the system around their averaged virial theorem value.

In conclusion we have to note that derivation of the equation of dynamical equilibrium and its solution for conservative and dissipative systems shows that dynamics of celestial bodies in their own force field puts forward a wide class of geophysical, astrophysical and geodetic problems which can be solved by the methods of celestial mechanics introducing the new physical concepts we have introduced (Giordano and Plastino, 1999).

Chapter 6
The Nature of Oscillation and Rotation of the Earth

We have presented the physical and theoretical fundamentals of the virial theory of dynamical equilibrium for study of the unperturbed and perturbed motion of a self-gravitating body. As it was noted, the condition of dynamical (oscillating) equilibrium of the body, which is determined by a functional relationship between the polar moment of inertia and the potential (kinetic) energy of any natural conservative and dissipative system in the form of the generalized virial theorem or Jacobi's virial equation, serve as the bases of the presented theory. It was shown in Chap. 3 that the outstanding property of Jacobi's virial equation is its ability to be both the equation of dynamical equilibrium and equation of motion simultaneously. Moreover, it is valid for all the known models of the dynamics of natural systems. The secret of universality of the equation is in its ability to describe the motion of a material system as a whole in its fundamental integral characteristics which are the energy and polar moment of inertia. It was shown in Chap. 2 that the functional relationship between the potential energy and the polar moment of inertia was revealed by means of analyzing the orbits of the artificial satellites. The potential energy, which is generated by interaction of the mass particles, is the force function of the body, i.e. the active component of its motion. The potential energy creates the inner and outer force field of the body. The kinetic energy is the reactive (inertial) constituent of the force field and is developed in the form of motion of the body's mass particles and its shells. In the equations of motion written, for instance, for continuous media in coordinates and velocities under the mass force ρF one understands the gravity force induced by the outer force field. As a result, formulation and solution of any geophysical problems including dynamics of the body and change in its form and structure appears to be physically incorrect. As to the problem of the orbital motion of the body in the central force field, then the dynamical effects of the interacting bodies are ignored because of their smallness.

Now we would like to apply the obtained results of the general solution of Jacobi's virial equation presented in Chaps. 3, 4 and 5 to study the Earth's dynamics and to obtain some quantitative data concerning the concrete elements of its non-perturbed and perturbed motion. As in celestial mechanics, by non-perturbed motion of the Earth we understand its motion under action of its own force field generated by interaction of its own masses. The perturbation motion of the planet

V. I. Ferronsky, S. V. Ferronsky, *Dynamics of the Earth*,
DOI 10.1007/978-90-481-8723-2_6, © Springer Science+Business Media B.V. 2010

is considered to be the effects developed by the outer force fields of the Sun and the Moon. The perturbations from other planets are not considered because of their indirect effect. The study of dynamics of the Earth in its own force field starts first of all from investigation of its basic forms, namely, the oscillation and rotation of the shells, which are developed by interaction of its own masses. That part of the energy which is emitted from the body's surface forms the outer force field of the planet. Its pressure through interaction with the pressure of the outer force field of the Sun provides dynamical equilibrium of the body in space and guaranties its orbital motion. The integral effect of the differential rotation of the shells determines changes in the slope of the Earth's axis rotation. Herein, the observed rotation of the planet as a rigid body in reality relates only to its upper shell. The other shells rotate more slowly. The effect of perturbation of the upper shell and the integral effect of rotation of all other shells result in rotation of the outer planet's force field which operates the motion of the Moon and artificial satellites along their orbits.

In order to find a quantitative solution of the task it is necessary to have data about the mass, radius, moment of inertia and radial density distribution of the planet's mass. We have sufficiently reliable data about the mass and radius of the Earth. The reliable mean value of the moment of inertia has also now been obtained by satellites. The radial density distribution data appear to be unreliable. This is because the existing methodology of interpretation of seismic data by the Williamson-Adams equation is based on the planet's hydrostatic equilibrium and needs to be reconsidered. That is why we shall search for such a law of density distribution which would satisfy the experimentally found moment of inertia and the condition of the density – differentiated masses on the shell that have been observed.

We start the study of dynamics of the Earth from its own oscillations. After that we move to determine some properties of the interacting masses which have not been earlier considered and which are needed. Among them are the structure of the potential and kinetic energy, the nature of the Archimedes and Coriolis forces and the electromagnetic component of the body's potential energy. They are the basis for consideration of the Earth's dynamical effects. Finally, we turn to the solution of the problem of the planet's shells rotation and other aspects of its dynamics.

6.1   The Problem of the Earth's Eigenoscillations

In order to demonstrate application of the new theory to study of the Earth's dynamics, we consider both traditional differential and proposed integral (dynamical) approaches. We show that, within the framework of the solution of the problem in terms of volumetric forces and moments, the eigenoscillations of the Earth is the natural integral effect of the interacting particles of the system.

6.1.1 The Differential Approach

We consider first the problem of radial oscillations of a gravitating elastic sphere within the framework of the traditional hydrostatic equilibrium approach that has been used to study the Earth's eigenoscillations.

The Euler equation of motion of a deformable body in the presence of the mass forces of the outer uniform force field is written in the form (Landau and Lifshitz, 1954)

$$\frac{\partial \sigma_{ik}}{\partial x_k} + \rho F_i = \rho \frac{\partial^2 u}{\partial t^2}, \tag{6.1}$$

where ρF_i is the i-th component of the mass force; u_i is the i-th component of the displacement vector; σ_{ik} is the stress vector; ρ is the mass density of the sphere.

We write the vector components of the mass force in the spherical system of co-ordinates as:

$$\rho F_0 = 0,$$

$$\rho F_\lambda = 0,$$

$$\rho F_r = -G\rho \frac{m(r)}{r^2},$$

where $m(r)$ is the Earth's mass within the sphere with radius r.

Because of radial deformation of the sphere only the radial component of the displacement vector differs from zero, i.e.

$$u_r = 0,$$

$$u_\theta = 0,$$

$$\frac{\partial^2 u}{\partial r^2} = \frac{\partial^2 r}{\partial t^2}.$$

For isotropic media and for small deformations the stress tensor σ_{ik} and the deformation tensor u_{ik} have the linear relationship, according to Hooke's law:

$$\sigma_{ik} = k u_{ik} \delta_{ik} + 2\mu \left(u_{ik} - \frac{1}{3} \delta_{ik} u_{ik} \right) = \lambda u_{ik} + 2\mu u_{ik}, \tag{6.2}$$

where k is the displacement modulus; μ is the shear modulus; $\lambda = k - 2/3\mu$ is the Lamé constant; $\sigma_{ik} = 0$ at $i \neq k$ and $\sigma_{ik} = 1$ at $i = k$.

In the case of radial deformations, the components of the deformation tensor are equal to

$$u_{rr} = \frac{du_r}{dr},$$

$$u_{\theta\theta} = u_{\varphi\varphi} \frac{u_r}{r},$$

$$u_{\theta\varphi} = u_{\lambda r} = u_{r\theta}.$$

(6.3)

The components of the stress tensor are:

$$\sigma_{rr} = (\lambda + 2\mu)\frac{du_r}{dr} + 2\lambda\frac{u_r}{r},$$

$$\sigma_{\theta\theta} = \sigma_{\varphi\varphi} = \lambda\frac{du_r}{dr} + (2\lambda + \mu)\frac{u_r}{r},$$

$$\sigma_{\theta\varphi} = \sigma_{\varphi r} = \sigma_{r\theta}.$$

(6.4)

The general equation (6.1) of motion now takes the form

$$\frac{d\sigma_{rr}}{dr} + \frac{1}{r}\left(2\sigma_{rr} - \sigma_{\theta\theta} - \sigma_{\varphi\varphi}\right) + \rho F_i = \rho\frac{d^2 u_r}{dt^2}.$$

(6.5)

Putting Eq. (6.4) into (6.5), we obtain an equation describing the radial displacement of matter in the sphere:

$$(\lambda + 2\mu)\left(\frac{d^2 u_r}{dr^2} + \frac{2}{r}\frac{du_r}{r^2}\right) + \left(\frac{du_r}{r^2} + 2\frac{u_r}{r}\right)\frac{d\lambda}{dr}$$

$$+ 2\frac{d\mu}{dr}\frac{du}{dr} + \rho F_r = \rho\frac{d^2 u_r}{dt^2}.$$

(6.6)

Equation (6.6) is used to study the problem of the radial oscillations of the Earth in the traditional differential (linear) approach. This equation is solved at boundary conditions of uniform radial displacement of matter or at uniform pressure over a spherical surface enveloping the body.

The normal stress at the sphere's surface which is formed by the outer layer of the body with radius r is

$$T_r = \overline{\Sigma} \cdot \frac{\bar{r}}{r}.$$

(6.7)

Where $\overline{\Sigma}$ is the stress tensor, the components of which are

$$\left|t_r, t_\theta, t_\varphi\right| = |100| \begin{vmatrix} \sigma_{rr} & 0 & 0 \\ 0 & \sigma_{\theta\theta} & 0 \\ 0 & 0 & \sigma_{\varphi\varphi} \end{vmatrix}.$$

(6.8)

Here only $t_r = \sigma_{rr} \neq 0$, but tensor T_r is the purely normal stress, which is

$$T_r = \left| (\lambda + 2\mu)\frac{du_r}{dr}\frac{du_r}{dr} + 2\lambda\frac{u_r}{r} \right|\frac{\bar{r}}{r}. \tag{6.9}$$

Let us consider a uniform sphere with $\lambda = $ const and $\mu = $ const. Then

$$\frac{d}{dt}\left(\frac{du_r}{dr} + 2\frac{u_r}{r}\right) = \frac{d\theta}{dr} = 0, \tag{6.10}$$

where $\theta = u_{rr} + u_{\theta\theta} + u_{\varphi\varphi}$ is the dilation of the body.
   The general solution of Eq. (6.10) is

$$u_r = Ar + \frac{B}{r^2}. \tag{6.11}$$

Constants A and B can be defined from the following boundary conditions: at the centre of the Earth ($r = 0$) the displacement $u_r = 0$ and the value $B = 0$; on the surface of the sphere with radius $r = a$ and $T_r = -p$.
   Then

$$\left| (\lambda + 2\mu)\frac{du_r}{dr} + 2\lambda\frac{u_r}{r} \right|_{r=A} = -p \tag{6.12}$$

from which it follows that

$$A = -\frac{p}{3\lambda + 2\mu}.$$

Now the general solution (6.11) takes the form

$$u_r = -\frac{pr}{3\lambda + 2\mu}$$

$$\theta = -\frac{3}{3\lambda + 2\mu} = -\frac{p}{\lambda + 2/3\mu} = -\frac{p}{k}. \tag{6.13}$$

Substituting the solution (6.13) into (6.4), we obtain the expression for the components of the stress tensor:

$$\sigma_{rr} = \sigma_{\theta\theta} = \sigma_{\varphi\varphi} = -p. \tag{6.14}$$

It is seen from (6.14) that the value of the stress components is reduced to the constant hydrostatic pressure of the body matter.
   To solve the problem of the eigenoscillations of a uniform spherical body, we assume

$$u_r = U(r)e^{i\omega t}, \tag{6.15}$$

where ω is the eigenoscillation frequency of the sphere.

Substituting (6.15) into Eq. (6.6), we obtain

$$(\lambda + 2\mu)\left(\frac{d^2U}{dr^2} + \frac{2}{r}\frac{dU}{dr} - \frac{2U}{r^2}\right) + \rho\omega^2 U + \frac{4}{3}\pi r G \rho^2 = 0. \qquad (6.16)$$

We introduce the new variable x:

$$x = \sqrt{\frac{p}{\lambda + 2\mu}}\,\omega r = hr. \qquad (6.17)$$

Then Eq. (6.16) can be rewritten as

$$\frac{d^2U}{dr^2} + \frac{2}{r}\frac{dU}{dr} - \frac{2U}{r^2} + \frac{x^2}{r^2}U = 0.$$

or

$$\frac{d^2U}{dx^2}\left(\frac{dx}{dr}\right)^2 + \frac{2}{r}\frac{dU}{dx}\left(\frac{dx}{dr}\right) - \frac{2U}{r^2} + \frac{x^2}{r^2}U = 0.$$

Considering that

$$\frac{dx}{dr} = \sqrt{\frac{p}{x + 2\mu}}\,\omega = \frac{x}{r},$$

we obtain

$$\frac{d^2U}{dx^2} + \frac{2}{r}\frac{dU}{dx} - \frac{2}{x^2}U = 0. \qquad (6.18)$$

Equation (6.18) is known as the Riccati equation. Its solution is

$$U(r) = \frac{d}{dx}\left(\frac{A\,\sin x + B\,\cos x}{x}\right). \qquad (6.19)$$

The arbitrary integration constants A and B can be found from the boundary conditions. In the centre of the Earth ($r = 0$, $x = 0$) we have $U(r) = 0$. Hence, $B = 0$. Moreover, the outer surface of the body should be in equilibrium. This means that the surface pressure should be equal to zero:

$$(\sigma_{rr})|_{r=A} = 0$$

and

$$(\lambda + 2\mu)\frac{dU}{dt}e^{i\omega t} + 2\lambda\frac{U}{r}e^{i\omega t} = 0,$$

or

$$(\lambda + 2\mu)\frac{d}{dx}\left|\frac{d}{dx}\left(A\frac{\sin x}{x}\right)\right| + 2\frac{\lambda}{r}\frac{d}{dx}\left(A\frac{\sin x}{x}\right) = 0. \qquad (6.20)$$

Differentiating this, we obtain

$$-(\lambda + 2\mu)x^2 \sin x - 4\mu x \cos x + 4\mu \sin x = 0. \qquad (6.21)$$

After dividing (6.21) by $4\mu \sin x$, we obtain

$$xctgx = 1 - \frac{\lambda + 2\mu}{4\mu}x^2.$$

The value x at the outer surface is equal to ah. Finally, we obtain the equation for determining the eigenoscillation values of the body:

$$ahctgah = 1 - \frac{\lambda + 2\mu}{4\mu}a^2h^2. \qquad (6.22)$$

6.1.2   The Dynamic Approach

Now we consider the problem of the eigenoscillations of the Earth as a uniform body on the basis of its dynamical equilibrium, i.e. under an action of its own internal force field or within the framework of the integral (dynamic) approach. For this purpose we use Eq. (6.5) assuming that the internal pressure in the body is isotropic, i.e.

$$\sigma_{rr} = \sigma_{\theta\theta} = \sigma_{\varphi\varphi} = -p.$$

Then we have from (6.5)

$$\frac{\partial^2 r}{\partial t^2} = -\frac{\partial p}{\partial r} + \rho F_r \qquad (6.23)$$

Now let us derive Jacobi's virial equation from Eq. (6.23) for the spherical symmetric model of the Earth. For this purpose we multiply the right-hand sides of Eq. (6.23) by $4\pi r^3 dr$ and integrate it with respect to dr from 0 to R:

$$\int_0^R 4\pi r^3 \rho(r)\frac{\partial^2 r}{\partial t^2}dr = -\int_0^R \frac{\partial p}{\partial r}4\pi r^3 dr - 4\pi G\int_0^R r\rho(r)m(r)dr. \qquad (6.24)$$

The left-hand side of Eq. (6.24) gives

$$
\int_0^R 4\pi r^3 \rho(r) \frac{\partial^2 r}{\partial t^2} dr = \frac{\partial^2}{\partial t^2} \left[\frac{1}{2} \int_0^R 4\pi r^4 \rho(r) dr - \int_0^R r\pi r^2 \left(\frac{\partial r}{\partial t} \right)^2 \rho(r) dr \right]
$$

$$
= \ddot{\Phi} - 2T, \tag{6.25}
$$

where Φ is the Jacobi function of the Earth, and T is the kinetic energy of the displacements of the matter.

The first term in the right-hand side of (6.24) is equal to

$$
\int_0^R \frac{dp}{dr} \frac{dr}{d\rho} d\rho 4\pi r^3 = - \int_{\rho_0}^{\rho_s} \frac{dp}{d\rho} 4\pi r^3 d\rho, \tag{6.26}
$$

where ρ_o and ρ_s are the mass densities in the centre and on the surface of the Earth respectively.

Taking into account that, within the framework of the model of an elastic medium, the system reaches its mechanical equilibrium faster than its thermal equilibrium, we assume that the entropy of the system is equal to the constant value, and therefore we write

$$
\frac{dp}{d\rho} = \left(\frac{dp}{d\rho} \right)_s = \frac{k}{\rho} = c_s^2, \tag{6.27}
$$

where c_s is the velocity of sound in elastic media and $c_s^2 = v_p^2 - 4/3v_g^2$; $v_p = \sqrt{(k + 4/3\mu)/\rho}$ is the velocity of the longitudinal waves in an elastic medium; $v_g = \sqrt{\mu/\rho}$ is the velocity of the transverse waves in the elastic medium.

Finally, Eq. (6.26) can be rewritten in the form

$$
- \int_0^R \frac{\partial p}{\partial r} 4\pi r^3 dr = - \int_0^R c_s^2 4\pi r^3 dr. \tag{6.28}
$$

If the velocity of sound does not depend on the radius of the body, then (6.28) is

$$
- \int_0^R \frac{\partial p}{\partial r} 4\pi r^3 dr = - \int_0^R c_s^2 4\pi r^3 dr = -\rho(r) c_s^2 4\pi r^3 \big|_0^R
$$

$$
+ \int_0^R 3x 4\pi r^3 \rho(r) c_s^2 dr = 3c_s^2 M, \tag{6.29}
$$

where $\rho(R) = 0$, $\rho(0) \neq \infty$.

In the general case when the velocity of sound depends on the radius, the expression for the energy of elastic deformations can be written by the expression for the velocity of sound as a mean value through the mass body, i.e.

$$2E_e = -\int_0^R \frac{\partial p}{\partial r} 4\pi r^3 dr = 3M\bar{c}_s^2, \tag{6.30}$$

where

$$\bar{c}_s^2 = -\frac{4\pi}{3M} \int_0^R c_c^2(r) r^3 d\rho(r). \tag{6.31}$$

The phenomenological parameter c_s takes into account the different aggregative states of the substance of a body, i.e. gaseous, liquid, solid, and plasma (Ferronsky et al., 1981a).

The second term in the right-hand side of Eq. (6.24) is the potential energy of the sphere:

$$4\pi G \int_0^R r\rho(r)m(r)dr = U. \tag{6.32}$$

Finally, Eq. (6.24) can be rewritten in the form of the Jacobi virial equation:

$$\ddot{\Phi} = 2T + 3M\bar{c}_s^2 + U = 2E - U, \tag{6.33}$$

where $E = T + 3/2M\bar{c}_s^2 + U$ is the total energy of the Earth.

One may see now that Equation (6.33) represents the generalized virial equation (2.31) obtained from Euler's equation of motion for a deformable body (6.1) by means of transformation of the kinetic energy of the mass interaction (6.25) through the polar moment of inertia (the Jacobi function Φ). The polar moment of inertia in (6.33) has a functional relation with the potential energy (6.32). Here the polar moment of inertia physically represents the body's structure and shows its changes with a change in the potential energy.

Averaging Jacobi's virial equation (6.33) with respect to a sufficiently long period of time gives the classical averaged virial theorem of the body and expresses the condition of its hydrostatic equilibrium in the outer force field:

$$2E = U,$$

or

$$-U = 3M\overline{c_s^2}. \tag{6.34}$$

Equation (6.34) follows also from the condition of hydrostatic equilibrium of the sphere matter when the left-hand side of the equation of motion (6.23) is equal to zero:

$$\frac{\partial p}{\partial r} = \rho F_r.$$

In accordance with (6.30), the left-hand side of this equation represents a double value of the total body's energy, the matter of which stays in hydrostatic equilibrium in the outer uniform force field. The right-hand side of the equation determines the potential energy of the matter interaction.

It follows from Eq. (6.34) that the velocity of sound in the elastic media determines not only the potential energy of the mass interaction, but also the velocity of propagation of the potential energy flux in the media. This relationship between the potential energy and the sound velocity is used now in seismic studies for determination of mass density. We also will apply it for interpretation of the radial density distribution.

To obtain the equation of virial oscillations from (6.33), we accept the following assumptions. We assume that the total energy E has a constant value and the relationship between the Jacobi function Φ and the gravitational potential energy U is held in the form

$$|U|\sqrt{\Phi} = B = const. \tag{6.35}$$

It follows from Eqs. (2.31) and (2.32) and Table 2.1 of Chap. 2 that, for a density-uniform sphere, the expression (6.35) is strictly sustained. Then, for the uniform Earth Eq. (6.33) with the help of (6.35) can be written in the form of a non-linear differential equation of the second order with respect to the variable Φ:

$$\ddot{\Phi} = -A + \frac{B}{\sqrt{\Phi}}. \tag{6.36}$$

Where $A = -2E = |U|$.

Expression (6.36) is the equation of virial oscillations of the uniform Earth. As it is shown in Chap. 4, the solution of Eq. (6.36) represents the periodic change of polar moment of inertia, i.e. oscillation of the interacting mass particles, and synchronously with this change of the potential (kinetic) energy. The solution of (6.36) is described by Eqs. (4.14) and (4.15) and we rewrite them in the same form:

$$\sqrt{\Phi_0} = \frac{B}{A}\left[1 - \varepsilon \cos(\lambda - \psi)\right], \tag{6.37}$$

$$t = \frac{4B}{(2A)^{3/2}}\left[\lambda - \varepsilon \sin(\lambda - \psi)\right]. \tag{6.38}$$

Here ε and ψ are the integration constants depending on the initial conditions of the Jacobi function Φ_0 and its first derivative at the first moment of time t_0.

Eqs. (6.37) and (6.38) and the integration constants after corresponding generalization were obtained in Chap. 4 in the explicit form:

$$\sqrt{\Phi_0} = a\,[1 - \varepsilon \cos \varphi], \tag{6.39}$$

$$\omega = \frac{2\pi}{T_v} = \frac{(2A)^{3/2}}{4B} = \frac{(2 \times 2E)^{3/2}}{4U\sqrt{\Phi}} = \sqrt{\frac{GM}{r^3}} = \sqrt{\frac{4}{3}\pi G\rho}, \tag{6.40}$$

$$a = \frac{B}{A}, \tag{6.41}$$

$$M_c = \varphi - \varepsilon \sin\varphi, \tag{6.42}$$

Where $M_c = n(t - \tau)$; T_v is the period of the virial oscillations; ω is the frequency of oscillations; ρ is the mass density of the body.

The physical meaning of Equations (6.39)–(6.42) is expressed by Kepler's motion laws, in particular, Eq. (6.39) describes the first and second laws, and Eq. (6.40) describes its third law. But now, due to the functional relationship between the potential energy and the Jacobi function (polar moment of inertia), the Kepler laws express dynamic but not static equilibrium of the planet. In fact, the Jacobi function $\sqrt{\Phi_0}$, which represents the polar moment of inertia, traces a second-order curve with period of oscillations T_v and frequency ω. The curve of the uniform body is the circle having the greatest semi-axis a and the eccentricity ε. The expression (6.42) represents the Kepler equation (1.1). Eqs. (6.39)–(6.42) as a whole describe an oscillating motion of the Earth in accordance with Kepler's laws. But their significance grows owing to the involved volumetric polar moment of inertia and its relationship with the potential energy of the body determining its dynamical effects. Figure 4.3 demonstrates the graphic picture of this effect and Eqs. (4.36)–(4.38) show the effect in the explicit form as the Lagrange series:

$$\sqrt{\Phi_0} = \frac{B}{A}\left[1 + \frac{\varepsilon^2}{2} + \left(-\varepsilon + \frac{3}{8}\varepsilon^3\right)\cos M_c - \frac{\varepsilon^2}{2}\cos 2M_c - \frac{3}{8}\varepsilon^3 \cos 3M_c + \ldots\right] \tag{6.43}$$

$$\Phi_0 = \frac{B^2}{A^2}\left[1 + \frac{3}{2}\varepsilon^2 + \left(-2\varepsilon + \frac{\varepsilon^3}{4}\right)\cos M_c - \frac{\varepsilon^2}{2}\cos 2M_c - \frac{\varepsilon^3}{4}\cos 3M_c + \ldots\right]. \tag{6.44}$$

Note that because the polar moment of inertia of the body has a functional relationship with the potential energy, Eqs. (6.43)–(6.44) and Fig. 4.3 express also the effect of a change of the potential energy in time. This fact is important for understanding of the nature and mechanism of the mass particle interaction, the result of which is generation of the potential energy. Later on we will return to this problem.

Quantitative values of the parameters of the Earth's virial oscillations are considered together with solution of the problem of oscillation and rotation of the non-uniform planet. In order to do this the potential and kinetic energies of a non-uniform body need to be to expanded and some other effects of its non-uniform structure have to be understood. It was noted in Sect. 2.5 that the non-uniformities play an important role in dynamical processes of the planet. Let us start consideration of the effect of the non-uniformities with separation of the potential and kinetic energies.

6.2 Separation of Potential and Kinetic Energies of the Non-uniform Earth

In fact, the Earth is not a uniform body. It has a shell structure and the shells themselves are also non-uniform elements of the body. It was shown in Sect. 2.2 that according to the artificial satellite data all the measured gravitational moments including tesseral ones have significant values. In geophysics this fact is interpreted as a deviation of the Earth from the hydrostatic equilibrium and attendance of the tangential forces which are continuously developed inside the body. From the point of view of the planet's dynamical equilibrium, the fact of the measured zonal and tesseral gravitational moments is a direct evidence of permanent development of the normal and tangential volumetric forces which are the components of the inner gravitational force field. In order to identify the above effects the inner force field of the Earth should be accordingly separated.

The expressions (2.46)–(2.49) in Chap. 2 indicate that the force function and the polar moment of a non-uniform self-gravitating sphere can be expanded with respect to their components related to the uniform mean density mass and its non-uniformities. In accordance with the superposition principle these components are responsible for the normal and tangential dynamical effects of a non-uniform body. Such a separation of the potential energy and polar moment of inertia through their dimensionless form-factors α^2 and β^2 was done by Garcia Lambas et al. (1985) with our interpretation (Ferronsky et al., 1996). Taking into account that the observed satellite irregularities are caused by the non-uniform distribution of the mass density, an auxiliary function relative to the radial density distribution was introduced for the separation:

$$\psi(s) = \int_0^s \frac{(\rho_r - \rho_0)}{\rho_0} x^2 dx, \tag{6.45}$$

where $s = r/R$ is the ratio of the running radius to the radius of the sphere R; ρ_o is the mean density of the sphere of radius r; ρ_r is the radial density; x is the running coordinate; the value $(\rho_r - \rho_o)$ satisfies $\int_0^R (\rho_r - \rho_o)r^2 dr = 0$ and the function $\Psi(1) = 0$.

The function $\Psi(s)$ expresses a radial change in the mass density of the non-uniform sphere relative to its mean value at the distance r/R. Now we can write expressions for the force function and the moment of inertia by using the structural form-factors α^2 and β^2 which were found in Sect. 2.6:

$$U = 4\pi G \int_0^R r\rho(r)m(r)dr = \alpha^2 \frac{GM^2}{R}, \tag{6.46}$$

$$I = 4\pi \int_0^R r^4 \rho(r)dr = \beta^2 MR^2. \tag{6.47}$$

By (6.45) we can do the corresponding change of variables in (6.46) and (6.47). As a result, the expressions for the potential energy U and polar moment of inertia I are found in the form of their components composed of their uniform and non-uniform constituents (Garcia Lambas et al., 1985; Ferronsky et al., 1996):

$$U = \alpha \frac{GM^2}{R} = \left[\frac{3}{5} + 3 \int_0^1 \psi x dx + \frac{9}{2} \int_0^1 \left(\frac{\psi}{x} \right)^2 dx \right] \frac{GM^2}{R}, \tag{6.48}$$

$$I = \beta^2 MR^2 = \left[\frac{3}{5} - 6 \int_0^1 \psi x dx \right] MR^2. \tag{6.49}$$

It is known that the moment of inertia multiplied by the square of the frequency ω of the oscillation-rotational motion of the mass is the kinetic energy of the body. Then Eq. (6.49) can be rewritten

$$K = I\omega^2 = \beta^2 MR^2 \omega^2 = \left[\frac{3}{5} - 6 \int_0^1 \psi x dx \right] MR^2 \omega^2. \tag{6.50}$$

Let us clarify the physical meaning of the terms in expressions (6.48) and (6.50) of the potential and kinetic energy.

As it follows from (2.46) and Table 2.1, the first terms in (6.48) and (6.50), numerically equal to 3/5, represent α_0^2 and β_0^2 being the structural coefficients of the uniform sphere with radius r, the density of which is equal to its mean value. The ratio of the potential and kinetic energies of such a sphere corresponds to the condition of the body's dynamical equilibrium when its kinetic energy is realized in the form of oscillations.

The second terms of the expressions can be rewritten in the form

$$3\int_0^1 \psi x dx \equiv 3\int_0^1 \left(\frac{\psi}{x}\right)x^2 dx, \tag{6.51}$$

$$-6\int_0^1 \psi x dx \equiv -6\int_0^1 \left(\frac{\psi}{x}\right)x^2 dx. \tag{6.52}$$

One can see that there are written here the additive parts of the potential and kinetic energies of the interacting masses of the non-uniformities of each sphere shell with the uniform sphere having a radius r of the sphere shell. Note that the structural coefficient β^2 of the kinetic energy is twice as high as the potential energy and has the minus sign. It is known from physics that interaction of mass particles, uniform and non-uniform with respect to density, is accompanied by their elastic and inelastic scattering of energy and appearance of a tangential component in their trajectories of motion. In this particular case the second terms in Eqs. (6.48) and (6.50) express the tangential (torque) component of the potential and kinetic energy of the body. Moreover, the rotational component of the kinetic energy is twice as much as the potential one.

The third term of Eq. (6.48) can be rewritten as

$$\frac{9}{2}\int_0^1 \left(\frac{\psi}{x}\right)^2 dx \equiv \frac{9}{2}\int_0^1 \left(\frac{\psi}{x^2}\right)^2 x^2 dx. \tag{6.53}$$

Here, there is another additive part of the potential energy of the interacting non-uniformities. It is the non-equilibrated part of the potential energy which does not have an appropriate part of the reactive kinetic energy and represents a dissipative component. Dissipative energy represents the electromagnetic energy which is emitted by the body and it determines the body's evolutionary effects. This energy forms the electromagnetic field of the body (see Chap. 8).

Thus, by expansion of the expression of the potential energy and the polar moment of inertia we obtained the components of both forms of energy which are responsible for oscillation and rotation of the non-uniform body. Applying the above results we can write separate conditions of the dynamical equilibrium for each form of the motion and separate virial equations of the dynamical equilibrium of their motion.

6.3 Conditions of Dynamical Equilibrium of Oscillation and Rotation of the Earth

Equations (6.48) and (6.50) can be written in the form

$$U = \left(\alpha_0^2 + \alpha_t^2 + \alpha_\gamma^2\right) \frac{GM^2}{R}, \tag{6.54}$$

$$K = \left(\beta_0^2 - 2\beta_t^2\right) MR^2\omega^2, \tag{6.55}$$

where $\alpha_0^2 = \beta_0^2$ and $2\alpha_t^2 = \beta_t^2$ and the subscripts o, t, γ define the radial, tangential, and dissipative components of the considered values.

Because the potential and kinetic energies of the uniform body are equal $(\alpha_0^2 = \beta_0^2 = 3/5)$ then from (6.48) and (6.50) one has

$$U_0 = K_0, \tag{6.56}$$

$$E_0 = U_0 + K_0 = 2U_0. \tag{6.57}$$

In order to express dynamical equilibrium between the potential and kinetic energies of the non-uniform interacting masses we can write, from (6.48) and (6.50),

$$2U_t = K_t, \tag{6.58}$$

$$E_t = U_t + K_t = 3U_t, \tag{6.59}$$

where $E_0, E_t, U_0, K_0, U_t, K_t$ are the total, potential and kinetic energies of oscillation and rotation accordingly. Note, that the energy is always a positive value.

Eqs. (6.56)–(6.59) present an expression for uniform and non-uniform components of an oscillating system which serves as the conditions of their dynamical equilibrium. Evidently, the potential energy U_γ of interaction between the non-uniformities, being irradiated from the body's outer shell, is irretrievably lost and provides a mechanism of the body's evolution.

In accordance with classical mechanics, for the above-considered non-uniform gravitating body, being a dissipative system, the torque N is not equal to zero, the angular momentum L of the sphere is not a conservative parameter, and its energy is continuously spent during the motion, i.e.

$$N = \frac{dL}{dt} > 0, \quad L \neq const., \quad E \neq const. > 0.$$

A system physically cannot be conservative if friction or other dissipation forces are present, because $F \times ds$ due to friction is always positive and an integral cannot vanish (Goldstein, 1980), i.e.:

$$\oint_{J} F \times ds > 0.$$

6.4 Equations of Oscillation and Rotation of the Earth and Their Solution

After we have found that the resultant of the body's gravitational field is not equal to zero and the system's dynamical equilibrium is maintained by the virial relationship between the potential and kinetic energies, the equations of a self-gravitating body motion can be written.

Earlier (Ferronsky et al., 1987) we used the obtained virial equation (6.33) for describing and studying the motion of both uniform and non-uniform self-gravitating spheres. Jacobi (1884) derived it from Newton's equations of motion of n mass points and reduced the n-body problem to the particular case of the one-body task with two independent variables, namely, the force function U and the polar moment of inertia Φ, in the form

$$\ddot{\Phi} = 2E - U. \tag{6.60}$$

Eq. (6.60) represents the energy conservation law and describes the system in scalar U and Φ volumetric characteristics. In Chap. 3 it was shown that Eq. (6.60) is also derived from Euler's equations for a continuous medium, and from the equations of Hamilton, Einstein, and quantum mechanics. Its time-averaged form gives the Clausius virial theorem. It was earlier mentioned that Clausius was deducing the theorem for application in thermodynamics and, in particular, as applied to assessing and designing of Carnot's machines. As the machines operate in the Earth's outer force field, Clausius introduced the coefficient 1/2 to the term of "living force" or kinetic energy, i.e.

$$K = \frac{1}{2} \sum_{i} m_i v_i^2.$$

As Jacobi has noted, the meaning of the introduced coefficient was to take into account only the kinetic energy generated by the machine, but not by the Earth's gravitational force. That was demonstrated by the work of a steam hammer for driving piles. The machine raises the hammer, but it falls down under the action of the force of the Earth's gravity. That is why the coefficient 1/2 of the kinetic energy of a uniform self-gravitating body in Eqs. (6.48)–(6.50) has disappeared. In its own force field the body moves due to release of its own energy.

Earlier by means of relation $U\sqrt{\Phi} \approx$ const, an approximate solution of Eq. (6.60) for a non-uniform body was obtained (Ferronsky et al., 1987). Now, after expansion of the force function and polar moment of inertia, at $U\gamma = 0$ and taking into account the conditions of the dynamical equilibrium ((6.57) and (6.59)), Eq. (6.60) can be written separately for the radial and tangential components in the form

$$\ddot{\Phi}_0 = \frac{1}{2}E_0 - U_0, \tag{6.61}$$

$$\ddot{\Phi}_t = \frac{1}{3}E_t - U_t. \tag{6.62}$$

Taking into account the functional relationship between the potential energy and the polar moment of inertia

$$|U|\sqrt{\Phi} = B = const,$$

and taking into account that the structural coefficients $\alpha_0^2 = \beta_0^2$ and $2\alpha_t = \beta_t^2$, both Eqs. (6.61) and (6.62) are reduced to an equation with one variable and have a rigorous solution:

$$\ddot{\Phi}_n = -A_n + \frac{B_n}{\sqrt{\Phi_n}}, \tag{6.63}$$

where A_n and B_n are the constant values and subscript n defines the non-uniform body.

The general solution of Eq. (6.62) is (4.14) and (4.15):

$$\sqrt{\Phi_n} = \frac{B_n}{A_n}[1 - \varepsilon \cos(\xi - \varphi)], \tag{6.64}$$

$$\omega = \frac{2\pi}{T_v} = \frac{4B}{(2A)^{3/2}}[\xi - \varepsilon \sin(\xi - \varphi)], \tag{6.65}$$

where ε and φ are, as previously, the integration constants depending on the initial values of Jacobi's function Φ_n and its first derivative $\dot{\Phi}_n$ at the time moment t_0 (the time here is an independent variable); T_v is the period of virial oscillations; ω is the oscillation frequency; ξ is the auxiliary independent variable; $A_n = A_0 - 1/2E_0 > 0$; $B_n = B_0 = U_0\sqrt{\Phi_0}$ for radial oscillations; $A = A_t = -1/3E_t > 0$; $B = B_t = U_t\sqrt{\Phi_t}$ for rotation of the body.

The expressions for the Jacobi function and its first derivative in an explicit form can be obtained after transforming them into the Lagrange series:

$$\sqrt{\Phi_n} = \frac{B}{A}\left[1 + \frac{\varepsilon^2}{2} + \left(-\varepsilon + \frac{3}{8}\varepsilon^3\right)\cos M_c - \frac{\varepsilon^2}{2}\cos 2M_c - \frac{3}{8}\varepsilon^3 \cos 3M_c + \dots\right],$$

$$\Phi_n = \frac{B^2}{A^2}\left[1 + \frac{3}{2}\varepsilon^2 + \left(-2\varepsilon + \frac{\varepsilon^3}{4}\right)\cos M_c - \frac{\varepsilon^2}{2}\cos 2M_c - \frac{\varepsilon^3}{4}\cos 3M_c + \dots\right],$$

$$\dot{\Phi}_n = \sqrt{\frac{2}{A}}\varepsilon B\left[\sin M_c + \frac{1}{2}\varepsilon \sin 2M_c + \frac{\varepsilon^2}{2}\sin M_c\left(2\cos^2 M_c - \sin^2 M_c\right) + \dots\right].$$

$$\tag{6.66}$$

Radial frequency of oscillation ω_{or} and angular velocity of rotation ω_{tr} of the shells of radius r can be rewritten from (6.40) as

$$\omega_{0r} = \frac{(2A_0)^{3/2}}{4B_0} = \sqrt{\frac{U_{or}}{J_{or}}} = \sqrt{\frac{\alpha_{or}^2 Gm_r}{\beta_{or}^2 r^3}} = \sqrt{\frac{4}{3}\pi G\rho_{0r}}, \qquad (6.67)$$

$$\omega_{tr} = \frac{(2A_t)^{3/2}}{4B_t} = \sqrt{\frac{2U_{tr}}{J_{tr}}} = \sqrt{\frac{2\alpha_{tr}^2 Gm_r}{\beta_{tr}^2 r^3}} = \sqrt{\frac{4}{3}\pi G\rho_{0r}k_{er}}, \qquad (6.68)$$

where U_{0r} and U_{tr} are the radial and tangential components of the force function (potential energy); J_{0r} and $J_{tr} = 2/3 J_{0r}$ are the polar and axial moment of inertia; $\rho_{0r} = \frac{1}{V_r}\int\limits_{V_r}\rho(r)dV_r$; $\rho(\mathrm{r})$ is the law of radial density distribution; ρ_{0r} is the mean density value of the sphere with a radius r; V_r is the sphere volume with a radius r; $2\alpha_{tr} = \beta_{tr}^2$; k_{er} is the dimensionless coefficient of the energy dissipation or tidal friction of the shells equal to the shell's oblateness.

The relations (6.64)–(6.65) represent Kepler's laws of body rotation in dynamical equilibrium. In the case of uniform mass density distribution the frequency (6.57) of oscillation of the sphere's shells with radius r is $\omega_{or} = \omega_o = const$. It means that here all the shells are oscillating with the same frequency. Thus, it appears that only non-uniform bodies are rotating systems.

Rotation of each body's shell depends on the effect of the potential energy scattering at the interaction of masses of different density. As a result, a tangential component of energy appears which is defined by the coefficient k_{er}. In geodynamics the coefficient is known as the geodynamical parameter. Its value is equal to the ratio of the radial oscillation frequency and the angular velocity of a shell and can be obtained from Eqs. (6.67)–(6.68), i.e.

$$k_e = \frac{\omega_t^2}{\omega_0^2} = \frac{\omega_t^2}{\frac{4}{3}\pi G\rho_0}. \qquad (6.69)$$

It was found that in the general case of a three-axial (a, b, c) ellipsoid with the ellipsoidal law of density distribution, the dimensionless coefficient $k_e \in [0.1]$ is equal to (Ferronsky et al., 1987)

$$k_r = \frac{F(\varphi,f)}{\sin\varphi} \bigg/ \frac{a^2 + b^2 + c^2}{3a^2},$$

where $\varphi = \arcsin\sqrt{\frac{a^2-c^2}{a^2}}, f = \sqrt{\frac{a^2-b^2}{a^2-c^2}}$, and $F(\varphi, f)$ is an incomplete elliptic integral of the first degree in the normal Legendre form.

Thus, in addition to the earlier obtained solution of the Earth's radial oscillations (Ferronsky et al., 1987), now we have a solution of its rotation. It is seen from expression (6.67) that the shell oscillations do not depend on the phase state of the body's mass and are determined by its density.

It follows from Eqs. (6.67) and (6.68) that in order to obtain the frequency of oscillation and angular velocity of rotation of the non-uniform Earth, the law of radial density distribution should be revealed. This problem will be considered later on. But before that the problem of the nature of the Earth's shells separation with respect to their density needs to be solved.

6.5 Application of Roche's Tidal Approach for Separation of the Earth's Shells

It is well known that the Earth has a quasi-spherical shell structure. This phenomenon has been confirmed by recording and interpretation of seismic longitudinal and transversal wave propagation during earthquakes. In order to understand the physics and mechanism of the Earth's mass differentiation with respect to its density, we apply Roche's tidal dynamics.

Newton's theorem of gravitational interaction between a material point and a spherical layer states that the layer does not affect a point located inside the layer. On the contrary, the outside-located material point is affected by the spherical layer. Roche's tidal dynamics is based on the above theorem. His approach is as follows (Ferronsky et al., 1996).

There are two bodies of masses M and m interacting in accordance with Newton's law (Fig. 6.1a).

Let $M \gg m$ and $R \gg r$, where r is the radius of the body m, and R is the distance between the bodies M and m. Assuming that the mass of the body M is uniformly distributed within a sphere of radius R, we can write the accelerations of the points A and B of the body m as

$$q_A = \frac{GM}{(R-r)^2} - \frac{Gm}{r^2}, \quad q_B = \frac{GM}{(R+r)^2} + \frac{Gm}{r^2}.$$

The relative tidal acceleration of the points A and B is

$$q_{AB} = G\left[\frac{M}{(R-r)^2} - \frac{M}{(R+r)^2} - \frac{2m}{r^2}\right]$$
$$= \frac{4\pi}{3}G\left[\rho_M R^3 \frac{4Rr}{(R^2-r^2)^2} - 2\rho_m r\right] \approx \frac{8\pi}{3}Gr(2\rho_M - \rho_m). \quad (6.70)$$

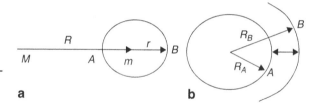

Fig. 6.1 The tidal gravitational stability of a sphere (**a**) and the sphere layer (**b**)

Here $\rho_M = M/\frac{4}{3}\pi R^3$ and $\rho_m = m/\frac{4}{3}\pi r^3$ are the mean density distributions for the spheres of radius R and r. Roche's criterion states that the body with mass m is stable against the tidal force disruption of the body M if the mean density of the body m is at least twice as high as that of the body M in the sphere with radius R. Roche considered the problem of the interaction between two spherical bodies without any interest in their creation history and in how the forces appeared. From the point of view of the origin of celestial bodies and of the interpretation of dynamical effects, we are interested in the tidal stability of separate envelopes of the same body. For this purpose we can apply Roche's tidal dynamics to study the stability of a non-uniform spherical envelope.

Let us assess the tidal stability of a spherical layer of radius R and thickness $r = R_B - R_A$ (Fig. 1.2b). The layer of mass m and mean density $\rho_m = m/4\pi\ R_A^2 r$ is affected at point A by the tidal force of the sphere of radius R_A. The mass of the sphere is M and its mean density $\rho_M = M/\frac{4}{3}\pi R_A^3$. The tidal force in point B is generated by the sphere of radius $R + r$ and mass $M + m$. Then the accelerations of the points A and B are

$$q_A = \frac{GM}{R_A^2} \quad \text{and} \quad q_B = \frac{G(M+m)}{(R_A+r)^2}.$$

The relative tidal acceleration of the points A and B is

$$q_{AB} = GM\left[\frac{1}{R_A^2} - \frac{1}{(R_A+r)^2}\right] - \frac{Gm}{(R_A+r)^2}$$

$$= \left(\frac{8}{3}\pi G\rho_M - 4\pi G\rho_m\right)r = 4\pi Gr\left(\frac{2}{3}\rho_M - \rho_m\right), (R \gg r). \quad (6.71)$$

Eqs. (6.70)–(6.71) give the possibility to understand the nature of the Earth's shell separation including some other dynamical effects.

6.6 Physical Meaning of Archimedes and Coriolis Forces and Separation of the Earth's Shells

The Archimedes principle states: *The apparent loss in weight of a body totally or partially immersed in a liquid is equal to the weight of the liquid displaced.* We saw in the previous section that the principle is described by Eqs. (6.70) and (6.71) and the forces that sink down or push out the body or the shell are of a gravitational nature. In fact, in the case of $\rho_n = \rho_M$ the body immersed in a liquid (or in any other medium) is kept in place due to equilibrium between the forces of the body's weight and the forces of the liquid reaction. In the case of $\rho_n > \rho_M$ or $\rho_n < \rho_M$ the body is sinking or floating up depending on the resultant of the above forces. Thus, the

Archimedes forces seem to have a gravity nature and are the radial component of the Earth's inner force field.

It is assumed that the Coriolis forces appeared as an effect of the body motion in the rotational system of co-ordinates relative to the inertial reference system. In this case rotation of the body is accepted as the inertial motion and the Coriolis forces appear to be the inertial ones. It follows from the solution of Equation (6.62) that the Coriolis' forces appear to be the tangential component of the Earth's inner force field, and the planet rotation is caused by the moment of those forces that are relative to the three-dimensional centre of inertia which also does not coincide with the three-dimensional gravity centre.

In accordance with Eq. (6.71) of the tidal acceleration of an outer non-uniform spherical layer at $\rho_M \neq \rho_m$, the mechanism of the gravitational density differentiation of masses is revealed. If $\rho_M < \rho_m$, then the shell immerses (is attracted) up to the level where $\rho_M = \rho_m$. At $\rho_M > \rho_m$ the shell floats up to the level where $\rho_M = \rho_m$ and at $\rho_M > 2/3\rho_m$ the shell becomes a self-gravitating one. Thus, in the case when the density increases towards the sphere's center, which is the Earth's case, then each overlying stratum appears to be in a suspended state due to repulsion by the Archimedes forces which, in fact, are a radial component of the gravitational interaction forces.

The effect of the gravitational differentiation of masses explains the nature of creation of the Earth's crust and also the ocean, geotectonic, orogenic and seismic processes, including earthquakes. All these phenomena appear to be a consequence of the continuous gravitational differentiation in density of the planet's masses. We assume that this effect was one of the dominate forces during creation of the Earth and the Solar system as a whole. For instance, the mean value of the Moon's density is less than 2/3 of the Earth's, i.e. $\rho_M < 2/3\rho_m$. If one assumes that this relation was maintained during the Moon's formation, then, in accordance with Eq. (6.71), this body separated at the earliest stage of the Earth's mass differentiation. Creation of the body from the separated shell should occur by means of the cyclonic eddy mechanism, which was proposed in due time by Descartes and which was unjustly rejected. If we take into account existence of the tangential forces in the non-uniform mass, then the above mechanism seems to be realistic.

6.7 Self-similarity Principle and the Radial Component of a Non-uniform Sphere

It follows from Eq. (6.71) that in the case of the uniform density distribution ($\rho_m = \rho_M$), all spherical layers of the gravitating sphere move to the centre with accelerations and velocities which are proportional to the distance from the centre. It means that such a sphere contracts without loss of its uniformity. This property of self-similarity of a dynamical system without any discrete scale is unique for a uniform body (Ferronsky et al., 1996).

A continuous system with a uniform density distribution is also ideal from the point of view of Roche's criterion of stability with respect to the tidal effect. That is why there is a deep physical meaning in separation of the first term of potential energy in expression (6.48). A uniform sphere is always similar in its structure in spite of the fact that it is a continuously contracting system. Here, we do not consider the Coulomb forces effect. For this case we have considered the specific proton and electron branches of the evolution of the body (see Sect. 8.5).

Note that in Newton's interpretation the potential energy has a non-additive category. It cannot be localized even in the simplest case of the interaction between two mass points. In our case of a gravitating sphere as a continuous body, for the interpretation of the additive component of the potential energy we can apply Hooke's concept. Namely, according to Hooke there is a linear relationship between the force and the caused displacement. Therefore the displacement is in square dependence on the potential energy. Hooke's energy belongs to the additive parameters. In the considered case of a gravitating sphere, the Newton force acting on each spherical layer is proportional to its distance from the centre. Thus, here from the physical point of view, the interpretations of Newton and Hooke are identical.

At the same time in the two approaches there is a principal difference even in the case of uniform distribution of the body density. According to Hooke the cause of displacement, relative to the system, is the action of the outer force. And if the total energy is equal to the potential energy, then equilibrium of the body is achieved. The potential energy plays here the role of elastic energy. The same uniform sphere with Newton's forces will be contracted. All the body's elementary shells will move without change of uniformity in the density distribution. But the first terms of Eqs. (6.48) and (6.50) show that the tidal effects of a uniform body restrict motion of the interacting shells towards the centre. In accordance with Newton's third law and the d'Alembert principle the attraction forces, under the action of which the shells move, should have equally and oppositely direct forces of Hooke's elastic counteraction. In the framework of the elastic gravitational interaction of shells, the dynamical equilibrium of a uniform sphere is achieved in the form of its elastic oscillations with equality between the potential and kinetic energy. The uniform sphere is dynamically stable relative to the tidal forces in all of its shells during the time of the system contraction. Because the potential and kinetic energies of a sphere are equal, then its total energy in the framework of the averaged virial theorem within one period of oscillation is accepted formally as equal to zero. Equality of the potential and kinetic energy of each shell means the equality of the centripetal (gravitational) and centrifugal (elastic constraint) accelerations. This guarantees the system remaining in dynamical equilibrium. On the contrary, all the spherical shells will be contracted towards the gravity centre which, in the case of the sphere, coincides with the inertia centre but does not coincide with the geometric centre of the masses (see Fig. 2.2). Because the gravitational forces are acting continuously, the elastic constraint forces of the body's shells are reacting also continuously. The physical meaning of the self-gravitation of a continuous body consists in the permanent work which applies the energy of the interacted shell masses on one side and the energy of the elastic reaction of the same masses in the form of oscillating motion on the other side. At dynamical equilibrium the body's equality of potential and kinetic energy means that the shell motion should be restricted by the elastic oscillation amplitude of the system. Such an oscillation is similar to the standing wave which appears without transfer of energy into outer space. In this case the radial forces of the shell's elastic interactions along the outer boundary sphere should have a dynamical equilibrium with the forces of the outer gravitational field. This is the condition of the system to be held in the outer force field of the mother's body. Because of this, while studying the dynamics of a conservative system, its rejected outer force field should be replaced by the corresponding equilibrated forces as they do, for instance, in Hooke's theory of elasticity.

Thus, from the point of view of dynamical equilibrium the first terms in Eqs. (6.48) and (6.50) represent the energy which provides the field of the radial forces in a non-uniform sphere. Here, the potential energy of the uniform component plays the role of the active force function, and the kinetic energy is the function of the elastic constraint forces.

6.8 Charges-like Motion of Non-uniformities and Tangential Component of the Force Function

Let us now discuss the tidal motion of non-uniformities due to their interactions with the uniform body. The potential and kinetic energies of these interactions are given by the second terms in Eqs. (6.48) and (6.50). In accordance with (6.71), the non-uniformity motion looks like the motion of electrical charges interacting on the background of a uniform sphere contraction. Spherical layers with densities exceeding those of the uniform body (positive anomalies) come together and move to the centre in elliptic trajectories. The layers with deficit of the density (negative anomalies) come together, but move from the centre on the parabolic path. Similar anomalies come together, but those with the opposite sign are dispersed with forces proportional to the layer radius. In general, the system tends to reach a uniform and equilibrium state by means of redistribution of its density up to the uniform limit. Both motions happen not relative to the empty space, but relative to the oscillating motion of the uniform sphere with a mean density. Separate consideration of motion of the uniform and non-uniform components of a heterogeneous sphere is justified by the superposition principle of the forces action which we keep here in mind. The considered motion of the non-uniformities looks like the motion of the positive and negative charges interacting on the background of the field of the uniformly dense sphere (Ferronsky et al., 1996). One can see here that in the case of gravitational interaction of mass particles of a continuous body, their motion is the consequence not only of mutual attraction, but also mutual repulsion by the same law $1/r^2$. In fact, in the case of a real natural non-uniform body it appears that the Newton and Coulomb laws are identical in details. Later on, while considering the Earth's by-density differentiated masses, the same picture of motion of the positive and negative anomalies will be seen.

If the sphere shells, in turn, include density non-uniformities, then by means of Roche's dynamics it is possible to show that the picture of the non-uniformity motion does not differ from that considered above.

In physics the process of interaction of particles with different masses without redistribution of their moments is called elastic scattering. The interaction process resulting in redistribution of their moments and change in the inner state or structure is called inelastic scattering. In classical mechanics while solving the problems of motion of the uniform conservative systems (like motion of the material point in the central field or motion of the rigid body), the effects of the energy scattering do not appear. In the problem of dynamics of the self-gravitating body, where interaction of the shells with different masses and densities are considered, the elastic and inelastic scattering of the energy becomes an evident fact following from consideration of the physical meaning of the expansion of the energy expressions in the form of (6.48) and (6.50). In particular, their second terms represent the potential and kinetic energies of gravitational interaction of masses having a non-uniform density with the uniform mass and express the effect of elastic scattering of density-different shells. Both terms differ only in the numeric coefficient and sign. The difference in the numerical coefficient evidences that the potential energy here is equal to half of the kinetic one ($U_t = 1/2K_t$). This part of the active and reactive force function characterizes the degree of the non-coincidence of the volumetric centre of inertia and that of the gravity centre of the system expressed by Eqs. (2.48)–(2.49). This effect is realized in the form of the angular momentum relative to the inertia centre.

Thus, we find that inelastic interaction of the non-uniformities with the uniform component of the system generates the tangential force field which is responsible for the system rotation. In other words, in the scalar force field of the by-density uniform body the vector component appears. In such a case, we can say that, by analogy with an electromagnetic field, in the gravitational scalar potential field of the non-uniform sphere $U(R, t)$ the vector potential $A(R, t)$ appears for which $U = rotA$ and the field $U(R, t)$ will be solenoidal. In this field the conditions for vortex motion of the masses are born, where $div A = 0$. This vector field, which in electrodynamics is called solenoidal, can be represented by the sum of the potential and vector fields. The fields, in addition to the energy, acquire moments and have a discrete-wave structure. In our case the source of the wave effects appears to be the interaction between the elementary shells of the masses by means of which we can construct a continuous body with a high symmetry of forms and properties. The source of the discrete effects can be represented by the interacting structural components of the shells, namely, atoms, molecules and their aggregates. We shall continue discussion about the nature of the gravitational and electromagnetic energy in Chap. 8.

6.9 Radial Distribution of Mass Density and the Earth's Inner Force Field

The existent idea about the radial mass density distribution of the planet is based on interpretation of transmission velocity of the longitudinal and transverse seismic waves. Figure 6.2 presents the classic curve of transmission velocities of the

Fig. 6.2 The curves of
transmission velocities of
the longitudinal (*1*) and
transverse (*2*) seismic waves,
density (*3*), and hydrostatic
pressure (*4*) in the Earth

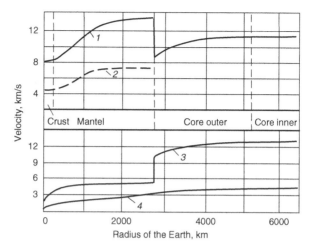

longitudinal and transverse seismic waves in the Earth plotted after generalization
of numerous experimental data (Jeffreys, 1970; Melchior, 1972; Zharkov, 1978).
The curves of the radial density and hydrostatic pressure distribution based on interpretation of the velocities of the longitudinal and transverse seismic waves are
also shown.

The picture of the transmission velocities of the seismic waves was obtained by
observations and therefore is realistic and correct. But interpretation of the obtained
data was based on the idea of hydrostatic equilibrium of the Earth. It leads to incred-
ibly high pressures in the core and high values of the mass density.

In accordance with Bullen's approach for interpretation of the seismic data,
the density distribution is characterized by the following values (Bullen, 1974;
Melchior, 1972; Zharkov, 1978). The density of the crust rocks is 2.7–2.8 g/cm^3 and
increases towards the centre by a certain curve up to ~13.0 g/cm^3 with jumps at the
Mohorovici discontinuity, between the upper and lower mantle, and on the border
of the outer core. Within the inner core the values of the transverse seismic waves
are equal to zero. Despite the jump of the longitudinal seismic wave velocity at the
outer core border dropping down, Bullen accepted that the density increases toward
the center. It was done after his unsuccessful attempt to approximate the seismic
data of the parabolic curve which gives a decrease of density in the core. Such a
tendency is not consistent with the idea of iron core content. Bullen certainly had
no idea that the radius of inertia and radius of gravity of the body do not coincide
with its geometric centre of mass and, therefore, the maximum value of density is
not located there. In accordance with our concept of the equilibrium condition of the
planet and its dynamical parameters, the approach to interpretation of the seismic
data related to the radial density and radial pressure distribution should be done on
a new basis.

Now, when we accept the concept of dynamical equilibrium of the Earth and refuse
its hydrostatic version, the basic idea to search for a solution of the problem seems

to be the found relationship between the polar moment of inertia and the potential (kinetic) energy. The value of the structural form-factor of the Earth's mean axial moment of inertia $\beta_\perp^2 = J_\perp/MR^2 = 0.3315$ found by artificial satellites (Zharkov, 1978) should be taken as a starting point. The mean polar moment of inertia of the assumed spherical non-uniform planet is equal to $\beta^2 = (3/2)\beta_\perp^2 = 0.49725$. We accept this value for development of the methodology.

Let us take as a basis the found mechanism of the shell separation with respect to the mass density which was presented in Sects. 6.5–6.8. The conditions and mechanism of the shell separation into radial and tangential components of the inner force field (by the Archimedes and Coriolis forces) represent continually acting effects and create physics for the Earth's structure formation. These effects explain the jumps between the shells observed by seismic data density. We take also into account the effect, expressed by Eq. (6.30), according to which the velocity of the sound recorded by the transmission velocity of the longitudinal and transverse seismic waves quantitatively characterize the energy of the elastic deformation of the media and velocity of its transmission there.

Applying the conception of Sect. 6.8, we accept that the non-uniformities of the spherical shells come together and, after their density becomes lower than that of the mean density of the inner sphere, move from the center by the parabolic law because they interact according to the law $1/r^2$. So, we can find a probable law of the radial density distribution in the form

$$\rho(r) = \rho_0(ax^2 + bx + c), \tag{6.72}$$

where $x = r/R$ is the ratio of the running and the final radius of the planet; ρ_0 is the body's mean density; a, b, c are the numerical coefficients.

The numerical coefficients were selected for different densities for the upper shell and in such a way that the planet's total mass M would be constant, i.e.

$$M = 4\pi \int_0^R r^2 \rho(r) dr = 4\pi \int_0^R r^2 \rho_0 \left(-a\frac{r^2}{R^2} + b\frac{r}{R} + c\right) dr$$

$$= \frac{4}{3}\pi R^3 \rho_0 \left(-\frac{3}{5}a + \frac{3}{4}b + c\right)$$

Here the term $(3/5)a + (3/4)b + c = 1$ in the right-hand side of the expression allows us to calculate and plot the distribution density curves in a dimensionless form.

We have selected three most typical parabolas (6.73) which satisfy the condition of equality of their moment of inertia, found by artificial satellite data, namely, the axial moment of inertia $J_\perp = \beta_\perp^2 MR^2 = 0.3315 MR^2$ or the polar moment of inertia $J = \beta^2 MR^2 = 0.4973 MR^2$. In addition, the first relation in (6.73) represents the straight line for which the surface mass density and that in the centre correspond to the present-day version and to the form-factor $\beta_\perp^2$. The fifth straight line represents the uniform spherical planet. The curve equations with selected numerical coefficients a, b, and c are as follows:

(1) $\rho(r) = \rho_0 \left(-2\dfrac{r}{R} + 2.495 \right)$, $\quad a = 0, \rho_s = 2.73$ g/cm^3;

(2) $\rho(r) = \rho_0 \left(-1.51\dfrac{r^2}{R^2} + 0.016\dfrac{r}{R} + 1.894 \right)$, $\quad \rho_s = 2.08$ g/cm^3;

(3) $\rho(r) = \rho_0 \left(-3.26\dfrac{r^2}{R^2} + 2.146\dfrac{r}{R} + 1.3465 \right)$, $\quad \rho_s = 1.28$ g/cm^3;   (6.73)

(4) $\rho(r) = \rho_0 \left(-5.24\dfrac{r^2}{R^2} + 5.132\dfrac{r}{R} + 0.295 \right)$, $\quad \rho_s = 1.03224$ g/cm^3;

(5) $\rho(r) = \rho_0 = $ const.

Figure 6.3 shows all the curves of (6.73). They intersect the straight line 5 of the mean density in the common point which corresponds to the value $r/R = 0.61475$.

Using Eqs. (6.73) and the found (by observations) form-factor $\beta_\perp^2 = 0.3315$, the main dynamical parameters were calculated for all four curves. The calculations were done by the known formulae of the attraction theory (Duboshin, 1975) and taking into account the relations of (6.48) and (6.50) obtained in Sect. 6.2. These calculations are presented below for equation (4), as an example.

The potential energy of the non-uniform sphere with the density distribution law $\rho(r)$ is found from the equation:

$$U = 4\pi G \int_0^R r\rho(r)m(r)dr,$$ (6.74)

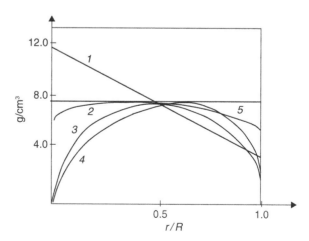

Fig. 6.3 Parabolic curves of radial density distribution calculated by Eq. (6.73)

where

$$\rho(r) = \rho_0 \left(a\frac{r^2}{R^2} + b\frac{r}{R} + c \right), \quad a = -5.24; \quad b = 5.132; \quad c = 0.295;$$

$$m(r) = 4\pi \int_0^r r^2 \rho(r)dr = 4\pi \int_0^r r^2 \rho_0 \left(a\frac{r^2}{R^2} + b\frac{r}{R} + c \right) dr$$

$$= \frac{4}{3}\pi r^3 \left(\frac{3}{5}a\frac{r^2}{R^2} + \frac{3}{4}b\frac{r}{R} + c \right).$$

Then

$$U = 4\pi G \int_0^R r\rho_0 \left(a\frac{r^2}{R^2} + b\frac{r}{R} + c \right) \frac{4}{3}\pi r^3 \left(\frac{3}{5}a\frac{r^2}{R^2} + \frac{3}{4}b\frac{r}{R} + c \right) dr$$

$$= \left(\frac{4}{3}\pi \rho_0 \right)^2 GR^5 \frac{R}{R} \left(\frac{1}{5}a^2 + \frac{81}{160}ab + \frac{9}{28}b^2 + \frac{24}{35}ac \right.$$

$$\left. + \frac{7}{8}bc + \frac{3}{5}c^2 = 0.660143\frac{GM^2}{R} \right).$$

The form-factor of the potential energy is $\alpha^2 = r_g^2/R^2 = 0.660143$, and the reduced radius of gravity is $r_g = \sqrt{0.660143R^2} = 0.8124918R$.

In accordance with Eq. (6.48), the potential energy of the non-uniform sphere is expanded into the components

$$U = U_0 + U_t + U_\gamma. \tag{6.75}$$

The potential energy of the uniform sphere is equal to

$$U_0 = \frac{3}{5}\frac{GM^2}{R}. \tag{6.76}$$

The form-factors of the potential and kinetic energy are equal to $\alpha_0^2 = 0.6$ and $\beta_0^2 = 0.6$ accordingly.

In accordance with the second term of the right–hand side of Eq. (6.48), the tangential component of the non-uniform sphere is written as

$$U_t = -\frac{1}{2}4\pi G \int_0^R r\rho_t(r)m_0(r)dr, \tag{6.77}$$

where

$$\rho_t(r) = \rho(r) - \rho_0 = \rho_0 \left(a\frac{r^2}{R^2} + b\frac{r}{R} + c \right) - \rho_0 = \rho_0 \left(a\frac{r^2}{R^2} + b\frac{r}{R} + c - 1 \right);$$

$$m_0(r) = 4\pi \int_0^r r^2 \rho_0 dr = \frac{4}{3}\pi \rho_0 r^3.$$

The coefficient $\frac{1}{2}$ in (6.77) is taken as the ratio of the second terms of the right–hand side of Eqs. (6.48) and (6.50), as in this particular case the tangential component of the potential energy is determined through the tangential component of the kinetic energy and is equal to half its value. Then

$$
\begin{aligned}
U_t &= -\frac{1}{2}4\frac{4}{3}(\pi \rho_0)^2 G \int_0^R r^4 \left(a\frac{r^2}{R^2} + b\frac{r}{R} + c - 1 \right) dr \\
&= -\frac{1}{2}\frac{GM^2}{R}\left(\frac{3}{7}a + \frac{1}{2}b + \frac{3}{5}c - \frac{3}{5} \right) = 0.0513571\frac{GM^2}{R}.
\end{aligned}
$$

The form-factors of the tangential components of the potential and kinetic energy are equal to $\alpha_t^2 = 0.051357$ and $\beta_t^2 = 2 \times 0.051357 = 0.102714$ accordingly.

In accordance with the third term in the right–hand side of Eq. (6.48), the dissipative component of the potential energy of the non-uniform sphere is

$$
U_\gamma = 4\pi G \int_0^R r\rho_t(r)m_t(r)dr, \tag{6.78}
$$

where

$$
\rho_t(r) = \rho(r) - \rho_0 = \rho_0 \left(a\frac{r^2}{R^2} + b\frac{r}{R} + c - 1 \right);
$$

$$
\begin{aligned}
m_0(r) &= 4\pi \int_0^r r^2 \rho_t(r)dr = 4\pi \int_0^r r^2 \rho_0 \left(a\frac{r^2}{R^2} + b\frac{r}{R} + c - 1 \right) dr \\
&= \frac{4}{3}\pi \rho_0 r^3 \left(\frac{3}{5}a\frac{r^2}{R^2} + \frac{3}{4}b\frac{r}{R} + c - 1 \right).
\end{aligned}
$$

Then

$$
\begin{aligned}
U_\gamma &= 4\frac{4}{3}(\pi \rho_0)^2 G \int_0^R r^4 \left(a\frac{r^2}{R^2} + b\frac{r}{R} + c - 1 \right)\left(\frac{3}{5}a\frac{r^2}{R^2} + \frac{3}{4}b\frac{r}{R} + c - 1 \right) dr \\
&= \frac{GM^2}{R}\left(\frac{1}{5}a^2 + \frac{81}{160}ab + \frac{24}{35}ac - \frac{24}{35}a + \frac{9}{28}b^2 \right.\\
&\quad \left. + \frac{7}{8}bc - \frac{7}{8}b + \frac{3}{5}c^2 - \frac{5}{6}c + \frac{3}{5} \right) \\
&= 0.008786\frac{GM^2}{R}. \tag{6.79}
\end{aligned}
$$

So, the value of the form-factor of the dissipative component is $\alpha_\gamma^2 = 0.008\,786$.

The radial distribution of the potential energy for interaction of a test mass point with the non-uniform sphere is

$$U(r) = \frac{4\pi G}{r} \int\limits_0^r r^2 \rho(r) dr + 4\pi G \int\limits_r^R r\rho(r) dr = \frac{4\pi G}{r} \int\limits_0^r r^2 \rho_0 \left(a\frac{r^2}{R^2} + b\frac{r}{R} + c \right) dr$$

$$+ 4\pi G \int\limits_r^R r\rho_0 \left(a\frac{r^2}{R^2} + b\frac{r}{R} + c \right) dr$$

$$= \frac{GMm_1}{R} \left(-\frac{3}{20} a\frac{r^4}{R^4} - \frac{1}{4} b\frac{r^3}{R^3} - \frac{1}{2} c\frac{r^2}{R^2} + \frac{3}{4a} + b + \frac{3}{2}c \right)$$

$$= \frac{GMm_1}{R} \left(0.786\frac{r^4}{R^4} - 1.283\frac{r^3}{R^3} - 0.1475\frac{r^2}{R^2} + 1,6445 \right) \tag{6.80}$$

At $r/R = 0$, then $\alpha_v^2(r) = 1.6445$; and at $r/R = 1$ then $\alpha_v^2(r) = 1$.

The radial distribution of the interaction force of the test mass point with the non-uniform sphere is

$$q(r) = -\frac{4\pi G}{r^2} \int\limits_0^r r^2 \rho(r) dr = -\frac{4\pi G}{r^2} \int\limits_0^r r^2 \rho_0 \left(a\frac{r^2}{R^2} + b\frac{r}{R} + c \right) dr$$

$$= -\frac{GMm_1}{R^2} \left(\frac{3}{5} a\frac{r^3}{R^3} + \frac{3}{4} b\frac{r^2}{R^2} + c\frac{r}{R} \right)$$

$$= -\frac{GMm_1}{R^2} \left(-3.144\frac{r^3}{R^3} + 3.849\frac{r^2}{R^2} + 0.295\frac{r}{R} \right). \tag{6.81}$$

At $r/R = 0$ then $\alpha_v^2(r) = 0$; and at $r/R = 1$ then $\alpha_v^2(r) = 1$.

Table 6.1 demonstrates the results of the calculated dynamical parameters for all the density curves (6.73) and Fig. 6.4 shows the curves of radial distribution of the potential energy and gravity force for the test mass point.

We wish to evaluate all four curves of mass density distribution in order to recognize which one is closer to the real Earth. In this case we keep in mind that the observed density jumps can be obtained for any curve by approximation of its continuous section with the mean value for each shell.

Figure 6.3 shows that the radial density values are substantially different for each curve. It refers, first of all, to the surface and centre of the body. At the same time Table 6.1 demonstrates the complete identity of the dynamical parameters of all the non-uniform spheres. It means that a fixed value of the polar moment of inertia permits us to have a multiplicity of curves of the radial density distribution with identical dynamical parameters of the body. The found property of the non-uniform self-gravitating sphere proves the rigor of the discovered functional relationship between the potential (kinetic) energy and the polar moment of inertia of the sphere. This property, in turn, is explained by the energy conservation law of a body during

Table 6.1 Physical and dynamical parameters of the earth for the density distribution presented by equation (6.73)

Equation N°	1	2	3	4
ρ_s, g/cm^3	2.76	2.08	1.65	1.03224
ρ_c, g/cm^3	13.8	10.455	6.315	1.6284
ρ_{max}, g/cm^3/km	13.8/0	10.455/0	8.26/2096	8.57/3122
$\beta^2_\perp$	0.3315	0.3315	0.3315	0.3315238
β^2	0.49725	0.49725	0.49725	0.49725858
β^2_t	0.10275	0.10275	0.102752	0.102714
α^2	0.660737	0.660737	0.660737	0.660143
α^2_t	0.051371	0.051371	0.0513714	0.0513571
α^2_γ	0.009366	0.009366	0.009366	0.0087859
r_g, km	5178.6	5178.7	5178.6	5176.4
r_m, km	4492.6	4492.6	4492.6	4492.7

Here ρ_s, ρ_c, ρ_{max} are the density on the sphere's surface, in the centre, and maximal accordingly; $\beta^2_\perp$, β^2, β^2_t are the form-factors of the axial, polar, and tangential components of the radius of inertia accordingly; α^2, α^2_t, α^2_γ are the form-factors of the radial, tangential, and dissipative components of the force function accordingly; r_g, r_m are the radiuses of the gravity and inertia.

its motion and evolution in the form of the dynamical equilibrium equation or generalized virial theorem.

If we accept the conditions of the mass density separation presented in Sects. 6.5–6.8, then the range of curves of the density distribution gives a principal picture of its evolutionary redistribution and can be applied for reconstruction of the Earth's history. It follows from Eq. (6.71) that the density value of each overlying shell of

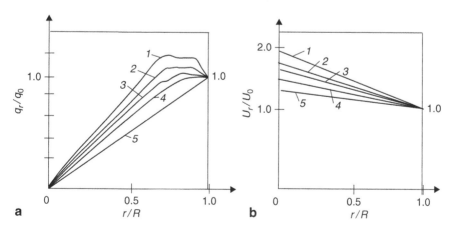

Fig. 6.4 The curves of the radial distribution of the potential energy (**a**) and gravity force (**b**) for the mass point test done by Eqs. (6.74) and (6.81)

the created Earth should be higher than the mean density of the inner mass. Otherwise, such a shell cannot be retained and should be dispersed by the tidal forces. It follows from this that the planet's formation process should be strictly operated by the dynamical laws of motion in the form of the virial oscillations and accompanied by differentiation of the non-uniform shells. The model of a cyclonic vortex which was proposed by Descartes is the most acceptable from the point of view of the considered ideas of planets' and satellites' creation from a common nebula. This problem needs a separate consideration. We only note here that from the presented curves of radial density distribution the parabola (4) more closely reflects the present-day planet's evolution as fixed by observations. In this case location of the Earth's reduced inertia radius falls on the lower mantle and the reduced gravity radius—on the upper mantle. The density maximum falls also on the lower mantle. Its value is found by ordinary means, namely, by taking the derivative from the density distribution law as equated to zero. From here $\rho_{max} = 8.57$ g/cm^3 is found to be at a distance of $r = 3122$ km. It means that the density maximum comes close to the border of the outer core where, as seismic observations show, the main density jump occurs. Curve (4) corrects the values of the radial density distribution in the mantle and changes its earlier interpretation in the outer and inner core. Because of zero values of the transverse velocities the matter of the inner core has a uniform density structure and, from the point of view of the equilibrium state, seems to be in a gaseous state at a pressure of 1–2 atmospheres. Taking into account the location of the maximum density value, there is a reason to assume that the outer core matter stays in the liquid or supercritical gaseous stage. In any case, the density and pressure of the inner and outer core are much lower and should have values corresponding to the seismic wave velocities. On the basis of the equation of mass density differentiation (6.71) we interpret the density jumps observed (by seismic data) near the Mohorovičich-Gutenberg and at the outer core borders as the borders of the shell's dynamical equilibrium. A shell which is found over that border appears in a suspended state due to action of the radial component of the gravitational pressure developed by the denser underlying shell. While the thickness of the suspended shell is growing it acquires its own equilibrium pressure (iceberg effect). The extremely high pressures in the Earth's interior, which follow from the hydrostatic equilibrium conditions, are impossible in its own force field.

The concept discussed above in relation to the Earth's density distribution is illustrated in Fig. 6.5.

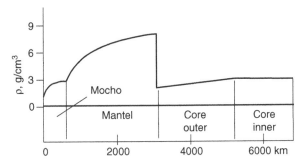

Fig. 6.5 Radial density distribution of the Earth by the authors' interpretation

6.10   Oscillation Frequency and Angular Velocity of the Earth's Shell Rotation

In order to determine numerical values of frequency of the virial oscillations and the angular velocities, which are the main dynamical parameters of the Earth's shells, we accept equation (4) of the density distribution (6.73) as the first approximation. All further relevant calculations can be made by applying this equation.

For calculations of the upper Earth's shell, the mean values of the planet's density $\rho_0 = 5.519$ g/cm^3 and the frequency of angular velocity $\omega_t = 7.29 \times 10^{-5}$ s^{-1} are known. Applying these values, the frequency and period of the virial oscillations, and the coefficient k_e of the tangential component of the inner forces, can be found. In accordance with Eq. (6.67) the frequency of the upper shell is equal to

$$\omega_0(r) = \sqrt{\frac{4}{3}\pi G \rho_0(r)} = \sqrt{\frac{4}{3}3.14 \times 6.67 \times 10^{-8} \times 5.519} = 1.24 \times 10^{-3} \text{s}^{-1}.$$

The period of oscillation is found from the expression

$$T_\omega = \frac{2\pi}{\omega_0(r)} = \frac{6.28}{1.24 \times 10^{-3}} = 5060.4 \text{ s} = 1.405 \text{ h.}$$

The product of the found frequency and the Earth's radius gives the value of the first cosmic velocity, the mean value of which is

$$v_1 = \omega_0(r) r_e = \left(1.2 \times 10^{-3}\right) \times 6370 = 7.9 \text{ km/s.}$$

Unlike the usual expression for the first cosmic velocity in the form of $v_1 = \sqrt{GM/r}$, we used here the physical condition of the dynamical equilibrium at the Earth's surface between the inner gravitational pressure of interacting masses and the outer background pressure including atmospheric pressure.

In the next chapter our own observation data on the near-surface atmospheric pressure and temperature oscillations at the near-surface layer and the results of the spectral analysis are given, which prove the above theoretical calculations of the planet's frequency of virial oscillations.

Now, applying the known mean value of the Earth's angular velocity $\omega_t = 7.29 \times 10^{-5}$s^{-1} and the known value of the frequency of virial oscillations for the upper shell $\omega_o = 1.24 \times 10^{-3}$s^{-1} by Eq. (6.69) the coefficient k_e can be found

$$k_e = \frac{\omega_t^2}{\omega_0^2} = \frac{\left(7.29 \times 10^{-5}\right)^2}{\left(1.24 \times 10^{-3}\right)^2} = \frac{1}{289.33} = 0.003456.$$

The coefficient k_e is known in geodynamics as a parameter that shows the ratio between the centrifugal force at the Earth's equator and the acceleration of the gravity force there equal to $k_e = 1/289.37$ (Melchior, 1972). The parameter is used to study the Earth's shape based on the Clairaut hydrostatic theory.

6.10.1 Thickness of the Upper Earth's Rotating Shell

It is known that the value of the mean linear velocity of the upper planet's shell is $v_e = 0.465$ km/c. We can find the thickness h_e at which the velocity v_e corresponds to the found frequency of radial oscillations of the shell $\omega_o = 1.24 \times 10^{-3}$ s^{-1}

$$h_e = \frac{v}{\omega_o(r)} = \frac{0.465}{1.24 \times 10^{-3}} = 375 \text{ km}. \tag{6.82}$$

Such is the thickness of the upper shell of the Earth which is rotating by forces in its own force field. It is assumed that the shell is found in the solid state. In reality it is known that the rigid shell has a thickness less than 50 km. The remaining more than 300 km-thick part of the shell has a viscous-plastic consistency, the density of which increases with depth. The border of the near-by Mohorovičich discontinuity has a decreased density because of the melted substance due to high friction and saturation by a gaseous component. The border plays a role of some sort of spherical hinge. Because the density of the Earth's crust is lower than that of the underlying matter, then it occurs in the suspended state. During the oscillating motion the crust shells are affected by the alternating-sign acceleration and the inertial isostatic effects.

6.10.2 Oscillation of the Earth's Shells

Let us obtain the expression of virial oscillations for the Earth's other shells by applying expression (4) of (6.73) for the radial density distribution. Write Eq. (6.67)

$$\omega_0(r) = \sqrt{\frac{4}{3}\pi G \rho_0(r)},$$

where

$$\rho_0(r) = \frac{m_0(r)}{\frac{4}{3}\pi r^3} = \frac{4\pi \int\limits_0^r r^2 \rho(r)dr}{\frac{4}{3}\pi r^3} = \frac{\frac{4}{3}\pi r^3 \rho_0 \left(\frac{3}{5}a\frac{r^2}{R^2} + \frac{3}{4}b\frac{r}{R} + c\right)}{\frac{4}{3}\pi r^3}$$

$$= \rho_0 \left(\frac{3}{5}a\frac{r^2}{R^2} + \frac{3}{4}b\frac{r}{R} + c\right).$$

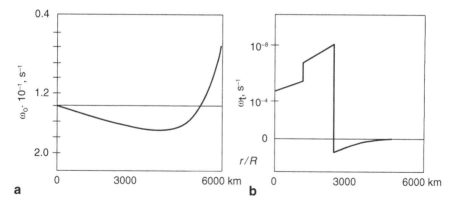

Fig. 6.6 Radial change in virial oscillation frequencies (**a**) and angular velocity of rotation (**b**) according to Eqs. (6.83) and (6.84)

Then

$$\omega_0(r) = \sqrt{\frac{4}{3}\pi G \rho_0(r)} = \sqrt{\frac{4}{3}\pi G \rho_0 \left(\frac{3}{5}a\frac{r^2}{R^2} + \frac{3}{4}b\frac{r}{R} + c\right)}$$

$$= 1.24 \times 10^{-3} \sqrt{\left(-3.144\frac{r^2}{R^2} + 3.849\frac{r}{R} + 0.295\right)}. \qquad (6.83)$$

At $r/R = 0$ then $\omega_o(r_o) = 0.6743 \times 10^{-3}$ s^{-1}; at $r/R = 1$ then $\omega_o(r_o) = 1.24 \times 10^{-3}$ s^{-1}; at $\rho_{max} = 8.57$ g/cm^3 $\omega_o(r_o) = 1.486 \times 10^{-3}$ s^{-1}, where $r/R = 0.49$.

Figure 6.6a shows a change in the virial oscillation frequencies of the Earth's shells by Eq. (6.83).

6.10.3 Angular Velocity of Shell Rotation

Angular velocity of the Earth's shell rotations is determined from Eq. (6.68)

$$\omega_t(r) = \sqrt{\frac{4}{3}\pi G \rho_t(r)} = \sqrt{\frac{4}{3}\pi G \rho_0(r) \left(\frac{3}{5}a\frac{r^2}{R^2} + \frac{3}{4}b\frac{r}{R} + c\right) k_e(r)}$$

$$= \omega_0(r) \sqrt{\left(\frac{3}{5}a\frac{r^2}{R^2} + \frac{3}{4}b\frac{r}{R} + c\right) k_e(r)}$$

$$= \omega_0(r) \sqrt{\left(-3.144\frac{r^2}{R^2} + 3.8475\frac{r}{R} + 0.295\right) k_e(r)}, \qquad (6.84)$$

where $\omega_t(r_o)$ is the angular velocity of the shell rotation; $\omega_o(r)$ is the shell oscillation frequency which is determined by Eq. (6.83).

The geodynamic parameter $k_e(r_o)$, which expresses the ratio of the tangential component of the force field and the gravity force acceleration for the upper shell, is approximated in the

$$k_e(r) = \frac{\omega_t^2(r)}{\omega_0^2(r)} = \frac{h_c^2}{R^2},$$

where h_c is the distance between the sphere's surface and the density jump; R is the sphere's radius.

At $r/R = 1$, $k_e(r_o) = 0.003456$; at $r/R = 0$, $k_e(r_o) = 1$, $\omega_t(0) = \omega_o(0)$, i.e. the virial oscillation frequency corresponds to the gravity pressure of the uniform density masses. In this particular case we are interested in a change in the angular velocity of rotation of the upper (1000 km) and lower (up to the core border) mantle (2900 km). Figure 6.6b shows the radial change of the angular velocity of rotation calculated by Eq. (6.84). It is seen that the angular velocity at the lower mantle–outer core is close to zero but changes its direction.

We emphasize once more that Equations (6.83) and (6.84) express the third Kepler law which determines radial distribution of both the virial oscillation frequencies and the angular velocities of rotation. Numerical values of these parameters are determined by the radial density distribution law. It also determines the density jumps which mark the effect of the shell's isostatic equilibrium.

6.11 Perturbation Effects in Dynamics of the Earth

The most noteworthy effects of dynamics of the Earth are the interrelated phenomena of the precession and nutation of the axis of rotation, tidal effects of the oceans, and atmosphere, the axial obliquity and declination of the plumb-line and the gravity change at each point of the planet. The present-day ideas about the nature of these phenomena were formed on the basis of the planet's hydrostatic equilibrium and since early times were considered as effects of perturbation from the Sun, Moon and other planets. All the above phenomena represent periodic processes and many observational and analytical works were done for their understanding and description. The present-day studies of these processes are still continuing to be specified and corrected. This is because such topical problems as correct time, ocean dynamics, short and long-term weather and climate changes and other environmental changes are important for every-day human life.

Now, after it was found that the conditions of the hydrostatic equilibrium are not acceptable for study of the Earth's dynamics, we reconsider the nature of the phenomena by applying the concept of the planet's dynamical equilibrium and developing a novel approach to solving the problem.

6.11.1 The Nature of Perturbations in the Framework of Hydrostatic Equilibrium

Phenomenon of precession. The phenomenon of the precession of equinoxes was already observed in the second century BP by the Greek astronomer and mathematician Hypparchos. His discovery was based on comparison of longitudes of the far stars with the longitudes of the same stars determined 150 years earlier by other astronomers.

The classical explanation of precession is based on the inertial rotation of a symmetrical rigid body with a fixed point. Such a motion of a body, presented in Fig. 6.7, includes its rotation with angular velocity Ω relative to the axis Oz, fixed in the body, and from rotation with angular velocity ω around the axis Oz_1. Here the axes x_1, y_1, z_1 are accepted to be immobile because motion of the body is considered relative to them. The straight line ON perpendicular to the plane z_1Oz is called the line of nodes, and angle $\psi = x_1ON$ is the precession angle. Together with precession, the body performs the nutation motions (axis wobbling) which cause changes in the nutation angle $\Theta = z_1Oz$.

Perturbation of the Earth's inertial rotation is considered as a result of the applied solar-moon force couple, the axis of which is at right angles to the rotation axis; the body turns around the third mutually perpendicular axis. In this model the Earth is accepted as a rigid body oblate along the rotary axis. Newton's idea was that the spherical body has an equatorial bulge that appears as the result of the planet's oblateness. In this case the Sun is more strongly attracted by the body's equatorial bulge and this tends to decrease the inclination of the Earth's equatorial plane to the ecliptic. The Moon has an analogous affect but is two times as powerful due

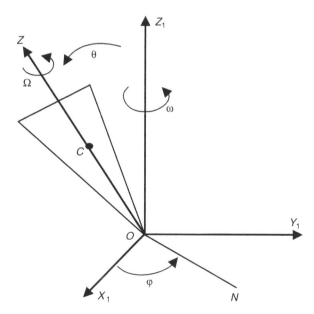

Fig. 6.7 Classical explanation of precession motion

to its closer distance. The common effect of the Sun and Moon on the equatorial excess of the rotating Earth's mass leads to the rotary axis precession. Because the induced precession forces are continuously varying due to changes in the Sun and Moon positions relative to the Earth, then additional nutations (wobble) of the axis are observed during translational motion of the planet. In addition to the lunisolar precession, the effect of the other planets of about a few tenths of an arc second is observed. The combined Earth precession rate is estimated to be equal to ~50.3"/ year or one complete rotation in ~26 000 years.

The theory of the precession and nutation of the Earth's axis of rotation based on hydrostatics was developed in the works of D'Alembert, Laplace and Euler. The precession values were calculated by Bessel and Struve and are undergoing verification even today. The physical basis of modern studies remains unchanged. The main emphasis in the studies is on consideration of the elastic and rheological properties of the planet, effects of dynamics of the atmosphere and the oceans and dynamics of the liquid core, the probability of which is assumed (Jeffreys, 1970; Munk and MacDonald, 1960; Melchior, 1972; Sabadini and Vermeersten, 2004; Molodensky, 1961; Magnitsky, 1965).

Tidal effects. The theory of the ocean tides was also presented first by Newton in his *Principia*, Proposition XXIV, Theorem XIX. He stated that the tides are caused by action of the Moon and the Sun. It follows from Corollaries IX and XX (Proposition LXVI, Book I), that the sea should rise and subside twice per every lunar and twice per every solar day, and the highest tide in the free and deep seas should appear less than six hours after the tide-body has passed the place meridian. And it exhibits the same behavior all along the East Atlantic and Pacific shores. The effects of both tide-bodies are summed up. At joining and opposing positions of the bodies their effects are summed up and provide the highest or lowest tide. Observation shows that the tide effect of the Moon is stronger than that of the Sun.

Modern studies in the theory of precession and nutation remain on the physical basis described by Newton. Besides, all the above phenomena are considered in close relationship and their amplitudes and periods are described by common equations which follow from the attraction theory (Melchior, 1972).

The modern physical picture for explanation of the tidal interaction is presented as follows (Pariysky, 1972). The tidal force is equal to the difference between any Moon-attracted particle on the Earth (including in the atmosphere, in the oceans, and in the solid body itself) and the same particle replaced to the center of the planet. (Fig. 6.8).

The normal tide forces are proportional to the mass of the Moon m and the distance to the center of the Earth r, and to the inverse cubic distance between the Moon and the Earth R, and the zenith distance of the Moon z. The vertical component of the tidal force per mass unit F changes the value of the gravity force into the value

$$F_v = 3G\frac{mr}{R^3}\left(\cos^2 z - \frac{1}{3}\right), \qquad (6.85)$$

where G is the gravity constant.

Fig. 6.8 Scheme of mass interaction between the Moon and the Earth for explanation of tidal effects. (By Pariysky, 1975)

To the Moon

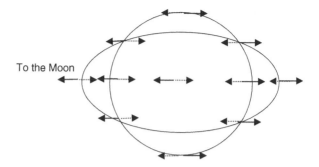

The gravity force decreases by 0.1 mgal or by 10^{-7} of its value on the Earth's surface when the Moon stays in zenith or nadir, and twice increases when the Moon rises or sets.

The horizontal component of the tidal force is equal to zero when the Moon stays zenith, nadir and on the horizon. Its maximum value reaches 0.08 mgal at zenith distance of the Moon equal to 45°:

$$F_h = \frac{3}{2} G \frac{mr}{R^3} \sin^2 z. \tag{6.86}$$

The tidal force of the Sun is formed analogously. But because of distance, its value is 2.16 times less than of the lunar one Due to rotational and orbital motion of the Earth, the Moon, and the Sun, the tide force of each point in the atmosphere, the oceans and the planet's surface continuously changes in time. Tables have been compiled of integral values of the tide forces in the form of the sums of periodic components (~500 terms or more) calculated by the theory of motion of the Moon round the Earth and the Earth round the Sun.

By estimation of many authors, the total tidal-slowing down of the Earth's rotation amounts to 3.5 msec in 100 years. By astronomic observation the Earth's rotation is accelerated by 1.5 msec per 100 years.

Note, that in the framework of the hydrostatic approach, the problems of the nature of the obliquity of the axis of the Earth's rotation to the ecliptic and the nature of the obliquity of axes of the Moon and the Sun to their orbit planes and their obliquity to the ecliptic are not discussed. These problems have as yet not even been formulated.

6.11.2 The Nature of Perturbations Based on Dynamic Equilibrium

To begin, let us consider the physical meaning of gravitational perturbation for interacting volumetric (but not point) body masses. Contrary to hydrostatics, where the measure of perturbation in the precession-nutation and the tidal phenomena is the perturbing force, in the dynamic approach that measure of perturbation is power pressure. In Chap. 2 we concluded that the mass points and the vector forces as a physical and mathematical instrument in the problem solution of dynamics of the Earth in its own force field are inapplicable. This is because the outer vector central force field of the interacting volumetric masses incorrectly expresses dynamical effects of their interaction. As a result, the kinetic effect of interaction of the mass particles, namely, the kinetic energy of their oscillation, is lost. And also the geometric center of a body is accepted as the gravity center and center of the inertia (reaction). In dynamics it leads to wrong results and conclusions. In this connection we found that in dynamics of the Earth as a self-gravitating body the effect of gravitational interaction of mass particles should be considered as the power pressure. In addition, in this case we are free in our choice of a reference system. Our conclusion does not contradict Newton's physical ideas which are presented in Book I of his *Principia* where he says:

> I approach to state a theory about the motion of bodies tending to each other with centripetal forces, although to express that physically it should be called more correct as pressure. But we are dealing now with mathematics and in order to be understandable for mathematicians let us leave aside physical discussion and apply the force as its usual name.

Accepting the power pressure as an effect of gravitational interaction, we come to an understanding that, in the considered problem of the mutual perturbations between the Earth, the Moon, and the Sun, the interaction results not between the body centers or shells along straight lines, but between the outer force fields of the bodies and between their inner force fields of the shells. Satellite observations show that the outer force field, induced by the Earth's mass, has 4π-outward direction of propagation and acquires a wave nature. We consider this outer wave force field as a physical media by which the bodies transmit their energy. Thus, the Earth and other planets are held and move on the orbits by the power of the outer wave field of the Sun. This power, in terms of its normal and equal to its tangential components, remains valid since the planets separation (see above Sects. 6.5 and 6.6). In order to demonstrate validity of the above conception let us calculate the mean values of velocity of the Earth and the Moon orbital motion from the frequencies of oscillation of the respective outer wave fields of their parents.

In accordance with Eq. (6.67), the frequency ω_s of oscillation of the self-gravitating Sun's outer force field at the mean distance R_e of the Earth's orbit is

$$\omega_s = \sqrt{\frac{GM_s}{R_{es}^3}} = \sqrt{\frac{6.67 \times 10^{-8} \times 1.99 \times 10^{33}}{\left(1.496 \times 10^{13}\right)^3}} = 1.9931 \times 10^{-7} \mathrm{s}^{-1},$$

where M_s is the Sun's mass; R_{es} is the mean distance between the Earth and the Sun.

In accordance with wave mechanics, the mean value of the Earth's orbital velocity is

$$v_e = \omega_s R_{es} = 1.9931 \times 10^{-7} \times 1.496 \times 10^{13} = 2.98 \times 10^6 \, \text{cm/s} = 29.9 \, \text{km/s}.$$

If we extend the outer force fields of the Sun and the Earth up to equality of their reduced densities, then in accordance with the same Eq. (6.67) the border between the two interacting fields can be found. By calculation the mean (nodal) value of the Earth's field border is found to extend up to 2.128×10^9 m and the Sun's border—to 1.478×10^{11} m.

Applying the same procedure, the orbital velocity of the Moon from the frequency oscillation ω_e of the Earth's outer force field at the distance R_{me} has value

$$\omega_e = \sqrt{\frac{GM_e}{R_{me}^3}} = \sqrt{\frac{6.67 \times 10^{-8} \times 5.976 \times 10^{27}}{\left(3.844 \times 10^{10}\right)^3}} = 2.64907 \times 10^{-6} \text{s}^{-1},$$

where M_e is the Earth's mass; R_{me} is the mean distance between the Moon and the Sun.

Then the Moon's orbital velocity is

$$v_m = \omega_e R_m = 2.64907 \times 10^{-6} \times 1.496 \times 10^{13} = 1.0183 \times 10^5 \, \text{cm/s}$$
$$= 1.0183 \, \text{km/s}.$$

The Moon's border of the outer force field in the nodal plane extends to the Earth up to 0.72×10^8 m and the Earth's border—to 3.724×10^8 m.

The obtained values of velocities, as well as the values for the pericenters and apocenters, are exactly the same as known from observation. It means that the observed ecliptic inclination relative to the equatorial plane of the Sun and inclination of the Moon's orbit relative to the ecliptic reflect asymmetric distribution of the solar and the planet's masses. It also means that the observed inclination of the Moon's orbit plane and the ecliptic are governed by asymmetric distribution of the Earth's and the Sun's force fields. The force fields of the Earth and the Moon, together with the bodies themselves being local "secondary" inclusions in the powerful force fields of the Sun and the Earth, are obliged to adjust their positions in order to be in dynamic equilibrium. The observed parameters of the orbits and their inclination relative to the plane diameters of the Sun, the Earth and the Moon give a general view of the asymmetric distribution of the body's masses. In particular, the northern hemisphere of the Earth is more massive than the southern one. So, in the perihelion the northern hemisphere is turned to the less massive hemisphere of the Sun. So that, the polar oblateness of each body controls the location of its pericenter and apocenter, and the equatorial oblateness of each body responds to location of its nodes. Thus, the body motion in the outer force field of its parent occurs under strict conditions of dynamic equilibrium which is also the main condition of its separation. It follows from the condition of dynamic equilibrium that the orbital motion of the Earth and the Moon reflects asymmetry in mass density distribution of the Sun, the Earth, and the Moon and asymmetry in the potential of the outer wave field distribution.

Only the structure of the Sun's outer wave field controls the Earth's trajectory at the orbital motion and the Earth's force field manages the orbital motion of the Moon, but not vice versa.

6.11.3 Change of the Outer Force Field and the Nature of Precession and Nutation

At the right time of motion of the bodies with the outer wave fields, their mutual perturbations are transferred not directly from each body to the other one or from their shells, but through the outer fields by means of the corresponding active and reactive wave pressure of the interacting fields. There is an important dynamic effect of all the perturbations. This is the continuous change in the outer wave field of each body which proceeds from its non-uniform radial distribution of the mass density. As it was earlier shown, the non-uniform radial distribution of mass density initiates the differential rotation of the body shells. And, in accordance with Eqs. (6.67)–(6.68) expressing the third Kepler's law, the reduced body shells' perturbing effects are transferred to the other body by means of the outer wave field. So that the Sun, for instance, through its outer wave field, continuously transfers to the Earth all the perturbations resulting during rotation of the interacting masses of the shells. The Earth, in the framework of the energy conservation law, demonstrates all the perturbations by changes in its orbit turns around the Sun (see below Fig. 6.9).

Earlier it was shown that in the case of non-uniform distribution of mass density the body's potential and kinetic energies have radial and tangential components which induce oscillation and rotation of the shells. It was defined by Eq. (6.82) that the observed daily rotation of the Earth concerns only the upper shell with thickness

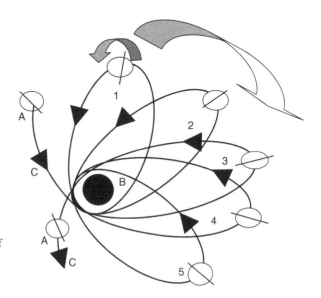

Fig. 6.9 Real picture of motion of a body *A* in the force field of a body *B*. Digits identify succession of turns of the body *A* moving around body *B* along the open orbit *C*

of ~375 km and reaches the near-by Mohorovicić discontinuity. By the same reasoning it is not difficult to find the thickness of the upper shells for the Sun and the Moon correspondingly equal to:

$$h_s = \frac{v_s}{\omega_{0s(R_s)}} \approx \frac{2}{6.23 \times 10^{-4}} \approx 3210 \text{ km,} \qquad (6.87)$$

$$h_m = \frac{v_m}{\omega_{0m(R_m)}} \approx \frac{4.56 \times 10^{-3}}{9.66 \times 10^{-4}} \approx 4.72 \text{ km.} \qquad (6.88)$$

We do not know real number and angular velocities of rotation for the inner shells of the three bodies. These velocities have a direct interrelation with the observed changes in parameters of the orbital motion of the Earth and the Moon including the retrograde motion of the orbital nodes and the apsidal line. In this connection let us try to understand first of all the nature of precession and nutation of the bodies from the point of view of the dynamic approach.

It was noted above that, in accordance with the hydrostatic approach, precession of the equinoxes of the Earth is an effect of the net torque of the Moon and the Sun on the equatorial "bulge" aroused from gravitational attraction. The torque aspires to diminish inclination of the equatorial belt with surplus mass relative to the ecliptic and induce the retrograde motion of the nodal line. In addition, because the ratio of distance between the interacted bodies is changed, then the relationship between the forces is also changed. In this connection the precession is accompanied by nutation (wobbling) motion of the axes of rotation.

Analysis of orbits of the artificial satellite motion around the Earth shows that, in spite of absence of the equatorial "bulge" of mass, the apparatus demonstrates the precession effect. Its orbital plane has a clockwise rotation with retrograde motion of the nodal line. But a new explanation of the phenomenon is given. It appears that the retrograde motion of the nodal line associates with the Earth equatorial and polar oblateness. The amplitude of the nodal line shift depends on the satellite orbit inclination to the Earth's equatorial plane. In the case of the poles' orbital plane the nodal line shift is completely absent. This is because the motion excludes both the polar and the equatorial oblatenesses of the Earth. The direction of motion of the apsidal line depends on the satellite's orbit inclination and is determined by the Lentz law.

It is also known that for the other free-of-satellite planets the retrograde motion of the nodal line is also a characteristic phenomenon called the "secular perihelion shift". It was found from observation of Mercury, Venus, Earth and Mars that their secular perihelion shifts are decreased from ~40″ through ~8.5″, ~5″ to ~1.5″ accordingly (Chebotarev, 1974).

All these facts imply that the explanation given for the satellites' precession depending on their orbital inclination to the ecliptic is correct. But the nature of this unique phenomenon, characteristic for all celestial bodies, are inconsistent with the hydrostatic approach and should be reconsidered, taking also into account the satellite observations.

6.11.4   Observed Picture of a Body Precession

The precession of the Earth, the Moon and the artificial satellites in the form of motion of an orbital plane toward the backward direction of the body's motion should be considered as a virtual explanation of the phenomenon. In fact, the orbit's plane is a geometric shape traced by the body. And there is no reason to consider its movement without the body itself. There is no difficulty to present the real body motion in space in two opposite directions synchronously. In particular, the actual picture of the Earth, the Moon and the satellite motion in counterclockwise direction and retrograde movement of the nodal line is shown in Fig. 6.9.

Here the satellite is moving in the counterclockwise direction along the unlocked elliptic orbit *1* in the continuously changing (perturbed by oblatenesses) planet's force field. Because of the counterclockwise rotation of the Earth's mass, the satellite in perigee started to move on the orbit *2* and makes a shift in retrograde direction in the ascending and descending nodes. At the same time the eccentricity of the orbit *2* changes by a proper value. Analogously the body passes on orbit *3, 4, 5* and so on. The theory of dynamic equilibrium of the Earth explains the physics of the observed phenomenon as follows.

6.11.5   The Nature of Precession and Nutation Based
on Dynamical Equilibrium

The dynamic equilibrium theory assumes that the Earth is a self-gravitating body, the interacting mass particles of which induce the inner and outer force fields. Separation of the planet's asymmetric shells results in an inner force field in accordance with the radial mass density distribution. The normal component of the body's power pressure provides oscillation, and the tangential component induces rotation of the shells having a different angular velocity. At the same time the mantle shells A and the outer shell of the core B may have the same (Fig. 6.10a) or opposite direction (Fig. 6.10b) of rotation depending on the radial mass density distribution.

The seismic data show that the inner core C has a uniform density distribution. Because of this, it does not rotate and its potential energy is realized in the form of oscillation of the interacting particles. The potential E of the outer force field is controlled by integral effect of the interacted masses of all the shells and presented by the reduced shell D having continuously changing power.

The energy of the Earth's outer force field is changed from the body surface in accordance with the $1/r$ law and at every r is continuously varied because of differences in the angular velocity of rotation of the shell's masses. This force field controls the direction and angular velocity of orbital motion of a satellite. Taking into account the non-uniform and asymmetric distribution of the masses of rotating shells, the change in the trajectory of the body motion is accompanied by a corresponding change in eccentricity of the orbit both at each and subsequent turns. Its maximum value is reached when the non-uniformities of the rotating masses coincide and the minimal value appears at the opposite position.

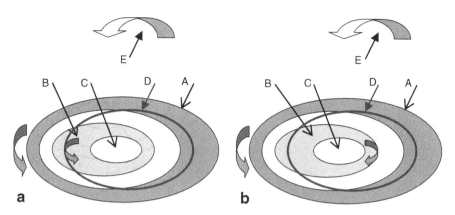

Fig. 6.10 Sketch of rotation of the Earth's shells by action of the inner force field: *A* is the mantle shells; *B* is the outer core; *C* is the inner core; *E* is the outer force field; *D* is the reduced shell of the inner force field of the planet

It is worth noting that because the effect of retrograde motion of the nodal line of the Earth, the Moon and artificial satellites appears to be a common phenomenon, then the conclusion follows that the Sun, in the outer force field of which the Earth and the other planets move, has the same effects in its shell structure and motion. It is obvious that the other planets with their satellites have the same character of structure and motion.

If one takes into account the effect of a planet's orbital plane inclination to the equatorial plane of the Sun, then the above changes are found to follow the law of $1/r$ This observable fact proves our conclusion that the changes in the outer force field of a body are controlled by rotation of its reduced inner force shell (see the force shell *D* on Fig. 6.10). It explains why Mercury has maximal value of the "secular perihelion shift" between the other planets.

Thus, the Earth's orbital motion and retrograde movement of its nodal line are controlled by the Sun's dynamics of the masses through the outer force field. The Earth plays the same role for the Moon and the artificial satellites. As to the nutation motion, then its nature is related to the same peculiarities in the structure and motion of the bodies but the effects of their perturbations are fixed by the axis wobbling.

6.11.6 The Nature of Possible Clockwise Rotation of the Outer Core of the Earth

The question arises why the outer planet's core may have a clockwise rotation. It was shown in Sect. 2.6 that the law of radial density distribution determines the direction of a body's shell rotation (Fig. 2.2).

It was found that in the case of uniform mass density distribution all energy of the mass interaction is realized in the form of oscillation of the interacting particles (Fig. 2.2a). If the density increases from the body's surface to the center, then there

Fig. 6.11 Dependence of shell rotation on the parabolic law of radial density distribution for the Earth. Here r_m and r_g are the reduced radiuses of inertia and gravitation

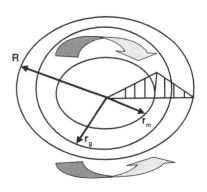

are oscillations and counterclockwise rotation of shells (Fig. 2.2b). Increase of mass density from center to surface leads to oscillation and clockwise rotation with different angular velocities of the body shells (Fig. 2.2c). Finally, the parabolic law of radial density distribution (Fig. 6.11), where the density increases from the surface and then it decreases, leads to oscillation and reverse directions of rotation. Namely, the upper shells have a counterclockwise and the central shells—clockwise rotation. Case of Fig. 6.11, obviously, is characteristic for a self-gravitating body.

Note that direction of the body rotation depends on radial density distribution and corresponds with the Lenz right-hand or right-screw rule, well known in electrodynamics. Taking into account the observed effect of the retrograde motion of the satellite nodal line, the gravitational induction of the inner and outer force fields of the Earth has a common nature with electromagnetic induction noted earlier. Just Fig. 6.11 may explain the nature of the retrograde motion of the nodal line of a satellite orbit related to change in the potential of the outer Earth's force field according to the induction law. The continuous and opposite-directed movement of the asymmetric mass density distribution of the mantle and the outer core (Fig. 6.10) seems to be the physical cause of precession, nutation and variation of the inner and outer force fields observed by satellites. This idea is proved by the satellite data about the retrograde motion of the nodal line depending on inclination of its orbital plane with respect to the planet's equatorial plane.

It is worth recalling, from the literature, that the idea of dynamical effects of the (probably) liquid core of the Earth has been discussed among geophysicists for a long time (Melchior, 1972).

6.11.7   The Nature of the Force Field Potential Change

It follows from the above discussion that in the frame of the considered dynamical approach, the variation of potential of the inner and outer force field relates to the non-uniform distribution of mass density of a self-gravitating body. Rotation of the outer and inner shells leads to the observed effects of precession, nutation and variation of their own force fields.

6.11.8   The Nature of the Earth's Orbit Plane Obliquity

Celestial mechanics does not discuss the problem of obliquity of the planet's and satellite's orbit planes and accepts it as an observable fact. The theory of dynamic equilibrium explains this phenomenon and the nature of apocenters and pericenters by asymmetric distribution of masses and by effect of rotation of asymmetric shells of self-gravitating bodies (see Fig. 6.11). In fact, if the mass of the Sun's shells has an asymmetric distribution, then the potential of the outer force field has the same asymmetry. This asymmetry determines inclination of the Earth's orbit plane relative to the plane of the Sun's rotation. Each point of the orbit reflects a condition of dynamical equilibrium of the interacting outer force fields of the planet and the Sun. The position of the Earth's aphelion and perihelion reflects the position of the reduced maximal and minimal concentration of the Sun's mass density in the shells. Because the Sun's asymmetric shells have different angular velocities of rotation, then amplitude of the nodal line will decrease with increasing distance between the mass anomalies and vise versa. The effect of variation of the nodal line is proved by observation. So, the present-day angle of ecliptic inclination to the plane of rotation of the Sun equal to ~7°15′ expresses the relation between maximum and minimum concentrations of the reduced mass density of the Sun's shells. An analogous effect is shown by inclination of the Moon's orbital plane to the plane of rotation of the Earth.

6.11.9   The Nature of Chandler's Effect of the Earth Pole Wobbling

As it was noticed, changes in the planet's inner force field are observed in the form of nutation or wobbling of the axis of rotation. The axis itself reflects the dynamics of the upper planet's shell, the thickness of which, by our estimate, is about 375 km. The Moon is rotating about the Sun in the force field of the Earth which is perturbed by its natural satellite. Its maximum yearly perturbation should be the Chandler effect. The Moon's yearly cycle seems to be the ratio of the Earth's to the Moon's month (in days). Then this cycle is $365(30.5/27) \approx 410$ days.

6.11.10   Change in Climate as an Effect of Rotation of the Earth's Shells

The above analysis of dynamical effects of the Earth's shells is based first of all on the data of satellite's orbit changes and measurements of the planet's force field. Unfortunately, a specific feature of an artificial satellite orbital motion is its artificial velocity which is ~16 times higher than the angular velocity of the upper Earth's

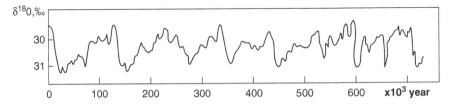

Fig. 6.12 Isotopic composition of oxygen in shells of mollusk *Globigerinoides Sacculifera* within time period 0–730 000 years. (Emiliani, 1978)

shell. In this connection all its parameters of satellite motion are unnatural. So, we cannot directly divide the natural component of its nodal retrograde shift in order to get the total picture of perturbations which propagate the Earth's inner shells. This is an experimental problem.

But there are also long term astronomical observations of the Earth's dynamics relative to the far stars, the results of which correspond to the presented ones. In addition, periodicity in rotation of asymmetric inner shells of the Sun can be fixed by climatic changes on the Earth over a long period of time. Such changes were being studied, for instance, by data of the oxygen isotopic composition in mollusk shells over a number of years. Figure 6.12 demonstrates the results of Emiliani (1978) who studied the core obtained during deep sea drilling in the Caribbean basin.

The author obtained the picture of climate change in the Pleistocene era over 730 000 years. It is seen that the periods of climate change vary from 50 000 to 120 000 years. It means that the pure period of rotation of the asymmetric mass shells of the Sun is absent and the orbital trajectory has not been locked into place during the studied time.

6.11.11   The Nature of Obliquity of the Earth's Equatorial Plane to the Ecliptic

It is obvious that the obliquity of the planet's equatorial plane is related to the polar and equatorial oblateness of the Earth's masses. It follows from Eq. (6.68) that the obliquity, in turn, is determined by the tangential component of the inner force pressure generated by the non-uniform radial mass density distribution. This tangential component of the inner force field induces the inner field of the rotary moments, the energy of which was discussed in Sect. 6.9 and presented in Table 6.1. The obliquity value can be obtained from the ratio of the potential energy of the uniform U_0 and non-uniform U_t body of the same mass. Accepting this physical idea and the data of Table 6.1, we write and obtain:

$$\cos\Theta = \frac{U_o}{U_t} = \frac{\alpha_o^2}{\alpha_t^2} = \frac{0.6}{0.66} = 0.909, \quad \Theta = 24.5°, \tag{6.89}$$

where α_o and α_t are the structural form-factors taken from Table 6.1.

The error obtained in calculation of obliquity by formula (6.89) equal to about $1°$ or $\Delta\alpha^2_t = 0.006$—can be explained by the accepted law of the continuous radial distribution of the planet's mass density.

Eq. (6.89) is an integral effect of the obliquity of the planet's equatorial plane which is observed on the surface of the upper rotating shell. It was shown earlier that the observed obliquity is really an integral dynamical effect of the Earth's mass including the upper shell up to the near-by Mohorovičich discontinuity. But being in a suspended state, relative to the other parts of the body, the upper shell is able to wobble as if on a hinge joint by perturbation from the Sun and the Moon. This effect of the upper shell wobbling gives an impression of the axial wobbling.

By the same cause the obliquity of the ecliptic with respect to the solar equator is determined by the Sun's polar and equatorial oblateness. The trajectory of the Earth's orbital motion at each point is controlled by the outer asymmetric solar force field in accordance with the dynamic equilibrium conditions. And only in the nodes, which are common points for equatorial oblateness of the Sun and the Earth, is the Huygens' effect of the innate initial conditions fixed by the third Kepler law.

6.11.12 Tidal Interaction of Two Bodies

Let us consider the mechanism and effects of interaction of the outer force pressure of two bodies being in dynamic equilibrium. Come back to the mechanism and conditions of separation of a body mass with respect to its density when a shell with light density is extruded to the surface. Rewrite Eq. (6.71) for acceleration of the gravity force in points A and B of the two body shells (Fig. 6.1b) and their densities ρ_M and ρ_m

$$q_{AB} = 4\pi\, Gr \left(\frac{2}{3}\rho_M - \rho_m \right), \qquad (6.90)$$

After the shell with density ρ_m appears on the outer surface of the body, the condition of its separation by Eq. (6.90) will be:

$$\rho_M > 2/3\rho_m. \qquad (6.91)$$

The gravitational pressure will replace the shell up to the radius $A + \delta A$, where the condition of its equilibrium reaches $\rho_M = \rho_m$. This condition is kept on the new border line between the body and its upper shell. Taking into account that the shell in any case has a thickness, then, by the Archimedes law, the body will be subject to its hydrostatic pressure. If the separated shell is non-uniform with respect to density, then a component of the tangential force pressure appears in it, and the secondary self-gravitating body-satellite is formed. The new body will be kept on the orbit by the normal and equal tangential components of the outer force pressure. In this case the reaction of the normal gravitational pressure will be local and non-uniform. If the upper shell is uniform with respect to density, then the reaction of the normal gravitational pressure along the whole surface of the body and the shell remains uniform in value. In this case the separated shell remains in the form of a uniform ring.

The above schematic description of the physical picture of the separation and creation of a secondary body can be used for construction of a mechanism of the tidal phenomena in the oceans, the atmosphere and the upper solid shell at interaction between the Earth and the Moon. The outer gravitational pressure of the Moon, due to which it maintains itself in equilibrium on the orbit, at the same time renders hydrostatic pressure on the Earth's atmosphere, oceans and upper solid shell through its outer force field. This effect determines the tidal wave in the oceans and takes active part in formation and motion of cyclonic and anti-cyclonic vortexes. In accordance with the Pascal law, the reaction of the Moon's hydrostatic pressure is propagated within the total mass of the ocean water and forms two tidal bulges. Because the upper shell of the Earth is faster-moving relative to motion of the Moon, the front tidal bulge appears ahead of the moving planet. Our perception of the ocean tides as an effect of attraction of the Moon appears to be speculative.

6.12   Earthquakes, Orogenesis and Volcanism

It is known from geological and geophysical observations that the Earth's crust is subjected to continuous vertical and horizontal movements. The physical mechanisms such as vertical rising and sinking of the mantle matter because of its heating up and density differentiation, horizontal shift of the crust and its plate-blocks as a result of the rotary convective flow of the planet's matter, are used for explanation of the tectonics hypotheses. Each of such hypotheses is proved by many facts.

In this work we have tried to solve those physical and analytical parts of the problem in which we give explanations for the nature of forces performing the work on transferring of enormous masses of matter with tectonic movements, as well as to determine the mechanism of action of such forces in time. We have shown that the forces of the Earth in the form of the force pressure are generated by the planet itself through interaction of its masses on the level of micro-particles. These forces form inner and outer fields of force pressure having a non-uniform structure because of heterogeneous distribution of the mass density. The gravitational force field includes radial, tangential and dissipative components and acts in an oscillating regime. The radial component is responsible for the radial transfer of the masses. The tangential component performs horizontal movement of the matter. And the dissipative component forms the inner and outer electromagnetic fields of the planet.

Let us discuss in general form the nature of such practically important phenomena as the Earth's crust oscillation, earthquakes, orogenesis and volcanism. From the point of view of the presented theory all the above phenomena have a common nature. It associates local, regional or global changes in the potential and kinetic energy of the Earth's shells. Therefore, in order to find the cause of one or another event, the perturbing energy link has to be first of all identified. Herein, one should proceed from these causes to the fact that the Earth's energy and its electromagnetic component are being continuously generated. The perturbations from the outer force fields of the Sun and the Moon propagate through the outer force field of the Earth and reciprocal perturbation of the planet's shells in the inner force field takes place. The observed diurnal rotation of the Earth is valid only for the upper shell of ~375 km thick and the total integral period of the planet's mass rotation is ~18.6 year. In addition, each planet's shell occurs in a suspended state and is subjected to an effect of the hydrostatic pressure.

6.12.1 Earth Crust Tremor and Earthquakes

The world net of seismic stations continuously records an innumerable amount of elastic, mainly weak, oscillations of the Earth's surface. Their total energy is estimated as 10^{19} J. Jeffreys assumed that the Earth is absolutely an elastic body and during its rotation the elastic stresses are developed in it. If the limit of the elastic deformations is exceeded they develop into a plastic flow and finally a disconnection of continuity and a break of the matter happen. At this moment a sharp local change in the stresses arises which leads to appearance and propagation of elastic waves in the body, causing an earthquake. This physical basis of earthquakes continues to exist. It is assumed that in the seismic center of the body volume, where the process is developed, the energy, accumulated during some time, is released. The release of energy is accompanied by ruptures of the geological structures and instantaneous displacement of the matter masses.

From the point of view of the considered theory for understanding the nature of the given phenomena, the following refinements and additions should be introduced. In accordance with the effects of differentiation of the planet's shells with respect to their radial masses density distribution (see Sect. 6.5 and Eq. (6.71)) by the action of gravitational energy, continuously generated in the auto-oscillating regime, and of its force pressure, the process of mass-transfer is constantly occurring. In the existing phase states of the Earth's masses the most effective process of mass-transfer by the force pressure towards the surface is preferential in the gaseous phase. The gases under the pressure would be accumulated in the shells with densities expressed by the relationship $\rho_M = 2/3\rho_m$, which corresponds to the jumps observed by seismic observations. The hydrostatic pressure of the accumulated gases increases with time and breaks through the overlying shell in the regime of explosion. The break of the pressurized gases can happen at any depth where they perform a wide spectrum of shallow and deep focus underground shocks and local breaks in mines and adits.

The catastrophic aftereffects of earthquakes on the oceanic bottom can be amplified by a disturbance in the dynamical equilibrium of water masses in the form of tsunami waves. The earthquake energy in the oceans can be amplified or attenuated by the resulting effect of the equilibrium state if wave processes such as tidal effects exist.

As to the weak oscillations, many of them have been identified through spectral analysis as periodic virial oscillations of the shells with periods from minutes to one hour.

6.12.2 Orogenesis

As observations show, orogenesis is the most powerful natural process having a relation with dynamical equilibrium of the planet as a whole. The term "orogenesis", in a wide meaning of the phenomenon, includes not only creation of the mountains on the continents, but also the continents themselves. From the position of dynamical equilibrium of the planet, the nature and mechanism of this phenomenon appear to be an effect of the mass density differentiation of the planet discussed in Sect. 6.5. Geologists long ago noticed that the Earth's crust and the underlying shell of the substratum represent a light (by density) sialic matter composed of silicon and aluminum. The matter of the deeper shells of the Earth represents heavier simatic matter composed of silicon and manganese. In addition, the fact of the suspended state of the crust is well documented by seismic records and by study of the Earth's gravitational field. It has also been determined that the continental mountains have no anomalies in the gravitational field and, thus, they are equilibrated on the surface with plain regions. It follows from the condition expressed by Eq. (6.71) that the continental mountains, together with the bottom topography of the oceans, were being formed and continue forming in the course of differentiation of the planet's shells with respect to density, under actions of both the radial component of the inner force field (the Archimedes forces) and its tangential component (Coriolis force). In other words, differentiation of the matter with respect to density occurs in the vertical and horizontal directions. Thus, both geological hypotheses of fixizm (vertical motion of rocks) and mobilizm (horizontal motion of the rocks), based on geological observations, are correct and represent two sides of the unique global evolutionary process resulting in the interior of the planet. The presented theory allows us to obtain quantitative estimations of this process.

6.12.3 Volcanism

As it was noticed, all geotectonic catastrophic phenomena have a common nature and are interrelated. The inner gravitational pressure and effect of mass density differentiation are the main physical cause of volcanism and volcanologists well understand the mechanism of action of the phenomenon. The subjects of discussion are the conditions and real mechanism of melting of the rocks up to their magmatic state. In order to answer the question we may turn again to the Earth's shells with jumps of their mass density. We found earlier that such shells should serve as the borders of the angular velocity of the shell rotation change. In Sect. 6.4 according

Eq. (6.69) it was found that the shell's angular velocity is equal to its frequency of oscillation multiplied by the coefficient of oblateness. In turn, the coefficient of oblateness is proportional to the coefficient of friction, which determines the temperature. Thus, the effect of volcanic activity of the Earth's interior is linked with development of the pressure of gases, being the real product of virial oscillations of the masses and their differentiation with respect to density. In turn, the shell friction at the density jumps is the real mechanism of rock melting where gases are accumulated during their migration to the surface. The gases burst in the weakest locations of the rock strata, accompanied by ejection of the melted products, is a probable mechanism of the volcanic process. The ejected eruptive and fumigate gases from a volcano in the form of H, H_2O, HCl, HF, CO, CO_2 and other volatile compounds are a weighty confirmation for the above described mechanism.

6.13   Earth's Mass in its Own Force Field

The conception of mass as a physical value of some volume of matter was introduced by Newton. In his mechanics the inert and gravity or heavy masses are distinguished. In definition of the movement ($p = mv$) and the force ($F = ma$) the mass plays the role of the coefficient of proportionality, i.e. constant value, characterizing a measure of inertia and dynamical ability of a body.

In Newton's gravitational theory the mass is represented as a source of the gravity field and at interaction with the mass of another body it feels its action also through the gravity field. In this case, the interacting fields possess the only property of attraction. On this basis the relationship for defining the weight of a body, being in the gravity field of the other body, has the form $P = mg$ (where $g = GM/R^2$ is the acceleration of gravity). Here the mass m represents the gravity mass of a body.

In relativity mechanics the mass is not an additive characteristic of a body. When two particles are divided or joined into a steady-state compound, an excess of energy is released. This energy is called the constraint energy or the mass defect. This effect is notably developed in nuclear reactions.

As it was mentioned in Chap. 1, Galileo found the constancy of acceleration of free fall of different bodies on the surface of the Earth in its gravity field. This effect is called the principle of equivalence. This principle certifies that the gravity and inert masses of a body in the outer force field are equivalent and at appropriate measuring units are equal to each other. Later on the principle was checked and proved experimentally by Eötvös, Dicke, Braginsky and other researchers. On the basis of this principle the value of mass of the Earth, the Moon and other bodies was determined relative to the mass of the Sun by analytical calculations applying the equations of their orbital motion and taking into account the perturbations from other planets. In the calculations, orbital motion is considered as the motion of the bodies as mass points in the central force field of the Sun. The Earth's mass, found by such "weighting", amounts to $\sim 5.976 \times 10^{24}$ kg, which we find adequate for practical use.

The given method of the Earth's mass determination ignores its own force field because of its negligible value in comparison with the Sun's force field. The same reason was given to ignore their own force field effect in the bodies studied in Galileo's and other researchers' experiments. But when our study concerns the dynamics of the Earth as a self-gravitating body, we cannot ignore its own force field and our experience has proved this fact. In particular, it appears that the gravitational and inert non-uniform masses are different. It means that the problem of a body in its own force field cannot be identified with the problem of a body in the outer central force field. It is worth noting that such a problem is not everywhere forgotten. The effects of their own force fields have been for a long time studied in non-linear systems. For instance, the non-linear generalization of Born and Infield in classical electrodynamics has resolved the problem of the energy of the Coulomb's field of the mass point, which was infinite. For mathematical convenience in field theory, the charge point is often considered as continuously distributed in some volume, and the definition of density of the charge is introduced, with its distribution written by means of Dirac's delta-function. Then, the total charge is defined by the density integration for an accepted volume, keeping, thus, the meaning of the charge point. One of the important discovered effects of the non-linear system is the exited non-damping oscillations due to transforming of the friction-dragging energy. The effects of inner energy release like this are discovered in our study of the self-gravitating Earth and the Sun, remaining at the same time in the framework of classical mechanics.

So, the mass of the Earth together with the mass of the Moon in the framework of hydrostatic equilibrium is determined by the data of astronomical observations of the planet's motion around the Sun expressing the results in the units of solar mass. We mark the sought-for mass in the solar and the Earth's reference system (Fig. 2.1) as M_0 and M. They will differ by a mass defect (bonding energy), released into the potential energy of the self-gravitating planet. In addition, because of non-uniformity of the mass density distribution, the gravitational (M_g) and inertial (M_m) masses of the planet, in turn, differ from their mean value M. In order to determine the three values M_g, M_m and M of the Earth's mass in its own force field the value M_0 (without the mass of the Moon) should be multiplied by the coefficients (see Table 6.1), i.e.

$$(1 - \beta_0^2) = (1 - 0.6) = 0.4,$$
$$(1 - \beta^2) = (1 - 0.49725) = 0.50275,$$
$$(1 - \alpha^2) = (1 - 0.66074) = 0.33926.$$

Then the values of the corresponding masses are equal to

$$M = 5.9765 \times 10^{24} \times 0.4 = 2.3906 \times 10^{24} \text{ kg},$$
$$M_m = 5.9765 \times 10^{24} \times 0.50275 = 3.004685 \times 10^{24} \text{ kg}, \qquad (6.92)$$
$$M_g = 5.9765 \times 10^{24} \times 0.33926 = 2.027587 \times 10^{24} \text{ kg},$$

Thus, the mass of a self-gravitating body implements simultaneously two functions. On one side, at interaction of its discrete components the bonding energy is realized in the form of potential energy of irradiation. This energy creates the inner and outer force fields and plays the role of an active component of the force function. On the other side, the same mass takes the reactive action of constraints of the irradiated energy and plays the role of a reactive component of the force function. This second function of the mass is observed in the form of oscillation, rotation or translation (e.g. a rocket). The direction of irradiation and reactive motion of an inert mass is determined by distribution of its mass. The effect of the reactive motion of the realized binding energy (the mass defect) in general is developed in the form of attraction or repulsion of the mass.

Thus, the active component of the force function of the interacting masses of a body forms through the energy irradiation the inner and outer force fields and determines the body's gravitational properties. The reactive component of the force function determines the inert properties and the character of the body's mass motion. The reactive component of the mass interaction represents the force function of inertia. The gravitational and the inertial masses of a self-gravitating body are not equal.

As to the experiments of Galileo, Eötvös and the conclusion about the equality of the gravitational and inertial masses, they remain correct for the study conditions, i.e. for the body's interaction in the uniform outer force field of the Earth.

Chapter 7
Dynamics of the Earth's Atmosphere and Oceans

The atmosphere and the oceans (collectively) are both upper shells of the Earth. The first of them occurs in a gaseous and the second one in a liquid phase. These shells exist in the solid Earth's outer force field but the atmosphere exists in dynamical equilibrium, and the oceans are in a weighted transition to a hydrostatic state. The atmosphere as a gaseous shell is totally "dissolved" in the Earth's outer force field in the form of atomic and molecular sub-layers, differentiated with respect to density, and these self-gravitating masses appear in a dynamical equilibrium state. Relatively homogeneous water masses of the oceans have too low a density (approximately 2/3) in comparison with the mineral crust to be an independent, with respect to its dynamics, self-gravitating shell. Therefore, it appears to be suspended in a semi-hydrostatic equilibrium relative to the crust's shell. Its inner gravitational pressure on the shell's surface is equilibrated with the atmospheric and outer gravitational pressure. The surface water is practically found to be in a limiting hydrostatic equilibrium. The small portion of the continuously pumping solar energy varies within 7% in the annual cycle and leads to the dynamical process of water transfer from its liquid to vapor, and vice verse, phase.

Dynamics of the main nitrogen and oxygen components with close densities (atomic weights) of the atmospheric masses in a non-perturbed state could be represented by its own virial oscillation and rotation about the solid Earth. But the water vapor which is continuously injected into the gaseous shell, being a component of very large-capacity energy (1 cm^3 of water generates more than 1000 cm^3 water vapor) appears to be the cause of the stormy dynamical perturbation within the near-surface shell of the atmosphere. The cyclonic activity of the atmospheric vapor has scientific, but mainly practical, interest. The weather and climate change by variation of the solar energy flux through the water vapor dynamics have a negative effect on the biosphere.

In turn, the oceans being in a balanced hydrostatic state, are continuously perturbed both by the inner gravity pressure of the planet related to the density differentiation of the shells and by perturbation from the Sun's and the Moon's outer force fields. All the above dynamical processes seem to be of interest for consideration from the point of view of dynamical equilibrium theory.

V. I. Ferronsky, S. V. Ferronsky, *Dynamics of the Earth*,
DOI 10.1007/978-90-481-8723-2_7, © Springer Science+Business Media B.V. 2010

In this chapter we search for a solution of Jacobi's virial equation for the non-perturbed atmosphere as the Earth's shell which is affected by both inner and outer perturbations. In order to justify applicability of the virial equation for the study of dynamics of the atmosphere in the framework of a model of a continuous medium, we derive this equation from the Euler equations.

7.1 Derivation of the Virial Equation for the Earth's Atmosphere

We consider the problem of global oscillations of the Earth's atmosphere as its shell. We accept that the atmosphere is found to be in dynamical equilibrium both in its own force field and in the outer force field of the Earth. Derivation of the virial equation is done in the framework of the continuous medium model. We accept also that the dynamics of the atmosphere is described by Euler's equations and the medium is composed of an ideal gas.

The Euler equations are written in the form (Landau and Lifshitz, 1954)

$$\rho\frac{\partial v}{\partial t} + \rho(v\nabla)v = -grad\ p + \rho F, \qquad (7.1)$$

where ρ is the gas density; p is the gas pressure; $\partial v/\partial t$ is the rate of velocity change at a fixed point of space; F is the density of mass forces.

In addition, the continuity equation

$$\frac{\partial p}{\partial t} + div\rho v = 0 \qquad (7.2)$$

holds for a gas medium.

Multiplying Eq. (7.2) by v and summing the product with Eq. (7.1), we obtain

$$\frac{\partial \rho}{\partial t}v = -\rho(v\nabla)v - vdiv\rho v - grad\ p + \rho F. \qquad (7.3)$$

We take the divergence of Eq. (7.3) and note that

$$\frac{\partial}{\partial t}[div\rho v] = \left[\frac{\partial}{\partial t}(div\rho v)\right] = -\frac{\partial^2 \rho}{\partial t^2}.$$

Then Eq. (7.3) can be rewritten in the form

$$\frac{\partial^2 \rho}{\partial t^2} = div\left[\rho(v\nabla)v + vdiv(\rho v) + grad\ p - \rho F\right]. \qquad (7.4)$$

Multiplying Eq. (7.4) by $r^2/2$ and integrating the obtained expression over the whole volume of the atmosphere, we have

$$\int\limits_{(V)} \frac{r^2}{2} \frac{\partial^2 \rho}{\partial t^2} dV = \int\limits_{(V)} \frac{r^2}{2} div\left[\rho(v\nabla)v + vdiv(\rho v) + grad\, p - \rho F\right] dV. \quad (7.5)$$

The left-hand side of Eq. (7.5) can be rewritten in the form

$$\int\limits_{(V)} \frac{r^2}{2} \frac{\partial^2 \rho}{\partial t^2} dV = \frac{\partial^2}{\partial t^2}\left(\frac{1}{2}\int\limits_{(V)} r^2 \rho dV\right) = \ddot{\Phi},$$

where Φ is the Jacobi function of the atmosphere.

Thus, we obtain Jacobi's equation for the Earth's atmosphere derived from the Eulerian equation (7.1) and the continuity equation (7.2) as follows:

$$\ddot{\Phi} = \int\limits_{(V)} \frac{r^2}{2} div\left[\rho(v\nabla)v + vdiv(\rho v) + grad\, p - \rho F\right] dV. \quad (7.6)$$

We assume that the Earth is a rigid spherical body with mass M and radius R and that the mass of the atmosphere is negligible in comparison with the earth's mass. We can transform the right-hand side of Eq. (7.6) as follows:

$$\int\limits_{(V)} \frac{r^2}{2} div\left[\rho(v\nabla)v + vdiv(\rho v) + grad\, p - \rho F\right] dV$$

$$= 2T + U + 3\int\limits_{(V)} pdV - 4\pi r^3 p \Big|_R^{R_{max}},$$

where

$$T = \frac{1}{2}\int\limits_{(V)} \rho v^2 dV$$

is the kinetic energy of the gas of the atmosphere in the Earth's gravitational field;

$$U = GM \int\limits_{(V)} \frac{\rho dV}{r}$$

is the potential energy of the atmosphere in the Earth's field;

$$3\int\limits_{(V)} pdV$$

is the internal energy of the gas atmosphere;

$$-2\pi r^3 p|_R^{R_{max}}$$

is the energy of the outer surface force pressure effecting the atmosphere.

In fact, we can write an expression for the mass forces, taking into account the spherical symmetry of the system considered:

$$F = -GM\frac{\mathbf{r}}{r^3}.$$

Then

$$-\int_{(V)}\frac{r^2}{2}\mathrm{div}(\rho F)\,dV = \int_{(V)}\frac{r^2}{2}\mathrm{div}\left(\rho GM\frac{\mathbf{r}}{r^3}\right)dV = \int_R^{R_{max}}\frac{r^2}{2}\frac{1}{r^2}\frac{\partial}{\partial r}\left(\frac{r^2\rho GM}{r^2}\right)4\pi r^2\,dr$$

$$= \frac{4\pi r^2}{2}\rho GM|_R^{R_{max}} - 4\pi GM\int_R^{R_{max}}\rho r\,dr = U + \frac{4\pi r^2}{2}\rho GM|_R^{R_{max}}.$$

Analogously, we have

$$\int_{(V)}\frac{r^2}{2}\mathrm{div}(\mathrm{grad}\,p)dV = 4\pi\int_R^{R_{max}}\frac{r^2}{2}\frac{1}{r^2}\frac{\partial}{\partial r}\left(r^2\frac{\partial p}{\partial r}\right)r^2\,dr$$

$$= \frac{4\pi r^2}{2}\frac{\partial p}{\partial r}\bigg|_R^{R_{max}} - 4\pi r^3\,p|_R^{R_{max}} + 3\times 4\pi\int_R^{R_{max}}r^2 p\,dr$$

$$= 3\int_{(V)}p\,dV - 4\pi r^3|_R^{R_{max}} + \frac{4\pi r^4}{2}\frac{\partial p}{\partial r}\bigg|_R^{R_{max}}.$$

It is easy to show that

$$\int_{(V)}\frac{r^2}{2}\mathrm{div}[\rho(v\nabla)v + v\mathrm{div}(\rho v)]\,dV = \int_{(V)}v^2\rho dV = 2T.$$

Owing to the immobility of the rigid Earth and the atmospheric boundary as well as the condition of continuity for the gas, the equilibrium hydrostatic condition holds along the normal to this border, which can be written in the form

$$\rho\frac{GM}{r^2}\bigg|_R = -\frac{\partial p}{\partial r}\bigg|_R.$$

If we assume

$$\rho r^2|_{R_{max}} = 0, \quad r^4 \frac{\partial p}{\partial r}|_{R_{max}} = 0,$$

then

$$\frac{4\pi r^2}{2} \rho GM|_R^{R_{max}} + \frac{4\pi r^4}{2} \frac{\partial p}{\partial r}|_R^{R_{max}} = 0.$$

It follows from this that Jacobi's virial equation (7.6) takes the form

$$\ddot{\Phi} = 2T + U - 4\pi r^3 p|_R + 3 \int\limits_{(V)} p dV. \tag{7.7}$$

Finally, using the energy conservation law for a continuum (Sedov, 1970) and the conservativity of a system, following from the accepted model (the Earth is a solid body and atmospheric gas is ideal), Eq. (7.7) can be written in the form of the standard Jacobi virial equation

$$\ddot{\Phi} = 2E - U, \tag{7.8}$$

where the total energy of the atmosphere is conserved and equal to

$$E = T + U - 2\pi r^3 p|_R + \frac{3}{2} \int\limits_{(V)} p dV. \tag{7.9}$$

Let us make some important notes concerning the study of the atmosphere by a conventional approach based on use of the virial theorem.

The standard solution of the problem using the same assumption at zero approximation (barometric height formula) expresses the equilibrium condition between the atmospheric gravity and the gas pressure. In our case this condition is satisfied if Jacobi's virial equation (7.8) is averaged with respect to time, i.e. it is reduced to the condition of the hydrostatic equilibrium. But in that case the kinetic energy of the particle interaction is excluded. Assuming that the motion of the gas particles is finite, and after time averaging Eq. (7.8) over a sufficiently large time interval, the validity of the virial theorem is easily shown to be

$$E = U/2. \tag{7.10}$$

At known parameters of the atmosphere and the Earth, the potential and total energies of the atmosphere can be estimated as

$$U = -\frac{GMm}{R} \approx -3.2 \times 10^{33} erg,$$

$$E \approx -1.6 \times 10^{33} erg.$$

7.2   Non-perturbed Oscillation of the Atmosphere

Let us now consider the solution of Eq. (7.8) for the spherical model of the atmosphere and find the dependence of its Jacobi function Φ (polar moment of inertia) on time in explicit form.

In the previous chapter it was shown that Eq. (7.8) is resolved both for the uniform medium and for the medium with radial density distribution by some law. In the last case the polar moment of inertia and potential energy are expanded on the uniform and tangential components and instead of Eq. (7.8) two Eqs. (6.61) and (6.62) are written.

Let us consider a solution for the uniform component of the atmosphere whose radial density distribution changes by the barometric equation. The uniform component of Eq. (7.8) has a solution when there is a relationship between the Jacobi function and the potential energy of the system in the form

$$|U|/\sqrt{\Phi} = const. \tag{7.11}$$

Assume that the Earth has a spherical shape with mass M and radius R and be enveloped by a uniform atmosphere with mass m which has thickness Δ. Then the potential energy of the shell U_a, which is in the gravitational field of the sphere, is

$$U_a = -GM\rho_a \int\limits_{R}^{R+\Delta} \frac{4\pi r^2}{r} dr = -2\pi GM\rho_a(2R\Delta + \Delta^2),$$

where ρ_a is the mass density of the shell.

The Jacobi function of the atmosphere is

$$\Phi_a = 4\pi\rho_a \frac{1}{2} \int\limits_{R}^{R+\Delta} r^4 = \frac{4\pi\rho_a}{2\times 5}\left[5R^4\Delta + 10R^3\Delta^2 + 10R^2\Delta^3 + 5R\Delta^4 + \Delta^5\right]. \tag{7.12}$$

Expressing the gas density ρ_a through its mass, we can write the relationship (7.11) in the form

$$B = |U|_a\sqrt{\Phi_a} = \frac{GMm}{\sqrt{2}} \frac{3(2+\lambda)}{2(3+3\lambda+\lambda^2)}\sqrt{\frac{3(5+10\lambda+10\lambda^2+5\lambda^3+\lambda^4)}{5(3+3\lambda+\lambda^4)}}, \tag{7.13}$$

where $\lambda = \Delta/R$.

Note that Eq. (7.13) depends only on the ratio of thickness of the shell to the radius of a central body and varies over limited ranges, while λ varies from 0 to ∞. At $\lambda = 0$

$$|U_a|\sqrt{\Phi_a} \to \frac{GMm^{3/2}}{R},$$

and at $\lambda \to \infty$

$$|U_a|\sqrt{\Phi_a} \to \frac{GMm^{3/2}}{R}\left(\frac{27}{20}\right)^{1/2}.$$

On the other hand, the Jacobi function (7.11) expressed through the mass of the shell, is written in the form

$$\Phi_a = \frac{3m}{10}R^2 \frac{3\left(5 + 10\lambda + 10\lambda^2 + 5\lambda^3 + \lambda^4\right)}{5\left(3 + 3\lambda + \lambda^4\right)}. \tag{7.14}$$

It follows from (7.14) that the Jacobi function of the atmosphere does not depend only on the value λ but also on the radius of the body R. Moreover, the value Φ_a varies over unlimited ranges when λ runs from 0 to ∞. The same can be said about the potential energy of the atmosphere:

$$U_a = \frac{GMm}{R}\frac{3}{2}\frac{2 + \lambda}{2\left(3 + 3\lambda + \lambda^2\right)}.$$

Accepting the mass of the atmosphere m $= 10^{21}$ kg, we can find the value B as

$$|U_a|\sqrt{\Phi_a} = \frac{GMm}{R}\left(\frac{mR^2}{2}\right)^{1/2} = 1.0374 \times 10^{53}[\text{g}^{3/2}\text{cm}^3\text{s}^{-2}]. \tag{7.15}$$

Taking (7.15) into account, Eq. (7.8) can be written in the form of an equation of virial oscillations of the atmosphere (subscript a at Φ is farther drop):

$$\ddot{\Phi} = -A + \frac{B}{\sqrt{\Phi}}, \tag{7.16}$$

where $A = -2E$; $E = -1.6 \times 10^{33}$ erg.

As shown in Chap. 4, at $A = const$ and $B = const$, Eq. (7.16) has two first integrals, as follows:

$$C = -2A\Phi + 4B\sqrt{\Phi} - \dot{\Phi}^2, \tag{7.17}$$

$$- arccos\frac{(A/B)\sqrt{\Phi} - 1}{\sqrt{1 - AC/2B^2}} - \sqrt{1 - \frac{AC}{2B^2}\left[1 - \left[\frac{\left(A/B\sqrt{\Phi} - 1\right)}{\sqrt{1 - AC/2B^2}}\right]^2\right.}$$

$$= \frac{(2A)^{3/2}}{4B}(t - t_0), \tag{7.18}$$

where C and t_0 are integration constants. The constant C has a dimension of square angular momentum.

The integrals (7.17) and (7.18) are solutions of the equation (7.16). Introducing new variables

$$E'' = -arccos\frac{\left(A/B\sqrt{\Phi} - 1\right)}{\sqrt{1 - AC/B^2}}, \quad e'' = \sqrt{1 - \frac{AC}{2B^2}},$$

and rewriting (7.18) in the form of Kepler's equation

$$E'' - e'' \sin E'' = \frac{(2A)^{3/2}}{4B}(t - t_0) = M'',$$

expressions can be obtained for $\Phi(t)$ and $\dot{\Phi}(t)$ from the first integrals (7.17) and (7.18) in explicit form, using Lagrange's series (Duboshin, 1975):

$$\Phi(t) = \frac{B^2}{A^2}\left[1 + 2e''\cos M'' - e''^2\left(1 - \frac{3}{2}\cos^2 M''\right)\right.$$
$$\left. - \frac{5}{2}e''^3\left(1 - \cos^2 M''\right)\cos M'' + \cdots\right] \qquad (7.19)$$

$$\dot{\Phi}(t) = \sqrt{\frac{2}{A}}\varepsilon B\left[\sin M + \frac{1}{2}\varepsilon\sin 2M + \frac{\varepsilon^2}{2}\sin M\left(2\cos^2 M - \sin^2 M\right) + \cdots\right],$$

where $M'' = (2A)^{3/2}(t - t_0'')/4B = n''(t - t_0'')$; n'' is the frequency of the atmospheres own virial oscillations; t_0 is the time moment when Φ acquires its maximal value.

At $0 < C < 2B^2/A$ the Jacobi function changes in time in accordance with Eq. (7.19). At $C = 2B^2/A$ the Jacobi function is equal to $\Phi = B^2/A^2$, which corresponds to the hydrostatic equilibrium of the system or to the virial theorem.

The solution represents the non-linear periodic pulsation of the Jacobi function of the atmosphere as a whole with period T_v. Using the numerical values of constants $B = 1.03 \times 10^{53}$ and $A = -U = 3.2 \times 10^{33}$ erg, the period of unperturbed virial oscillations of the Earth atmosphere is equal to

$$T_v = \frac{8\pi B}{(2A)^{3/2}} = 2\pi\sqrt{\frac{R^3}{Gm}} = 5060.7\,\text{s} = 1.4057\,\text{h}. \qquad (7.20)$$

Note that the expression (7.20) for the period of virial oscillations T_v includes three fundamental constants: the body M, radius R and gravity constant G. The simplest combination of these three constants, which gives the dimension of time, coincides with (7.20). In this case the nature of the virial oscillations of the atmosphere can be explained by the change in time of the gravitational potential of the solid Earth. The period (7.20) coincides with the period of revolution of a satellite along a circular orbit with first cosmic velocity $v = 7.9$ km/s, radius of the Earth and with a mathematical pendulum the length of whose filament is equal to the radius of the

Earth. This is because the parameters considered are defined by the same constants g, m, and R.

7.3 Perturbed Oscillations

We now consider a general approach to solving the problem of perturbed virial oscillations of the atmosphere, taking as an example the perturbations caused by the variation throughout the year of the solar energy flux owing to the ellipticity of the Earth's orbit. We assume that all the dissipative processes that occur during the interaction between the atmosphere and the hydrosphere and in the atmosphere itself are compensated by solar energy. We note, however, that the value of the flux is evidently dependent on time. Assuming also that the eccentricity of the Earth's orbit and the mean total energy of the atmosphere are known and remain unchanged in time, the total energy of the Earth's atmosphere is proportional to the power of the solar energy flux $L(x)$ which reaches the atmosphere at a given point of the orbit. Then

$$L(t) = L_0 \frac{a^2}{r^2}, \tag{7.21}$$

where L_0 is the mean energy flux reaching the Earth's atmosphere; r is the radius of the Earth's orbit; a is the semi-major axis of the Earth's orbit.

Using the property of the elliptical motion, we obtain

$$L(t) = L_0(1 - 2e' \cos E' + e'^2 \cos^2 E')^{-1}, \tag{7.22}$$

where e' is the eccentricity of the Earth's orbit; E' is the eccentric anomaly that characterizes the location of the Earth on the orbit and is linked with time by the Keplerian equation

$$E' - e' \sin E' = n'(t - t_0) = M_0', \tag{7.23}$$

where $n = 2\pi/\tau$ is the cyclic frequency of the Earth's revolution round the Sun; τ is the period of revolution and is equal to one year; t_0 is the moment of time required by the Earth to pass through the orbit's perihelion; M' is the mean of the anomaly.

If the eccentricity $e' \leq \bar{e} = 0.6627...$, which is the Laplacian limit, then, using the Lagrangian series, we can obtain expressions for E' and $\varphi(E')$ in the form of an absolute convergent infinite series expanded by entire positive powers of e'.

Note that in order to obtain the expression for $(1 - e' \cos E')$ we can write the equality which follows from the Keplerian equation:

$$\frac{1}{(1 - e' \cos E')^2} = \left(\frac{dE'}{dM'} \right)^2.$$

Expanding the eccentric anomaly E' in the Lagrangian series by the power of eccentricity e' (Duboshin, 1975), we obtain

$$E' = \sum_{k=0}^{\infty} \frac{e'^k}{k!} \frac{d^{k-1}\left(\sin^k M'\right)}{dM'^{k-1}}. \tag{7.24}$$

This can be rewritten in a more convenient form:

$$E' = \sum_{k=0}^{\infty} e'^k E_k\left(M'\right),$$

where

$$E_k\left(M'\right) = \sum_{k=0}^{\infty} \frac{1}{k!} \frac{d^{k-1}\left(\sin^k M'\right)}{dM'^{k-1}}.$$

As the absolute convergent series can be differentiated term by term, we obtain

$$\frac{a}{r} = \frac{dE'}{dM'} = \sum_{k=0}^{\infty} e'^k \bar{R}_k\left(M'\right), \tag{7.25}$$

where

$$\bar{R}_k\left(M'\right) = \frac{dE_k\left(M'\right)}{dM'}.$$

Multiplying the series (7.25) by itself, we obtain

$$\frac{a^2}{r^2} = \sum_{k=0}^{\infty} e'^k \bar{R}_k^{(2)}\left(M'\right), \tag{7.26}$$

where

$$\bar{R}_k^{(2)}\left(M'\right) = \sum_{k=0}^{k} \bar{R}_s\left(M'\right) \bar{R}_{k-s}\left(M'\right).$$

Then the expression (7.21) for the solar energy flux reaching the atmosphere can be rewritten:

$$L(t) = L_0 \sum_{k=0}^{\infty} e'^k \bar{R}_k^{(2)}\left(M'\right). \tag{7.27}$$

In agreement with (7.27), the change of the total energy of the Earth's atmosphere is proportional to the change of the solar energy flux $L(t) - L_0$. Then the expression for the total energy of the atmosphere is

$$E(t) = E + k[L(t) - L_0],$$

where k is a proportionality factor.

Thus, in our problem of virial oscillations of the atmosphere perturbed by the solar energy flux varying during the motion of the Earth along the orbit, the equation can be written as

$$\ddot{\Phi} = -A + \frac{B}{\sqrt{\Phi}} + X(e', M'), \qquad (7.28)$$

where $X(e', M')$ is the perturbation function, which has the form

$$X(e', M') = 2k[L(t) - L_0]\left[\sum_{k=0}^{\infty} e'^k R_K^{(2)}(M') - 1\right]. \qquad (7.29)$$

To estimate the geophysical effect of the variation of the solar energy within a time period of one year, we introduce the perturbation function (7.29) into equation (7.28) of perturbed oscillations, to an accuracy of squared eccentricity, i.e.,

$$X(M') = kL_0\left[2e'\cos M' + \frac{5}{2}e'^2\cos 2M\right], \qquad (7.30)$$

where $kL_o \approx 3 \times 10^{31}$ erg.

In this case, the expressions for the Jacobi function of the atmosphere and its first derivative to an accuracy of e have the form

$$\Phi(M) = \Phi_0(1 - 2e\cos M), \qquad (7.31)$$

$$\dot{\Phi}(M) = 2\Phi_0 n\sin M. \qquad (7.32)$$

Then, differentiating the expression for the eccentricity of the perturbed oscillations $e = (1 - AC/2B^2)^{1/2}$ with respect to time and using Lagrange's method of varying arbitrary constants, we obtain

$$\frac{de}{dt} = \frac{A\dot{\Phi}(M)X(M')}{2B^2 e} = \frac{A\Phi_0 kL_0 ne'}{B^2}\left[\sin(M + M') + \sin(M - M')\right.$$
$$\left. + \frac{5}{2}e'\sin(M + 2M') + \frac{5}{2}e'\sin(M - 2M')\right]. \qquad (7.33)$$

Integrating (7.33) with respect to time, we obtain the law of variation of the virial oscillation eccentricity as a first approximation of the perturbation theory:

$$
e(t) = \bar{e} - \frac{A\Phi_0 ne'}{B^2} \left[\frac{cos(M + M')}{n + n'} + \frac{cos(M - M')}{n - n'} \right.
$$
$$
\left. + \frac{5}{2}e' \frac{cos(M + 2M')}{n + 2n'} + \frac{5}{2}e' \frac{cos(M - 2M')}{n - 2n'} \right]. \tag{7.34}
$$

Finally, putting the expression for the eccentricity of the perturbed oscillations into (7.31), we obtain

$$
\Phi(t) = \Phi_0 + \frac{1}{2}\alpha_1 \left[cos(2M + M') + cos M' \right] + \frac{1}{2}\alpha_2 \left[cos(2M - M') + cos M' \right]
$$
$$
+ \frac{1}{2}\alpha_3 \left[cos(2M + 2M') + cos 2M' \right] + \frac{1}{2}\alpha_4 \left[cos(2M - 2M') + cos 2M' \right], \tag{7.35}
$$

where

$$
\alpha_1 = \frac{2A\Phi_0^2 kL_0 ne'}{B^2(n + n')}, \qquad \alpha_2 = \frac{2A\Phi_0^2 kL_0 ne'^2}{B^2(n - n')},
$$

$$
\alpha_3 = \frac{5A\Phi_0^2 kL_0 ne'^2}{B^2(n + 2n')}, \qquad \alpha_4 = \frac{5A\Phi_0^2 kL_0 ne'^2}{B^2(n - 2n')},
$$

$$
M = n(t - t_0), \qquad\qquad M' = n'(t - t_0'),
$$

$$
n = 1.2 \times 10^{-3} s^{-1}, \qquad n = 1.2 \times 10^{-7} s^{-1}.
$$

It follows from (7.34) that a contribution of the long-periodic part of the first-order perturbations to the variation of the Jacobi function of the Earth's atmosphere is

$$
\Phi_a(t) = const + 1.44 \times 10^{-3} \Phi_0 cos M' + 1.44 \times 10^{-3} e' \Phi_0 cos 2M', \tag{7.36}
$$

where $\Phi_0 = 1.06 \times 10^{39}$ g·cm^2; $\Phi_a/I_a = \frac{3}{4}$.

Assuming in first approximation that the rotations of the Earth and the atmosphere are synchronous, and using the law of conservation of angular momentum for the Earth-atmosphere system as $(I_\oplus + I_a)\omega_\oplus = const$, we obtain

$$
\frac{\Delta I_a}{I_\oplus} = -\frac{\Delta\omega}{\omega_\oplus} = \frac{\Delta T}{T_0}, \tag{7.37}
$$

where ω_0 is the angular velocity of the Earth's daily rotation; $T = 8.64 \times 10^4$ s; $I_\oplus = 8.04 \times 10^{44}$ g·cm^2.

It is easy to show that the Earth's rate of rotation has annual variations with daily amplitude of variation of about 2 *ms* duration and can be approximated in our estimate by the sum of two harmonics with a period of one year and half a year respectively. This estimate is in good agreement with the observed data of the seasonal variation of the Earth's angular velocity.

7.4 Resonance Oscillation

We now consider the solution for identification of the resonance frequencies of the perturbed oscillations of the atmosphere due to the change of solar energy flux during the earth's motion along an elliptic orbit.

Let us assume that the Earth's atmosphere satisfies all the conditions needed for writing the equation (7.16) of unperturbed virial oscillations. We can solve this equation because its two first integrals of motion (7.17) and (7.18) are known. We accept from the solution of Eq. (7.16) that the Jacobi function Φ changes in time with the period $\tau'' = 5060.7$ s^{-1} and the frequency $n'' = 2\pi/\tau'' = 0.00124$ s^{-1}. Assuming also that the perturbation of the Earth's atmosphere is affected only by the change of power of the solar radiation flux during the year owing to the ellipticity of the Earth's orbit, then the equation of the perturbed virial oscillations should have the form of (7.28) and the perturbation function $X(e', M')$, represented by (7.29), should be a periodic function of time with period $\tau' = 31\,556\,929.9747$ s and frequency $n' = 2\pi/\tau' = 1.9910638$ s^{-1}.

We use the Picard method to obtain the solution of Eq. (7.28), which in this case is written as

$$\ddot{\Phi} = -A + \frac{B}{\sqrt{\Phi}} + 2kL_0\left[\sum_{k=0}^{\infty} e'^2 R_k^{(2)}(M') - 1\right], \tag{7.38}$$

where $A = -2E_0$; $B = |U_a|\sqrt{\Phi_a}$; $e' = 0.014$ is the eccentricity of the Earth's orbit; $M' = n'(t - t_0) = E' - e'\sin E'$ is the mean anomaly determined from the Kepler equation; E' is the eccentric anomaly.

Using Lagrange's method of variation of the arbitrary constants C and t, which determine the solution of Eq. (7.38) in the form

$$\Phi = \Phi\left(C, t_0'', t\right), \tag{7.39}$$

$$\dot{\Phi} = \dot{\Phi}\left(C, t_0'', t\right), \tag{7.40}$$

we can write the system of differential equations (5.19) and (5.20) (see Chap. 5) determining the change of C и t_0'' in time:

$$\frac{dC}{dt} = -2\dot{\Phi}X\left(M'\right), \tag{7.41}$$

$$\frac{dt_0''}{dt} = -2\dot{\Phi}X\left(M'\right)\Psi(\Phi, C), \tag{7.42}$$

where the function $\Psi(\Phi, C)$ was defined earlier, and $X(M')$ is a periodic function of the argument M' with the period 2π.

Substituting the expressions for Φ and $\dot{\Phi}$ from (7.39) and (7.40) into (7.41) и (7.42), we obtain the system of two differential equations:

$$\frac{dC}{dt} = F_1(M', M'', C, t_0''), \tag{7.43}$$

$$\frac{dt_0''}{dt} = F_2(M', M'', C, t_0''), \tag{7.44}$$

where F_1 and F_2 are periodic functions of the arguments M' and M'' with the period 2π.

Owing to the periodicity of the right-hand side of the system of equations (7.43) and (7.44), they can be expanded into a double Fourier series. In this case the system (7.43) and (7.44) can be written:

$$\frac{dC}{dt} = \left\{ A_{00} + \sum_{k',k''=-\infty}^{\infty} \left[A_{k',k''} \cos(k'M' + k''M'') + B_{k',k''} \sin(k'M' + k''M'') \right] \right\}, \tag{7.45}$$

$$\frac{dC}{dt} = \left\{ a_{00} + \sum_{k',k''=-\infty}^{\infty} \left[a_{k',k''} \cos(k'M' + k''M'') + b_{k',k''} \sin(k'M' + k''M'') \right] \right\}. \tag{7.46}$$

Here the coefficients $A_{00}, A_{k'k''}, B_{k'k''}, a_{00}, a_{k'k''}, b_{k'k''}$ do not depend on M' and M'', but are functions of the unknown quantities C and t_0''.

Using the Picard procedure, we determine $C^{(1)}$ and $t_0''^{(1)}$ in the first approximation by substituting the constant values $C^{(0)}$ and $t_0''^{(0)}$ into the expressions for F_1 and F_2. The values $C^{(0)}$ and $t_0''^{(0)}$ could be found through the initial conditions of Φ_0 and $\dot{\Phi}_0$ using corresponding formulas (7.17) and (7.18) which describe the unperturbed virial oscillations.

After integration of the system (7.45) and (7.46) with respect to time, we have

$$C^{(1)} = C^{(0)} + A_{00}^{(0)}(t - t_0) + \sum_{k',k''\to-\infty}^{\infty} \frac{1}{k'n' + k''n''} \{ A_{k',k''}^{(0)} [\cos(k'M' + k''M'')$$
$$- \cos(k'M_0' + k''M_0'')] - B_{k',k''}^{(0)} [\sin(k'M' + k''M'')$$
$$- \sin(k'M_0' + k''M_0'')] \}, \tag{7.47}$$

$$t_0''^{(1)} = t_0''^{(0)} + a_{00}^{(0)}(t - t_0) + \sum_{k',k''\to-\infty}^{\infty} \frac{1}{k'n' + k''n''} \{ a_{k',k''}^{(0)} [\cos(k'M' + k''M'')$$
$$- \cos(k'M_0' + k''M_0'')] - b_{k',k''}^{(0)} [\sin(k'M' + k''M'')$$
$$- \sin(k'M_0' + k''M_0'')] \}, \tag{7.48}$$

where $C^{(1)}$ and $t_0''^{(1)}$ are the arbitrary constants which determine solution of the equation (7.28); $A_{k',k''}^{(0)}, B_{k',k''}^{(0)}, a_{00}^{(0)}, a_{k',k''}^{(0)}, b_{k',k''}^{(0)}$ are corresponding coefficients of the system (7.45)–(7.46) after replacing C и t_0''. by $C^{(0)}$ and $t_0''^{(0)}$.

Thus we have obtained the analytical structure of the solutions (7.47) and (7.48) known in general perturbation theory which have three classes of terms: constant, periodic and secular. Of the periodic terms, the most important are the resonance terms, i.e. those quantities $n_s = k'n' + k''n''$ which are substantially less then both n'' and n'. These terms give a series of long periodic inequalities (their number is infinite) and they allow prediction of the development of the natural processes within relatively long intervals of time.

Let us calculate, as an example, such lower resonance frequencies which have climatic significance:

$$n_{2:12471} = (2 \times 12415 - 12471 \times 1.9910638)10^{-7} = 0.556 \times 10^{-7}\,\mathrm{s}^{-1},$$

$$n_{3:18706} = (3 \times 12415 - 18706 \times 1.9910638)10^{-7} = 0.161 \times 10^{-7}\,\mathrm{s}^{-1},$$

$$n_{11:68589} = (11 \times 12415 - 68589 \times 1.9910638)10^{-7} = 0.07 \times 10^{-7}\,\mathrm{s}^{-1}$$

and so on, and corresponding to those frequencies the periods are:

$$\tau_{2:12471} = 3.6 \text{ years}; \quad \tau_{3:18706} = 12.36 \text{ years}, \quad \text{and} \quad \tau_{11:68598} \approx 28 \text{ years}.$$

It should, however be kept in mind that the first approximation obtained in the framework of perturbation theory is in good agreement with observations within short (not cosmogenic) intervals of time.

Note that one can find analogously the resonance frequencies and the periods of virial oscillations which occur owing to other perturbations, such as, for example, diurnal perturbations, because of the rotation of the Earth around its axis and its latitudinal perturbations, etc. As is well known, the mean solar day has period $\tau \approx 8.64 \times 10^4$ s and frequency $n_s = 2\pi/\tau = 7.27 \times 10^{-5}$ s^{-1}. Then calculating the resonance frequency $n_s = k'n' + k''n'' \ll n', n''$, we obtain:

$$n_{1:17} = (1 \times 124.17 - 17 \times 7.27)10^{-5} = 0.58 \times 10^{-5}\,\mathrm{s}^{-1},$$
$$n_{2:24} = (2 \times 124.17 - 24 \times 7.27)10^{-5} = 1.16 \times 10^{-5}\,\mathrm{s}^{-1}$$

and so on, and corresponding periods:

$$\tau_{1:17} = 12.5\,d; \quad \tau_{2:24} = 6.24\,d \text{ and so on.}$$

For the monthly Moon perturbations, when $\tau = 2.352\,672 \times 10^6$ s and $n_s = 2.67 \times 10^{-6}$ s^{-1}, we have

$$n_{1:465} = (1 \times 1241.7 - 2.67 \times 465)10^{-6} = 0.15 \times 10^{-6}\,\mathrm{s}^{-1},$$
$$n_{2:930} = (2 \times 1241.7 - 2.67 \times 930)10^{-6} = 0.3 \times 10^{-6}\,\mathrm{s}^{-1}$$

and so on, and corresponding periods:

$$\tau_{1:465} = 1.3 \text{ years}; \quad \tau_{2:930} = 0.66 \text{ years}.$$

7.5   Observation of the Virial Eigenoscillations of the Earth's Atmosphere

It was predicted in Sect. 7.2, by means of the solution of Jacobi's virial equation, that the eigenoscillations of the Earth's atmosphere with a period of $T_v = 1^h.4$ and frequency $\omega = 1.24 \times 10^{-3}$ s^{-1} exists. This solution describes the periodic change of the Jacobi function of the Earth's atmosphere in time and can be expressed in the form of a trigonometric Fourier series expanded by entire multiple values of argument M related to t as

$$M = \frac{2\pi}{T_v}(t - t'_0),$$

where t'_0 is the moment of time defining the phase of the virial oscillations.

The first four terms of this series are

$$\Phi(t) = \Phi_0 \left[1 + \frac{3}{2}e^2 + \left(-2e + \frac{e^3}{4} \right) \cos M - \frac{e^2}{2} \cos 2M - \frac{e^3}{3} \cos 3M + \cdots \right],$$

$$(7.49)$$

where Φ_0 is the mean value of the Jacobi function, determined by the virial theorem; e is the parameter of the virial oscillations of the atmosphere which characterizes the amplitude of the Jacobi function change; $e \ll 1$.

It has been shown theoretically that the period of chance of the Jacobi function of the atmosphere depends on the value of its total energy, and in the case of non-perturbed atmosphere is equal to $1^h.4$. In the framework of the model considered with spherically symmetric atmosphere, the change of Jacobi function takes place owing to the change of the radial mass density distribution of the atmosphere having the same period. Direct experimental test of this statement is difficult because the value of the Jacobi function cannot be measured directly to prove the expression (7.49). But, as it was shown in Sects. 2.2 and 2.3, permanent changes in the Earth's Jacobi function (polar moment of inertia) is fixed by artificial satellites and earthquake measurements. Moreover, the process of virial oscillations is accompanied by synchronous changes of pressure, temperature, air moisture, magnetic field intensity and other measurable geophysical parameters at the Earth's surface. In addition, from the condition of a global scale of the virial oscillations, it follows that all the geophysical parameters are pulsating with the same period and are coherent both within the considerable interval of time and in space over all the Earth's surface as well as vertically. The expression for the virial pulsations of the atmospheric pressure $p(t)$ and temperature $T(t)$ can be expanded into a Fourier series (7.49) as follows:

$$p(t) = p_0 \left[1 + \frac{1}{2}e^2 + \left(2e + \frac{3}{4}e^3 \right) \cos M + \frac{5}{2}e^2 \cos 2M + \cdots \right], \quad (7.50)$$

$$T(t) = T_0 \left[1 + \frac{1}{2}e^2 + \left(2e + \frac{3}{4}e^3 \right) \cos M + \frac{5}{2}e^2 \cos 2M + \cdots \right], \quad (7.51)$$

where p_0 and T_0 are the mean values of the atmospheric pressure and temperature averaged over all the Earth's surface and through the mass of the atmosphere respectively.

We shall use for this expansion the analysis and interpretation of the experimental data.

As is well known, regular observation of the atmospheric pressure and temperature at various points of the Earth prevents the discovery of a rigorous periodicity in changes of atmospheric parameters, especially for short periods. There are a number of reasons for this, including their variability which defines the dynamics of air masses at any point of observation. The parameters recognized and studied until now are seasonal, diurnal and semidiurnal periodicity, in addition to variation of the atmospheric parameters connected with the motion of the planet along the elliptical orbit around the Sun, with the obliquity of the axis of rotation to the ecliptic, and with perturbations caused by the Moon.

In our case, in order to prove that the predicted oscillations of various geophysical parameters with period $1^h.4$ exist, we used the applicable spectral analysis of experimental data.

7.5.1 Oscillation of the Temperature

We now describe the results of spectral analysis of temperature data recorded by our colleagues in the Central Hydrometeorological Observatory at the Ostankino TV Tower, Moscow.

Let us consider two sets X_1 and Y_1 representing regular records of the air temperature variation of 34 hours duration, each of which were obtained simultaneously in July 1971 at heights of 503 m and 83 m (Fig. 7.1). The discreteness interval of the numerical record of the temperature was 120 s, the interval of the gauge was ~60 s, and the sensitivity was 0.1°C. The sets X_1 and Y_1 contain 1024 discrete values of recorded temperature starting at 10 h 34 m on 17 July through 21 h 34 m on 18 July. Spectral analysis of the data received was carried out by computer using the method of quick Fourier transformation with the program developed by A.B. Leybo and V.Yu. Semenov.

Fig. 7.1 Power spectra of
sets of temperature variation
at heights of 503 m and 83 m
(**a**); functions of mutual
coherence of the sets X_1 and
Y_1 (*1*), X_2 and Y_2 (*2*), X_3 and
Y_3 (*3*) (**b**); functions of phase
difference of the sets X_1 and
Y_1 (*1*), X_2 and Y_2 (*2*), X_3 and
Y_3 (*3*) (**c**). (*Arrows* show the
range of 95% of the confi-
dence intervals)

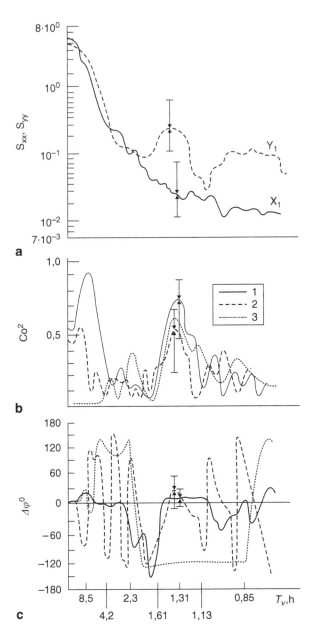

Figure 7.1a shows the recorded power spectra S_{xx} and S_{yy} of the sets X_1 and Y_1.
With the help of this initial information, we calculated the function of the mutual
spectral density $S_{xx} = S'_{xy} + iS''_{xy}$. Then the function of the mutual coherence was
found to be

$$Co^2 = \frac{|S_{xy}|^2}{|S_{xx}||S_{yy}|} \tag{7.52}$$

and the function of the phase difference was defined as

$$\Delta\varphi = atctg\frac{S''_{xy}}{S'_{xy}}. \tag{7.53}$$

They are both plotted in Figs. 6.12b and c.

It is known (Bendat and Pearson, 1971) that the range of the confidence intervals for estimating the phase difference $\Delta\varphi$ tends to zero as Co^2 runs to unity. Hence the higher the value of the function of the mutual coherence, the higher the stability of the phase difference of the harmonics of the two processes X and Y. Therefore, the relationship between the harmonics is probable for those frequencies where the value of Co^2 is close to unity. At the same values of Co^2 the meaning of the phase difference is lost because of the wide range of the confidence interval.

We can see from Fig. 7.1b that in the vicinity of the period $T_v = 1^h.4$, predicted by theory, the function of the mutual coherence of the sets X_1 and Y_1 is significant, i.e. has high probability of not being equal to zero, and sometimes of being even greater than 0.7. Note also that the phase difference function in the neighborhood of the period of oscillations T_v is equal to zero, which indicates that the harmonic constituents of the temperature variations with the period T_v within the considered interval of time are coherent at heights of 503 and 83 m.

Twice as much extended time for sets X_1 and Y_1 does not change the general character of the discovered regularity. Figures 7.1b and 7.1c demonstrate values of the mutual coherence function Co^2 and the phase difference $\Delta\varphi$ of two sets X_2 and Y_2 each containing 2048 experimental points and recorded synchronously with the same discreteness interval. Recording was continued for three days starting at 10 h 34 m on 17 July through 6 h 54 m on 20 July 1971. However, reduction of time for the sets leads to an increase in values of the mutual coherence function on the low frequencies, but the values of the frequencies corresponding to the period T_v do not increase. This proves the theoretically predicted conclusion concerning the existence of a coherent harmonic with the period T_v.

Figure 7.1b plots the mutual coherence function for two sets of temperature variations recorded at 17 h 2 m in each case. These sets were recorded at a height of 503 m with the same discreteness interval as discussed above, but in different years. Set X_3 covers a time interval from 10 h 34 m on 17 July until 3 h 36 m on 18 July 1971, and Y_3 was recorded from 9 h 14 m 30 July to 2 h 16 m on 31 July 1971. We can see from the plot that in this case the mutual coherence function acquires a value equal to 0.6 at $T = 1h.31$, which proves the theory of the steady state virial oscillations within sufficiently long intervals of time.

Let us estimate the amplitude of temperature virial oscillations in the neat surface layer of the atmosphere. For this purpose we recall that the value of the power spectrum of a process at a given value of the frequency is proportional to the square of the amplitude value of the harmonics with the same frequency in Fourier analysis of the process. We plot the power spectrum of a sinusoid of some known amplitude, for example, equal to 1°C with period $1^h.4$ and a given discreteness $\tau \approx 120$ s. The value of the power spectrum of the sinusoid for the frequency related to period $T_v = 1^h.4$ was found equal to 3.4 grad^2 per hertz, and the value of the power spectrum of temperature micro-fluctuation at heights of 83 m to 503 m in July 1971, according to the Observatory recording, was 0.3 grad^2 per hertz. It is easy to obtain the estimated value of the amplitude of the temperature virial oscillations, which is 0.3°C.

7.5.2   Oscillation of the Pressure

We now consider our results of the spectrum analysis of the atmospheric pressure records made by means of a microbarograph designed by V.N. Bobrov in the Institute of Earth Magnetism and Radio Wave Propagation, Russian Academy of Sciences. Analyzing the records, we took the pressure ordinates correct to 1 mm at amplitude of pressure oscillation within several cm (at microbarograph sensitivity equal to 0.02 mb/mm). The spectral analysis was done using the methodology described above.

Figure 7.2a shows the power spectrum of two sets X_4 and Y_4 containing 1024 of the pressure ordinates taken with discreteness interval 76.5 s covering the time interval from 18 h on 31 August through 15 h 30 m of 1 September 1977 and from 15 h 00 m on 1 September through 12 h 30 m on 2 September 1977.

The experimental data of micro-fluctuations of the atmospheric pressure were obtained at a near surface layer of the northeast coastal area of the Caspian Sea by our colleagues from the Institute of Earth Magnetism and Radio Wave Propagation. Using the function of mutual spectral density S_{xx}, we calculated the functions of mutual coherence Co^2 and phase difference $\Delta\varphi$, shown in Figs. 7.2b and 7.2c.

The spectral densities were analyzed by averaging the values of seven experimental points (the number of degrees of freedom equal to 14). The graph of spectral densities was plotted in relative units and on logarithmic scale (in order to have the same range of confidence intervals for any value of the spectral density).

Figure 7.2b shows that in the vicinity of the frequency value corresponding to the period $T_v = 1^h.4$, the coherence coefficient is significant and is equal to $Co^2 = 0.6$. It is also important to note that, for the same frequency, the phase difference is stable and close to zero (see Fig. 7.2c). These facts prove the theory of the existence of coherent oscillations of the Earth's atmosphere with period close to T_v.

Figures 7.2b and 7.2c also show the curves of mutual coherence and phase difference for the other two sets X_5 and Y_5 plotted on a base of 512 experimental points of atmospheric pressure recorded within time intervals starting at 16 h 56 m on 28 August and at 2 h 45 m on 29 August and at 18 h 00 m 31 August through 3 h 49 m on 1 September 1977, respectively. The mutual coherence function has a value equal to 0.5 in the vicinity of T_v, and the function of the phase difference has a value close to zero. This also proves the conclusion discussed above.

Fig. 7.2 Power spectra of the sets X_4 and Y_4 of atmospheric pressure micro-fluctuations northeast of the coastal area of the Caspian Sea (**a**); functions of mutual coherence of the sets X_4 and Y_4 (*1*) and X_5 and Y_5 (*2*) (**b**); functions of the phase difference of the sets X_4 and Y_4 (*1*) and X_5 and Y_5 (*2*) (**c**). (*Arrows* show the range of 95% of the confidence intervals)

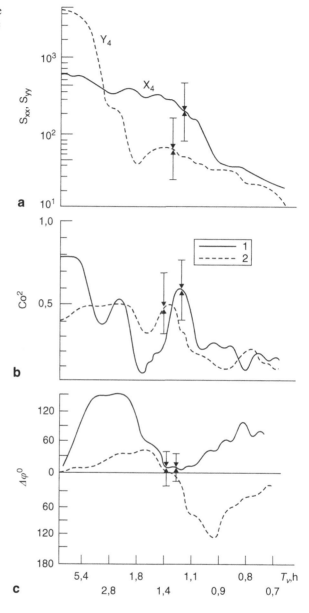

We can also show that virial pulsation of the Earth's atmosphere with period $T_v = 1^h.4$ is observed in both mid and low latitudes. For this purpose we studied records of micro-fluctuations of the atmospheric pressure obtained by the same researchers during their expedition to Cuba.

Let us analyze two sets of experimental data X_6 and Y_6 taken from their records of micro-fluctuations of the atmospheric pressure with discreteness interval equal to

Fig. 7.3 Power spectra of the sets X_6 and Y_6 of atmospheric pressure variations in Cuba (**a**); functions of mutual coherence of the sets X_6 and Y_6 (*1*) and X_7 and Y_7 (*2*) and X_8 and Y_8 (*3*) (**b**); functions of the phase difference of the sets X_6 and Y_6 (*1*) and X_7 and Y_7 (*2*) (**c**). (*Arrows* show the range of 95% of the confidence intervals)

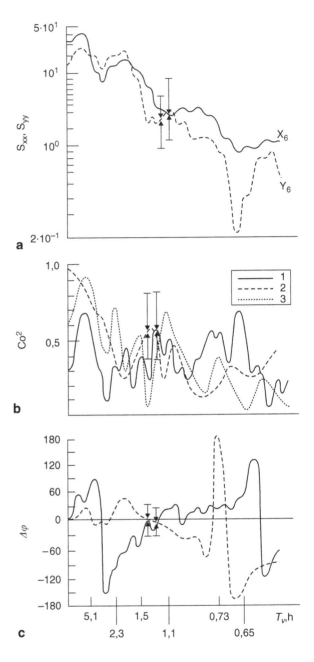

180 s. The process X_6 covers a time interval starting at 19 h 00 m on 7 May through 20 h 36 m on 8 May 1976 and the process Y_6 covers an interval from 21 h 54 m on 7 May through 23 h 30 m on 9 May 1976. Figure 7.3a shows the power spectra of the processes X_6 and Y_6. The functions of mutual coherence and phase difference

which, in the vicinity of the period T_v, have values $Co^2 = 0.56$ and $\Delta\varphi = 0°$ respectively, are shown in Figs. 7.3b and 7.3c. The same figures also show functions of mutual coherence and phase difference for two other sets, X_7 and Y_7, representing 256 experimental points each and analyzed with the same discreteness intervals. They cover time intervals starting at 21 h 37 m on 25 April through 10 h 23 m on 26 April and on 20 h 00 m on 28 April on 9 h 14 m on 29 April 1976. The value of function $Co^2 = 0.6$ and $\Delta\varphi = 0°$ relative to frequencies with period $T_v = 1^h.4$.

Our estimate of the amplitude of virial oscillations of the atmospheric pressure made by using the above procedure gives the value of 0.1 mbar.

Figure 7.3b shows the function of mutual coherence of two sets of atmospheric pressure X_8 and Y_8 taken with discreteness interval $\Delta t = 6$ m and recorded in various years and at different points of the globe: (a) in Cuba within a time interval starting at 19 h 00 m on 7 May through 2 h 30 m 8 May 1976; (b) on northeastern shore of the Caspian Sea at 19 h 00 m 4 September through 20 h 30 m on 5 September 1977. We also observed here that the function Co^2 takes a value of about 0.65 in the vicinity of the period $T_v = 1^h.4$. This also proves our hypothesis.

Analogous results were obtained on analysis of the experimental data of microfluctuation of the atmospheric pressure recorded at the Black Sea. The analyses were made with discreteness interval equal to 120 s.

We have shown the existence of harmonics with period close to $T_v = 1^h.4$, coherent within long time intervals for micro-fluctuations of atmospheric pressure and temperatures in the near surface layer of the atmosphere, obtained at various points of the globe and in different seasons of the year. The resulting data from analyses of experimental records of micro-fluctuations of pressure and temperature as well as the geomagnetic field of the Earth are given in Table 7.1.

7.6   The Nature of the Oceans

It follows from equation (6.71) that the world ocean is found to be in a suspended state because the density value of its water is by far less (2/3) than the mean density of the solid Earth. Accepting this criterion, we may assume that in the earlier stage of creation of the planet its hydrosphere remained in the gaseous phase. Only after irradiation of the corresponding part of the potential energy has the water vapor condensed into the liquid phase (see Chap. 9). Taking into account the observed geographic distribution of the Atlantic and Pacific oceanic basins, we also may assume that their floors were formed as a consequence of the planet's equatorial oblateness during formation. Applying the same argument and geography of location, one may conclude that the Indian oceanic floor was formed on the same base at the Earth's polar flattening and due to the asymmetric deformation of the southern hemisphere (see Chaps. 1 and 2).

The oceans have their own potential energy value which is equal to $U \approx 2 \times 10^{32}$ erg. This value is by four degrees less of their oscillating and rotating kinetic

Table 7.1 Experimental data after reciprocal spectral analysis

Place of observation and parameters	Data and hours of observation	Frequency points (h)	Discreteness (Δt, s)	Time (h)	$Co^2/\Delta\varphi°$
Caspian Sea, pressure	X: 31.08. (18-00)–01.09.1977 (05-05) Y: 28.08. (17-00)–29.08.1977 (04-05)	512/11	78.5	1.39	0.5/13
Caspian Sea, pressure	X: 31.08. (18-00)–01.09.1977 (15-38) Y: 01.09. (15-38)–02.09.1977 (12-41)	1024/22	78.5	1.36	0.56/6
Caspian Sea, pressure	X: 28.08. (16-45)–29.08.1977 (04-26) Y: 09.09. (22-00)–10.08.1978 (09-30)	512/11	78.5	1.39	0.47/0
Black Sea, pressure	X: 13.08. (22-30)–14.08.1979 (11-00) Y: 14.08. (11-00)–14.08.1979 (23-30)	512/12.5	87.7	1.39	0.76/–1.4
Moscow, temper. (503 m)	X: 24.01. (08-14)–25.01.1973 (01-16) Y: 25.01. (02-30)–25.01.1973 (19-32)	512/17	120	1.42	0.60/0
Moscow, temper. (50, 83 m)	X: 17.07. (10-34)–18.07.1971 (03-36) Y: 17.07. (10-34)–18.07.1971 (03-36)	512/17	120	1.42	0.67/–30
Moscow, temper. (503, 83 m)	X: 17.07. (10-34)–18.07.1971 (20-36) Y: 17.07. (10-34)–18.07.1971 (03-36)	1024/17	120	1.26	0.72/4
Moscow, temper. (503, 83 m)	X: 17.07. (10-34)–20.07.1971 (06-54) Y: 17.07. (10-34)–20.07.1971 (06-54)	2048/68	120	1.28	0.55/10
Moscow, temper. (503, 83 m)	X: 17.07. (10-34)–18.07.1971 (03-36) Y: 30.07. (09-14)–31.07.1977 (02-16)	512/17	120	1.31	0.62/0
Cuba, Pressure	X: 25.04. (21-37)–26.04.1976 (10-23) Y: 28.04. (20-00)–29.04.1976 (09-14)	256/13	180	1.42	0.57/10
Cuba, Pressure	X: 07.05. (19-00)–08.05.1976 (20-36) Y: 08.05. (21-54)–09.00.1976 (23-30)	512/26	180	1.28	0.59/1.7
Cuba, Pressure	X: 16.11. (12-00)–18.11.1976 (15-12) Y: 07.05. (19-00)–09.05.1976 (22-15)	1024/51	180	1.42	0.50/–17
Cuba, magnetic field pressure	X: 16.02. (12-00)–18.02.1976 (15-12) Y: 16.02. (12-00)–18.02.1976 (15-12)	1024/51	180	1.65 1.38	0.88/120 0.50/160
Cuba, Caspian Sea, pressure	X: 07.05. (19-00)–08.05.1976 (20-30) Y: 14.08. (11-00)–14.08.1979 (23-30)	256/26	360	1.1	0.65/–40

energy. It means that the oceans stay in hydrostatic equilibrium in the outer force field of the solid Earth which provides rotational motion of this shell.

This suspended stay of equilibrium of the oceans determines their dynamics. According to equation (6.71), angular velocity of the rotating oceanic water, because of low density, is less than that of the solid Earth. Therefore, the oceanic water in its rotary motion has lower angular velocity with respect to the solid Earth. Encountering the continents on their way, the ocean waters form the latitudinal currents along both American and Asian continents. The observed multiply disordered regional and local linear and eddy currents are the consequences of different perturbations, which dominate over the regular currents due to the above energy effects.

The regular currents, which move along the east shores of the continents, play an important role as heating system in formation of weather and climate processes in the middle and high latitudes.

7.7 The Nature of the Weather and Climate Changes

The atmosphere and oceans comprise a common natural system which controls the weather and climate on the planet. Both shells, being uniform in density and staying in semi-hydrostatic and semi-dynamic equilibrium in the outer force field of the solid Earth, are affected by virial oscillations of the planet and rotate with a corresponding angular velocity. However, irregular seasonal pumping of solar energy, because of ellipticity of the Earth's orbit and the precession and nutation perturbation of the upper solid shell, leads to continuous redistribution of solar energy in the latitude and longitude directions. Those perturbations cause permanent change in balance of the evaporated water from the ocean surface and lead to changes in baric topography and in trajectories of the cyclonic vortexes which carry clouds of moisture. From point of view of the considered theory, those are the perturbations that appear to be the cause of weather and climate change.

The principles of perturbation and resonance oscillation of the atmosphere presented in this chapter could be used as a basis for development of an analytical solution of the problem.

Chapter 8
The Nature of the Earth's Electromagnetic Field and Mechanism of its Energy Generation

We assume that the Earth's electromagnetic field is generated by the magneto-hydro-dynamo action provided by the planet's liquid metal core. Its essence is in the motion of the conducting liquid core where self-excitation of the electric and magnetic poloidal (meridional) and toroidal (parallel) fields occurs. During rotation of the inner planet's shells with different angular velocities, in the case of asymmetric thermal convection of the shell mass, the intensity of the fields is increased. This condition for the Earth is achieved because the rotation and magnetic axes do not coincide and thermal convection supposedly takes place. But a physically justified theory of the Earth's phenomenon of an electromagnetic field is absent. There is no explanation of the mechanism of generation of the energy of this field except for the general physical principle of mass and charge interaction. Also, ideas or hypotheses about the source of replenishment of the Earth's energy that is spent for gravitational and thermal irradiation are absent. The only source of solar and star irradiated energy is accepted to be interior nuclear fusion.

In order to find a solution of the problem, in this chapter we discuss a novel idea based on the innate capacity of a body's energy for performing motion. As it was shown in Chap. 3, energy is a measure of the motion and interaction of particles of any kind of a body's matter. The various forms of energy are interconvertible and its sum for a system remains constant. The above unique properties of energy, with its oscillating mode of motion in our dynamics, make it possible to consider the nature of the electromagnetic and gravitational effects of the Earth and other celestial bodies as interconnected events.

It was shown in Chap. 6 that the Earth's gravitational (potential) energy results in volumetric pulsations of the planet's shells, having an oscillating regime, frequencies of which depend on the mass density. In our consideration the Earth is accepted as a self-gravitating body. Its dynamics is based on its own internal force field and the potential and kinetic energies are controlled by the energy of oscillation of the polar moment of inertia, i.e. by interaction of the body's elementary particles.

Applying the dynamical approach and the results obtained, we show below that the nature of creation of the electromagnetic field and mechanism of its energy generation appears to be an effect of the volumetric gravitational oscillation of the body's masses. This effect is also characteristic for any celestial body.

V. I. Ferronsky, S. V. Ferronsky, *Dynamics of the Earth*,
DOI 10.1007/978-90-481-8723-2_8, © Springer Science+Business Media B.V. 2010

8.1   Electromagnetic Component of Interacting Masses

It was shown in Sect. 6.2 that electromagnetic energy is a component of the expanded analytical expression of potential energy. The expansion was done by means of an auxiliary function of the density variation relative to its mean value. The expression of the body's potential gravitational energy in the expanded form (6.48) was found as

$$U = \alpha^2 \frac{GM^2}{R} = \left[\frac{3}{5} + 3 \int\limits_0^1 \psi x dx + \frac{9}{2} \int\limits_0^1 \left(\frac{\psi}{x} \right)^2 dx \right] \frac{GM^2}{R}, \qquad (8.1)$$

where U is the potential energy of the gravitational interaction; α^2 is the form-factor of the force function; G is the gravity constant; M is the body mass; R is its radius; $\Psi(s)$ is the auxiliary function of radial density distribution relative to its mean value.

We have considered and applied the two first right-hand side terms of Eq. (8.1). The third term in dimensionless form represents an additive part of the potential energy of the interaction of the non-uniformities between themselves, which was written as

$$\frac{9}{2} \int\limits_0^1 \left(\frac{\psi}{x} \right)^2 dx \equiv \frac{9}{2} \int\limits_0^1 \left(\frac{\psi}{x^2} \right)^2 x^2 dx. \qquad (8.2)$$

The non-uniformities are determined as the difference between the given density of a spherical layer and the mean density of the body within the radius of the considered layer. We apply for interpretation of the third term the analogy of electrodynamics (Ferronsky et al., 1996). Each particle there generates an external field, which determines its energy. The energies of some other interacting particles and their own charges are determined by this field. If the potential of the field is expressed by means of the Poisson equation through the density of charge in the same point, then the total energy can be presented in additive form through application of the squared field potential. If the body mass is considered as a moving system, then the Maxwell radiation field applies. In our solution the dimensionless third term of the field energy is written as:

$$\frac{9}{2} \int\limits_0^1 \left(\frac{\psi}{x} \right)^2 dx \equiv \frac{9}{2} \int\limits_0^1 \left(\frac{\psi}{x^2} \right)^2 x^2 dx \equiv \frac{9}{2} \int\limits_0^1 E^2 dV, \qquad (8.3)$$

where $E = \Psi/x^2$ is a dimensionless form of the electromagnetic field potential which is a part of the gravitational potential; Ψ plays role of the charge; $dV = x^2 dx$ is the volume element in dimensionless form.

Table 8.1 Observational parameters of equilibrium nebulae

Parameters	Visible dark nebulae			
	Small globula	Large globula	Intermediate cloud	Large cloud
M/M_{Sun}	>0.1	3	8×10^2	1.8×10^4
R (pc)	0.03	0.25	100	20
n (n/cm^3)	>4×10^4	1.6×10^3	100	20
$M/\pi R^2$ (g/cm^2)	>10^{-2}	3×10^{-3}	3×10^{-3}	3×10^{-3}

In order to determine the numerical value of λ, which is integral of (8.3), calculations were made for a sphere with different laws of radial density distribution including the politropic model (Ferronsky et al., 1996). The results of the calculations show that, for physically significant models of the density distribution in celestial bodies, the parameter λ is equal to 0.022. There is also an observational confirmation of this conclusion. Spitzer (1968) demonstrates observational results of nebulae of different mass and size in Table 3.2 of his book, which we reproduce here in Table 8.1.

For the accepted parabolic low of the Earth's radial density distribution (see Chap. 6, Table 6.1) the third dimensionless term in Eq. (8.1) is $a_\gamma^2 = 0.01$ and the total field energy should be

$$U_\gamma = \left(\frac{9}{2} \int_0^1 E^2 dV \right) \frac{GM^2}{R} = 0.01 \frac{GM^2}{R}. \tag{8.4}$$

Thus, the virial approach to the problem solution of the Earth's global dynamics gives a novel idea about the nature of the planet's electromagnetic field. The energy of this field appears to be the component of the potential energy of the interacting masses. The question arises about the mechanism of the body's energy generation, which provides radiation in a wide range of the wave spectrum from radio through thermal and optical to x and γ rays.

8.2 Potential Energy of the Coulomb Interaction of Mass Particles

With the help of a model solution, we can show that for the Coulomb interactions of the charged particles, constituting a celestial body, the relationship between the potential energy of a self-gravitating system and its Jacobi function holds (Ferronsky et al., 1981a), i.e.

$$-U_c\sqrt{\Phi} = const, \tag{8.5}$$

where U_c is the potential energy of the Coulomb interactions.

Derivation of the expression for the potential energy of the Coulomb interactions of a celestial body is based on the concept of an atom following, for example, from the Tomas-Fermi model (Flügge, 1971). In our problem this approach does not result in limited conclusions since the expression for the potential energy, which we write, will be correct within a constant factor.

Let us consider a one-component, ionized, quasi-neutral and gravitating gaseous cloud with a spherical symmetrical mass distribution and radius of the sphere R. We shall not consider here the problem of its stability, assuming that the potential energy of interaction of charged particles is represented by the Coulomb energy. Therefore, in order to prove relationship (8.5) it is necessary to obtain the energy of the Coulomb interactions of positively charged ions with their electron clouds.

Assume that each ion of the gaseous cloud has the mass number A_i and the order number Z and the function $\rho(r)$ expresses the law of mass distribution inside the gaseous cloud. The mass of the ion will be $A_i m_p$ (where $m_p = 4.8 \times 10^{-24}$ g is the mass of the proton) and its total charge will be $+Ze$ (where $e = 4.8 \times 10^{-10}$ GCSE is an elementary charge). Then, let the total charge of the electron cloud, which is equal to $-Ze$, be distributed around the ion in the spherically symmetrical volume of radius r_i with charge density $q_e(r_e)$, $r_e \in [0, r_i]$. Radius r_i of the effective volume of the ion may be expressed through the mass density distribution $\rho(r)$ by the relation

$$\frac{4}{3}\pi r_i^3 = \frac{A_i m_p}{\rho(r)}. \tag{8.6}$$

Then

$$r_i = \sqrt[3]{\frac{3A_i m_p}{4\pi \rho(r)}}. \tag{8.7}$$

Let us calculate the Coulomb energy U_c' per ion, using relation (8.7) Assuming that the charge distribution law in the effective volume of radius r_i is given, we may write U_c' in the form

$$U_c' = U_c^{(+)} + U_c^{(-)}, \tag{8.8}$$

where $U_c^{(-)}$ is the potential energy of the Coulomb repulsion of electrons inside the effective volume radius r_i; $U_c^{(+)}$ is the potential energy of attraction of the electron cloud to the positive ion.

Let the charge distribution law $q_e(r_e) = q_0(r_e)$ inside the electron cloud be given. Then normalization of the electron charge of the cloud, surrounding the ion, may be written in the form

$$-Ze = \int\limits_0^{r_1} 4\pi q_e(r_e)r_e^2 dr_e. \tag{8.9}$$

From expression (8.9) we may obtain the normalization constant q_0, which will depend on the given law of charge distribution, as

$$q_0 = -\frac{Ze}{4\pi \int\limits_0^{r_1} r_e^2 f(r_e)dr_e}. \tag{8.10}$$

Now it is easy to obtain expressions for $U_c^{(-)}$ and $U_c^{(+)}$ in the form

$$U_c^{(-)} = (4\pi)^2 q_0^2 \int\limits_0^{r_1} r_e f(r_e)dr_e \int\limits_0^{r_1} (r_e')^2 f(r_e')dr_e', \tag{8.11}$$

$$U_c^{(+)} = 4\pi Zeq_0 \int\limits_0^{r_1} r_e f(r_e)dr_e. \tag{8.12}$$

Finally, expression (8.8) for the potential energy U_c' corresponding to one ion may be rewritten using (8.10)–(8.12) in the form

$$U_c' = -e^2 Z^2 \left[\frac{\int\limits_0^{r_1} r_e f(r_e)dr_e}{\int\limits_0^{r_1} r_e^2 f(r_e)dr_e} - \frac{\int\limits_0^{r_e} r_e f(r_e)dr_e \int\limits_0^{r_e} (r_e')f(r_e')dr_e'}{\left(\int\limits_0^{r_1} r_e^2 f(r_e)dr_e\right)^2} \right]. \tag{8.13}$$

It is easy to see that in the right-hand side of equation (8.13), the expression enclosed in brackets determines the inverse value of some effective diameter of the electron cloud, which may be expressed through the form-factor α_i^2 of the ion radius r_i, i.e.

$$\frac{\int\limits_0^{r_1} r_e f(r_e)dr_e}{\int\limits_0^{r_1} r_e^2 f(r_e)dr_e} - \frac{\int\limits_0^{r_e} r_e f(r_e)dr_e \int\limits_0^{r_e} (r_e')f(r_e')dr_e'}{\left(\int\limits_0^{r_1} r_e^2 f(r_e)dr_e\right)^2} = -\frac{\alpha_i^2}{r_i^2} \tag{8.14}$$

Thus, expression (8.13), using (8.14), yields

$$-U_c' = \alpha^2 \frac{e^2 Z^2}{r_i^2}. \tag{8.15}$$

The numerical values of the form-factor α_i^2 depending on the charge distribution $q_e(r_e)$ inside the electron cloud are given in Table 8.2 and their calculations were given in our work (Ferronsky et al., 1981a).

Table 8.2 Numerical values of the form-factors α_i^2 for different radial distribution of charge of the electron cloud around the ion

The law of charge distribution*	α_i^2
$q_e(r_e) = q_o(r_e/r_i)$	0.76
$q_e(r_e) = q_o = \text{const}$	0.9
$q_e(r_e) = q_o(1 - r_e/r_i)$	1.257
$q_e(r_e) = q_o(1 - r_e/r_i)^n$	$\dfrac{(n+3)\left(11n^2 + 41n + 36\right)}{8(2n+3)(2n+5)}$
$q_e(r_e) = q_o(r_e/r_i)^n$	$\dfrac{(n+3)^2}{(n+2)(2n+5)}$
The same for $n \rightarrow \infty$	$\alpha_i^2 \rightarrow 1/2$

*Here q_o is the charge value in the centre of the sphere; r_e is the parameter of radius, $r_e \in [0, r_i]$; n is an arbitrary number, $n = 0, 1, 2, \ldots$

Using expression (8.15), the total energy of the Coulomb interaction of particles may be written as:

$$-U_c = 4\pi \int\limits_0^R \frac{\rho(r)}{A_i m_p} U_c' r^2 dr = \frac{3\alpha_i^2 e^2 Z^2}{R} \int\limits_0^R R r^2 \left(\frac{4\pi\rho(r)}{3A_i m_p}\right)^{4/3} dr. \qquad (8.16)$$

Introducing in expression (8.16) the form-factor of the Coulomb energy α_i^2, depending on the mass distribution in the gaseous cloud and on the charge distribution inside the effective volume of the ion, we obtain:

$$-U_c = \alpha_c^2 \frac{e^2 Z^2}{r_i^2} \left(\frac{m}{A_i m_p}\right)^{4/3}, \qquad (8.17)$$

where

$$\alpha_c^2 = \frac{3\alpha_i^2 \int\limits_0^R [(4\pi/3)\rho(r)]^{4/3} R r^2 dr}{m^{4/3}},$$

$$m = \sum_{i=1}^N m_i = 4\pi \int\limits_0^R r^2 \rho(r) dr.$$

Since the total number of ions N in the gaseous cloud is equal to

$$N = \frac{m}{A_i m_p},$$

and the relation between the radius of the cloud and the radius of the ion may be obtained from the relationship of the corresponding volumes

$$\frac{4}{3}\pi R^3 = N\frac{4}{3}\pi r_i^3,$$

then the expression (8.17) may be rewritten in the following form:

$$-U_c = \alpha_c^2 \frac{N^{4/3}e^2Z^2}{R^2} = \alpha_c^2 N\frac{e^2Z^2}{r_i^2}. \tag{8.18}$$

Hence, the form-factor entering the expression for the potential energy of the Coulomb interaction acquires the same physical meaning that it has in the expression for the potential energy of the gravitational interaction of the masses considered in Sect. 2.6. It represents the effective shell to which the charges in the sphere are reduced, i.e.

$$\alpha_c^2 = \frac{r_i^2}{r_{ei}^2}. \tag{8.19}$$

Taking into account that the moment of inertia of the body is

$$I = \frac{8\pi}{3}\int_0^R r^4 \rho(r)dr = \beta^2 mR^2, \tag{8.20}$$

then the relation (8.5) can be written in the form

$$-U_c\sqrt{I} = \alpha_c^2 N^{4/3}\frac{e^2Z^2}{R}\sqrt{\beta^2 mR^2} = \alpha_c^2\beta^2 N^{4/3}m^{1/2}e^2Z^2 = const. \tag{8.21}$$

The numerical values of the form-factors α_c^2 and β^2 calculated by Eq. (8.21) are given in our work (Ferronsky et al., 1981a).

The considered task about the potential energy of the Coulomb interactions of the charged particles proves the legitimacy of solution of the virial equation of dynamical equilibrium for study of the Earth's electromagnetic effects.

8.3 Emission of Electromagnetic Energy by a Celestial Body as an Electric Dipole

In Chap. 5 we considered the solution of the virial equation of dynamical equilibrium for dissipative systems written in the form

$$\ddot{\Phi} = -A_0\left[1 + q(t)\right] + \frac{B}{\sqrt{\Phi}}. \tag{8.22}$$

Here the function of the energy emission $[1 - q(t)]$ was accepted on the basis of the Stefan-Boltzmann law without explanation of the nature of the radiation process.

Now, after an analysis of the relationship between the potential energy and the polar moment of inertia, considered in the previous section, and taking into account the observed relationship by artificial satellites, we try to obtain the same relation for the celestial body as an oscillating electric dipole (Ferronsky et al., 1987).

Equation (8.22) for a celestial body as a dissipative system can be rewritten as

$$\ddot{\Phi} = -A_0 + \frac{B}{\sqrt{\Phi}} + X(t - t_0),$$

where $X(t - t_o)$ is the perturbation function sought, expressing the electromagnetic energy radiation of the body as

$$X(t - t_0) = E_\gamma(t - t_0).$$

The electromagnetic field formed by the body is described by Maxwell's equations, which can be derived from Einstein's equations written for the energy-momentum tensor of electromagnetic energy. In this case only the general property of the curvature tensor in the form of Bianchi's contracted identity is used. We recall briefly this derivation sketch (Misner et al., 1975).

Let us write Einstein's equation in geometric form:

$$G = 8\pi T, \tag{8.23}$$

where G is an Einstein tensor and T is an energy-momentum tensor.

In the absence of mass, the energy-momentum tensor of the electromagnetic field can be written in arbitrary co-ordinates in the

$$4\pi T^{\mu\nu} = F^{\mu\alpha} F^{\nu\beta} g_{\alpha\beta} - \frac{1}{4} g^{\mu\nu} F_{\sigma\tau} F^{\sigma\tau}, \tag{8.24}$$

where $g_{\alpha\beta}$ is the metric tensor in co-ordinates, and $F^{\mu\nu}$ the tensor of the electromagnetic field.

From Bianchi's identity

$$\nabla G \equiv 0, \tag{8.25}$$

where ∇ is a covariant 4-delta, follows the equation expressing the energy-momentum conservation law:

$$\nabla T \equiv 0. \tag{8.26}$$

In the component form, the equation is

$$F^{\mu\alpha}{}_{;\sigma} g_{\alpha\tau} F^{\sigma\tau} + F^{\mu\alpha}{}_{;\tau} g_{\alpha\sigma} F^{\tau\alpha} = g^{\mu\nu} \left(F_{\nu\tau;\sigma} + F_{\sigma\nu;\tau} \right) F^{\sigma\tau}. \tag{8.27}$$

After a series of simple transformations, we finally have

$$F^{\beta u}{}_{;\nu} = 0. \tag{8.28}$$

Here and above, the symbol ";" defines covariant differentiation.

To obtain the total power of the electromagnetic energy emitted by the body, Maxwell's equations should be solved, taking into account the motion of the charges constituting the body. In the general case, the expressions for the scalar and vector potentials are

$$4\pi\phi = \int\limits_{(V)} \frac{[\rho]dV}{R},$$ (8.29)

$$4\pi\bar{A} = \int\limits_{(V)} \frac{[j]dV}{R},$$ (8.30)

where ρ and j are densities of charge and current; $[j]$ denotes the retarding effect (i.e. the value of function j at the time moment $t - R/c$); R is the distance between the point of integration and that of observation, and c the velocity of light.

In this case, however, it seems more convenient to use the Hertz vector Z of the retarded dipole $p(t - R/c)$ (Tamm, 1976). The Hertz vector is defined as

$$4\pi Z = \frac{1}{R}\rho\left(t - \frac{R}{c}\right).$$ (8.31)

Electromagnetic field potentials of the Hertz dipole can be determined from the expressions

$$\phi = -divZ,$$ (8.32)

$$\bar{A} = \frac{1}{c}\frac{dZ}{dt}.$$ (8.33)

Moreover, the Hertz vector satisfies the equation

$$\Box Z \equiv \left(\nabla^2 - \frac{1}{c^2}\frac{\partial^2}{\partial t^2}\right)Z = 0,$$ (8.34)

where $\Box$ is the d'Alembertian operator.

The intensities of the electric and magnetic fields E and H are expressed in terms $\bar{Z}$ by means of the equations

$$\bar{H} = rot\dot{\bar{Z}},$$ (8.35)

$$\bar{E} = grad\ div\bar{Z} - \frac{1}{c}\ddot{\bar{Z}}.$$ (8.36)

The radiation of the system can be described with the help of the Hertz vector of the dipole $\bar{p} = q\bar{r}$, where q is the charge and r the distance of the vector from the charge $(+q)$ to $(-q)$.

From the sense of the retardation of the dipole $p(t - R/c)$ we can write the following relations:

$$\frac{d\bar{p}}{\partial R} = -\frac{1}{c}\dot{\bar{p}}; \qquad \frac{d^2\bar{p}}{dR^2} = \frac{1}{c}\ddot{\bar{p}}.$$

Then the components of the fields $\bar{E}$ and $\bar{H}$ of the dipole are as follows:

$$H_\varphi = \frac{\sin\theta}{c^2 R}\ddot{\bar{p}}\left(t - \frac{R}{c}\right), \qquad (8.37)$$

$$E_\theta = \frac{\sin\theta}{c^2 R}\ddot{\bar{p}}\left(t - \frac{R}{c}\right), \qquad (8.38)$$

where θ is the angle between p and $\bar{R}$; $H_\varphi \perp E_\theta$ and $\perp R$; the other components of E and H in the wave zone are tending to zero quicker than $1/R$ in the limit $R \rightarrow \infty$.

The flax of energy (per unit area) is equal to

$$S = \frac{c}{4\pi}E_\theta H_\varphi = \frac{1}{4\pi c^2}\frac{\sin\theta}{R^2}(\ddot{p})^2. \qquad (8.39)$$

The total energy radiated per unit time is given by

$$\oiint S d\sigma = \frac{2}{3c^3}(\ddot{p})^2. \qquad (8.40)$$

Thus, transforming the dissipative system to an electric dipole by means of the Hertz vector, we have reduced the task of a celestial body model construction to the determination of the dipole charges $+Q$ and $-Q$ through the effective parameters of the body.

This problem can be solved by equating expression (8.40) for the total radiation of a celestial body as an oscillating electric dipole. In addition, the relation for the black body radiation expressed through effective parameters is presented below in Sect. 8.5.

The expression (8.40) for the total rate of the electromagnetic radiation J of the electric dipole can be written in the form (Landau and Lifshitz, 1973b)

$$J = \frac{2}{3}\frac{Q^2}{c^3}(\ddot{r}), \qquad (8.41)$$

where Q is the absolute value of each of the dipole charges, and r is the vector distance between the polar charges of the dipole. Its length in our case is equal to the effective radius of the body.

In our elliptic motion model of the two equal masses the vector $\bar{r}$ satisfies the equation

$$\ddot{\bar{r}} = -Gm\frac{\bar{r}}{r^3}.\tag{8.42}$$

Thus, the total rate of the electromagnetic radiation of the dipole is

$$J = \frac{2}{3}\frac{Q^2}{c^3}\frac{(Gm)^2}{r^4}.\tag{8.43}$$

In order to obtain the average flux of electromagnetic energy radiation, the calculated value of the factor $1/r^4$ should be averaged during the time period of one oscillation. Using the angular momentum conservation law, we can replace the time-averaging by angular averaging, taking into consideration that

$$dt = \frac{mr^2}{2M}d\varphi,\tag{8.44}$$

where M is angular momentum, and φ is the polar angle.

The equation of the elliptical motion is

$$\frac{1}{r} = \frac{1}{a\left(1-\varepsilon^2\right)}\left(1+\varepsilon\cos\varphi\right),\tag{8.45}$$

where a is the semi-major axis, and ε is the eccentricity of the elliptical orbit.

The value of $1/r^4$ can be found by integration. In our case of small eccentricities, we neglect the value of ε^2 and write

$$\overline{\left(\frac{1}{r^4}\right)} = \frac{1}{a^4}.\tag{8.46}$$

Finally we obtain

$$\bar{J} = \frac{2}{3}\frac{Q^2}{c^3}\frac{Gm^2}{a^4}.\tag{8.47}$$

Earlier it was shown (Ferronsky et al., 1987) that

$$\bar{J} = 4\pi\sigma\frac{1}{a^2}A_c^4,\tag{8.48}$$

where σ is the Stefan-Boltzmann constant; $A_e = Gm\mu_e/3k$ is the electron branch constant; μ_e is the electron mass; and k is the Boltzmann constant.

Equating relations (8.47) and (8.48), we find an expression for the effective charge Q as follows:

$$Q = \sqrt{6\pi\sigma}\frac{A_e^2}{cr_g},\tag{8.49}$$

where $r_g = Gm/c^2$ is the gravitation radius of the body.

We have thus demonstrated that it is possible to construct a simple model of the radiation emitted by a celestial body, using only the effective radius and the charge of the body. Moreover, a practical method was shown for determining the effective charge using the body's temperature from observed data.

The logical question about mechanism of generation of the wide spectrum energy emitted by the bodies arises. Let us consider this important problem at least in first approximation.

8.4 Quantum Effects of Generated Electromagnetic Energy

The problem of energy generation technology for human practical use has been solved long ago. In the beginning it was understood how to transfer the energy from wind and fire into the energy of translational and rotary motion. Later on scientists and engineers learned how to produce electric and atomic energy. Technology of thermo-nuclear fusion energy generation is the next step. It is assumed that the Sun replenishes its emitted energy by thermo-nuclear fusion of hydrogen, helium and carbon. The Earth's thermal energy loss is considered to be filled up by convection of its masses and thermal conductivity. But the source of energy for convection of the masses is not known.

The obtained solution of the problem of volumetric pulsations for a self-gravitating body based on their dynamical equilibrium creates a real physical basis on which to formulate and solve the problem. In fact, if a body performs gravitational pulsations (mechanical motions of masses) with strict parameters of contraction and expansion of any arbitrarily small volume of the mass, then such a body, like a quantum generator, should generate electromagnetic energy by means of its transformation from a mechanical form, through forced energy level transitions and their inversion, on both the atomic and nuclear levels. In short, the considered process represents transfer of mechanical energy of mass pulsation to the energy of an electromagnetic field (Fig. 8.1).

An interpretation of the process can be presented as follows. While pulsating and acting in a regime of the quantum generator, the body should generate and emit coherent electromagnetic radiation. Its intensity and wave spectrum should depend

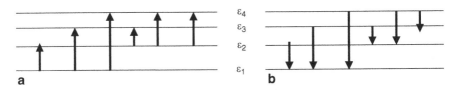

Fig. 8.1 Quantum transition of energy levels at contraction phase of the body mass (**a**) and inversion at the phase of its expansion (**b**); ε_1, ε_2, ε_3, ε_4 are levels of energy

on the body mass, its radial density distribution and chemical (atomic) content. As it was shown in Sect. 6.4, a body with uniform density and atomic content provides pulsations of uniform frequency within the entire volume. In this case, the energy generated during the contraction phase will be completely absorbed at the expansion phase. The radiation appearing at the body's boundary surface must be in equilibrium with the outer flux of radiation. A phenomenon like this seems to be characteristic for the equilibrated galaxy nebulae and for the Earth's water vapor in anti-cyclonic atmosphere.

The pulsation frequencies of the shell-structured bodies are different but steady for each shell density. In the case of density increase to the body's center, the radiation generated at the contraction phase will be partially absorbed by an over-lying stratum at the expansion phase. The other part of radiation will be summed up and transferred to the body's surface. That part of the radiation forms an outer electromagnetic field and is equilibrated by interaction with the outer radiation flux. The rest of the non-equilibrated and more energetic particles in the spectrum of radiation moves to space. The coherent radiation which reaches the boundary surface has a pertinent potential and wave spectrum depending on mass and atomic content of the interacting shells in accordance with Moseley's law. The Earth emits infrared thermal radiation in an optical short wave range of the spectrum. The Sun and other stars cover the spectrum of electromagnetic radiation from radio- through optical, x and gamma rays of wave ranges. The observed spectra of star radiation show that the total mass of a body takes part in generation and formation of surface radiation. According to the accepted parabolic law of density distribution of the Earth, it has maximum density value near the lower mantle boundary. The value of the outer core density has jump-like fall and the inner core density seems to be uniform up to the body's center. The mechanism of energy generation that we have discussed is justified by the observed seismic data of density distribution. It is assumed that the excess of electromagnetic energy generated from the outer core comes to the inner core and keeps there the pressure of dynamical equilibrium at the body's pulsation during the entire time of the evolution. The parabolic distribution of density seems to be characteristic for most of the celestial bodies.

In connection with the discussed problem it is worth considering the equilibrium conditions between radiation and matter on the body's boundary surface.

8.5 Equilibrium Conditions on the Body's Boundary Surface

Any conclusion about the quantum electromagnetic nature of energy in maintaining equilibrium of a self-gravitating body needs to analyze the conditions on its boundary surface. Electromagnetic energy has discrete-wave structure, therefore we shall search for the particle mass which satisfies equilibrium between the matter and the radiation on the border with the outer flux of energy. This question was considered earlier (Ferronsky et al., 1987; 1996) and here we continue the analysis.

We can write the equation of hydrodynamic equilibrium for the flux of particles that are blocked by gravity forces on the body's boundary surface

$$\frac{GM\mu}{R^2} = \frac{\mu v^2}{R},$$ (8.50)

where μ is the mean value of the particle mass; v is the velocity of thermal motion of the particle running from the gravitational field.

The heat velocity v depends on the boundary temperature T_0 as

$$\mu v^2 = 3kT_0,$$ (8.51)

where k is the Boltzmann constant.

Now using the equilibrium condition for retaining particle flow can be rewritten as

$$\frac{GM\mu}{3k} = T_0 R.$$ (8.52)

By means of Eq. (8.52) we can calculate numerical values of particle masses, which are "locked" by the gravitational forces at different stages of the system's evolution. With the help of the same equation we identified proton $T_p R_p$ and electron $T_e R_e$ phases in the evolution history of proto-solar nebula (Ferronsky et al., 1987). They are as follows:

$$T_p R_p = GM\mu_p/3k = 5 \times 10^{17} \text{cm} \cdot K,$$
$$T_e R_e = GM\mu_e/3k = 2.73 \times 10^{14} \text{cm} \cdot K.$$

It means, that for the contemporary Sun with radius $R = 7 \times 10^{10}$ cm and surface temperature $T_0 = 5000$ K the boundary equilibrium is maintained by the mass of the electrons. At the same temperature of the system its boundary equilibrium is "locked" by the mass of the protons on radius $R = 7 \times 10^{14}$ cm, which corresponds to the orbit of Jupiter.

Thus, we may assume that evolution of the proto-solar nebula has passed at least two phases. First was the proton phase AB (Fig. 8.2) when the planetary mass differentiation and separation took place. The second electron phase CD is characteristic for the present-day equilibrium state when the Sun, having a non-uniform or even a shell structure, is pulsating in the regime of an optic quantum generator. The process of mass density differentiation in the interior is definitely continuing and the light mass component is emerging on the surface in the form of so-called sunspots.

A more general consideration of the problem related to the particle mass, which determines radiation equilibrium on the boundary surface of a body at its separation from a gaseous nebula, can be obtained from solution of the equation suggested by Chandrasekhar and Fermi (1953).

Fig. 8.2 Proton AB and electron CD phases of the Sun's evolution

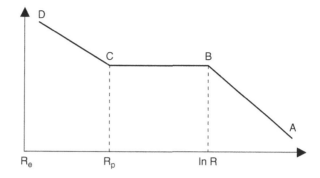

8.6   Solution of the Chandrasekhar-Fermi Equation

The following equation for a self-gravitating body, which is separated from gaseous nebula as a result of equilibrium between gravitational and electromagnetic forces, was proposed by Chandrasekhar and Fermi (1953)

$$\int\limits_{(V)}\left[\rho v^2 + 3p + \frac{H^2 + E^2}{8\pi} - \frac{(\nabla U)^2}{8\pi G}\right]dV = 0; \qquad (8.53)$$

here ρ is the mean density of the body; p is the pressure; v is the mean velocity of the particle; H and E are the components of the electromagnetic field; G is the gravitational constant; V is the volume of the system; ∇U is the gradient of the gravitational field.

At the moment of separation of the self-gravitating body, its kinetic effects compared to the mass term are small. In this case Eq. (8.53) can be written as (Ferronsky et al., 1981b; 1996)

$$\int\limits_{(V)}\left[3p - \frac{(\nabla U)^2}{8\pi G}\right]dV \approx 0,$$

or

$$\int\limits_{(V)} 3p\,dV \approx 0.1\frac{GM^2}{R}, \qquad (8.54)$$

where the coefficient 0.1 represents the electromagnetic component in expansion of the potential energy (8.1) found by means of data of the astronomical observations of equilibrated nebulae (Ferronsky et al., 1996).

The left-hand side of equation (8.54) is proportional to the Coulomb energy of interaction of the charged particles (electrons, protons, ionized atoms and molecules). The right-hand side of the equation is proportional to the gravitational energy of the mass particles' interaction. Then, assuming that the separated secondary body has mass m, radius R and mean value of mass of the constituting particles μ, equation (8.54) can be rewritten in the form of equality of energies of the gravitational and Coulomb interactions. For the purpose of estimating the order of magnitude we rewrote the expression for the energy of the Coulomb interaction in equation (8.54), through the Madelung energy, in the form

$$0.1\frac{GM^2}{R} \approx \frac{M}{\mu}\frac{e^2}{R\sqrt[3]{\mu/M}}. \tag{8.55}$$

where $e = 4.8 \times 10^{10}$ e.s.u., the electron charge. Then from equation (8.55) one has

$$M\mu^2 = 2 \times 10^{-16}\text{g}^3. \tag{8.56}$$

Eq. (8.56) allows estimation of the mass of the equilibrated particle of the solar nebula at the time of its separation from the galactic one, which is

$$\mu_s = (2 \times 10^{16}/M_c)^{1/2} = 3 \times 10^{-24}\text{g}.$$

For the Earth the mass of the particle is $\mu_e = (2 \times 10^{16}/M_e)^{1/2} = 1.8 \times 10^{-22}$ g, for Jupiter it is $\mu_j = 10^{-23}$ g and for the Universe, taking its mass to be $M \sim 10^{56}$ g, one has that $\mu_g = 10^{-36}$ g.

The maximal average mass of particles in cosmic space can be determined from the condition $\mu = M$. Then $\mu_{Max} = \sqrt[3]{2 \times 10^{-16}} = 0.6 \times 10^{-5}$ g. The obtained value is close to the Planck mass or the mass of the "cold" plasma.

8.7 The Nature of the Star Emitted Radiation Spectrum

We assume that the Novas and Supernovas after explosion and collapse fracture into neutron stars, white dwarfs, quasars, black holes and other exotic creatures which emit electromagnetic radiation in different ranges of the wave spectrum. The effects discussed in the book are based on dynamical equilibrium evolution of self-gravitating celestial bodies, allowing the exotic stars to be interpreted from a new position. We consider the observed explosions of stars as a natural logical step of evolution related to their mass differentiation with respect to the density. The process is completed by separation of the upper "light" shell. At the same time the wave parameters of the generated energy of the star after shell separation are changed because of changes in density and atomic contents. As a result, the frequency intensity and spectrum of the coherent electromagnetic radiation on the boundary surface are changing. For example, instead of radiation in the optical range, coherent emission is observed in the x or gamma ray range. But the body's dynamical equilibrium should remain during all the time of evolution. The loss of the upper body shell leads to decrease of the angular velocity and increase of the oscillation frequency.

The idea of a star's gravitational collapse seems to be an effect of the hydrostatic equilibrium theory.

As to the high temperature on the body surface, the order from the Rayleigh-Jeans equation is 10^7 K or more. In our interpretation wherein we apply Eq. (8.52) for evolution of a star of solar mass at the electron phase (Fig. 8.2), the limiting temperate $T_0 \rightarrow \mu_e c^2/3k$ or (Ferronsky et al., 1996)

$$3kT_0 \rightarrow \mu_e c^2 \approx 0.5\,MeV,$$
$$T \approx 5 \times 10^9\,K.$$

This means that on the body's surface the temperature of the gas approaches that of the electrons because the velocity of its oscillating motion runs to c.

The total energy is a quantitative measure of interaction and motion of all the forms of the matter being studied. In accordance with the law of conservation, energy neither disappears nor appears of its own volition. It only passes from one form to another. For a self-gravitating body the energy of mechanical oscillations, induced by gravitational interactions, passes to electromagnetic energy of the radiation emission and vice versa. The process results from the induced quantum transition of the energy levels and their inversion. Here transition of the gravitational energy into electromagnetic energy, and vice versa, results in a self-oscillating regime. In the outer space of the body's border, the emitted radiation energy forms an equilibrium electromagnetic field. The non-equilibrium part of the energy in the corresponding wave range of the spectrum is irradiated to outer space. The irreversible loss of the emitted energy is compensated by means of the binding energy (mass defect) at the fission and fusion of molecules, atoms and nuclei. The body works in the regime of a quantum generator. Those are conclusions that follow from the theory based on a body's dynamical equilibrium.

In conclusion we wish to stress that discovery of the artificial satellites' relationship between the gravitational field (potential energy) and the polar moment of inertia of the Earth leads to understanding the nature of the mechanism of the planet's energy generation as the force function of all the dynamical processes released in the form of oscillation and rotation of matter. Through the nature of energy we understand the unity of forms of gravitational and electromagnetic interactions which, in fact, are the two sides of the same natural effect.

Chapter 9
Observable Facts Related to Creation and Evolution of the Earth

During the second half of the twentieth century, many advances have been made in the field of cosmogenics. New facts and rich general information in physics of space near the Earth, the Moon, and nearby planets Mars, Venus, Mercury, and Jupiter were obtained by means of cosmic apparatus and techniques. Lunar rock samples brought back to the Earth by astronauts and automatic space devices have provided much data on the physical-mechanical properties of those rocks and on that basis a number of new cosmophysical and cosmochemical conclusions have been drawn. Due to new techniques available for studying natural substances, much knowledge was gained on the chemical and isotopic composition and the absolute ages of meteorites, lunar and terrestrial rocks, water, and gases.

The programme of deep sea drilling and ultra-deep inland drilling initiated the most important studies of the century concerning the structure of the ocean and the inland crust, and have already provided many important facts.

Finally, the most valuable studies of dynamical properties of the Earth's gravitational field were conducted by artificial geodetic satellites. All those results and the novel approach to dynamics of the Earth considered in this work, allow us to return with added insight to discussion of the oldest scientific problem of creation and evolution of our planet.

9.1 A Selection of Existing Approaches to Solution of the Problem

Let us briefly review the existing hypotheses that are available for research into the nature of the Solar system. At present the most common theories of formation of the Earth and the other planets of the Solar system are based upon the idea of accretion of substances from a proto-planetary gas-dust cloud. There are several options on the origin of the cloud itself. Some scientists believe that the proto-planetary cloud was captured from interstellar nebula by the existing Sun during its motion through interstellar space (Schmidt, 1957). Others consider it to be the product of evolution

V. I. Ferronsky, S. V. Ferronsky, *Dynamics of the Earth,*
DOI 10.1007/978-90-481-8723-2_9, © Springer Science+Business Media B.V. 2010

of a more massive cloud from which the Sun itself originated (Cameron and Pine, 1973; Cameron, 1973).

Before the 'accretion' theory was advanced, there were a lot of other ideas and theories concerning the origin of the solar system which have been discussed repeatedly in great detail. (Spenser, 1956). But the majority of these theories were rejected, being unable to explain a number of observed astronomical facts and aspects of celestial mechanics and cosmochemistry. However, in the light of effects of the dynamic equilibrium approach presented in this book, some of them, such as the theories of Laplace and Descartes deserve rehabilitation.

According to the accretion hypothesis, the planets and satellites of the solar system were formed as a result of successive accumulation of dust particles. The subsequent growth of these accumulations into planetary and satellite bodies has been considered on the basis of the mechanism of gravitational instability and the amalgamation of small bodies with larger ones.

The majority of geological and geochemical reconstructions concerning the formation of the Earth and its shells, based on the accretion hypothesis, postulate the initial existence of a chemically homogeneous substance. It is assumed that at the initial stage of the Earth's formation the temperature of the condensing matter was low and water was retained, along with carbon, nitrogen, and other volatile constituents (Urey, 1957). Subsequently, due to the gravitational heat generated during accretion and the energy produced by radioactive decay, the matter constituting the Earth underwent complete or partial melting. During the melting of this matter, the water and the volatile components degassed to the surface producing the hydrosphere and the atmosphere.

Rubey (1964) notes that the CO_2 concentration in the atmosphere and the oceans remains more or less constant over a considerable period of geologic time. In his opinion, if only $1/100$ of the carbonates in sedimentary rocks were transferred to the oceans and the atmosphere, life on Earth would not be possible. From analytical computations based on the geochemical balance of the volatile substances contained in old sedimentary rocks and in the present hydrosphere, atmosphere, and biosphere on the one hand, and in disintegrated igneous rocks on the other, he obtained the very interesting data shown in Table 9.1.

As can be seen from Table 9.1, there is an enormous excess of volatile substances in the surface zone of the Earth which could not have been produced by the disintegration of igneous rock material and its subsequent transfer into sediments and solutions. Analyzing possible reasons for the excess, Rubey concludes that the present-day oceans formed gradually through the emergence of water together with other volatile substances from the deep interior of the Earth or, more specifically, from what he called hydrothermal sources. Therefore, in Rubey's opinion, the present hydrosphere is of juvenile origin.

Although satisfactory explanations exist for a number of observed facts in the framework of the cold origin hypothesis, many authors have noted certain fundamental contradictions. For example, Ringwood (1966; 1979) has drawn attention to the extremely high concentrations of iron and nickel oxides in ultramafic basic

Table 9.1 The content of volatile components near the surface of the earth and in fractured rocks (Rubey, 1964)

Object	Volatiles ($\times 10^{20}$ g)					
	H_2O	Carbon in CO_2	CI	N	S	H, B, Br and others
Present atmosphere, hydrosphere, biosphere	14 600	1.5	276	39	13	1.7
Ancient sedimentary rocks	2100	920	30	4	15	15
Total	16 700	921.5	306	43	28	16.7
Fractured igneous rocks	130	11	5	0.6	6	3.5
Excess of volatile components	16 570	910.5	301	42.4	22	13.2

rocks and basalts and to very large amounts of CO and CO_2 which would have been released through the reduction of the iron constituting the planet's core. If the theory were correct, these gases would have had a mass half that of the entire core, which is hardly possible. The mechanism of formation of the core and the other shells of the Earth also remains problematic.

Cosmochemical facts, which have been obtained in recent years by many researchers studying meteorites, the Moon, and the planets, indicate that other ideas concerning the formation of the bodies in the solar system should be employed. These new ideas have shown that the processes of condensation of the chemical elements and compounds that might have occurred in the protoplanetary substance have actually taken place. Due to these processes the bodies of the solar system could have become inhomogeneous in chemical composition and differentiated into shells. The development of these ideas is closely related to the problem of the original formation and evolution of the Earth and its shells and are therefore considered in detail below, including the effects following from our theory of the planet's motion based on its dynamical equilibrium. We consider that the main effect of planet body formation is differentiation of the cloud's mass in density and their separation due to development of the normal and the tangential components of the force function at gravitational interaction of the non-uniform masses as discussed in Chap. 6.

The fact of existence of gaseous clouds and nebulae is proved by observation. The evolutionary process inevitably leads to their density separation. The study of the isotopic composition of a substance is one of the most efficient ways of learning about the origin of that substance. We carried out analysis of vast experimental data on isotopic composition of the hydrogen and oxygen of water on the Earth in order to solve the problem of origin of the hydrosphere by means of its separation during formation. Analogous analysis was done on the basis of isotopic composition of the carbon and sulfur in natural objects of organic and inorganic origin. The results prove the fact of the separation process which has taken place in accordance with the chosen theory. We present our analysis below (Ferronsky, 1974; Ferronsky and Polyakov, 1982; 1983).

9.2   Separation of Hydrogen and Oxygen Isotopes
in Natural Objects

At present we definitely know the physical ground for the separation of isotopes, and much experimental data concerning the isotope composition of water in different objects has been obtained. These objects are: atmospheric moisture, water of the oceans and their sediments, inland surface and ground waters, liquid inclusions in rocks and minerals, deuterium and oxygen-18 contents in sedimentary and igneous rocks and minerals, in meteorites, and in lunar samples. The available data permit us to discuss the problem of origin of the Earth's hydrosphere as the planet's shell.

When considering the evolution of the isotopic composition of any element through what is assumed to be a wide range of temperatures, the isotopic separation can be expected to occur only within strictly defined temperature and pressure limits corresponding to the phase transition of the element to and from the gaseous, liquid and solid state. The isotopes of a given element in the gaseous state will be distributed uniformly throughout the gas volume. If, on the other hand, an element in the gas phase reacts with other substances on reaching an appropriate temperature for the formation of new compounds constituting a condensed phase, then the separation effect will begin to manifest itself as a depletion of the gas phase in these same isotopes. The transition of the compounds in question to each successive phase state will be accompanied by enrichment of the new phase in heavy isotopes.

In a single-component system, the isotopic composition of the condensed and gas phases under equilibrium conditions is determined by the vapour pressure of the isotopic varieties of the molecules forming the condensed phase, and the enrichment of the low-temperature phase in heavy isotopes occurring within a certain temperature range. In a multi-component system, where the element of interest is present in various chemical compounds, the isotopic composition of each compound will be determined by the isotopic exchange constants, which are in turn a function of the free energy of the system.

Close examination of this process from the thermodynamic point of view shows that, with increasing system temperature, isotopic separation among the various compounds decreases. Under certain equilibrium conditions the system components or phase will be depleted or enriched in one or another isotope according to the isotopic exchange constant, and as the isotopic exchange constant is a function of temperature, changes in temperature must lead to change in isotopic composition of the components or phases through isotopic exchange reactions. For example, CO_2 gas, being in isotopic equilibrium with water in the liquid phase, will be enriched in ^{18}O by approximately 40% relative to the oxygen constituting H_2O molecules at 25°C.

It should be borne in mind that during the Earth's formation and subsequent evolution, isotopic separation of the elements also occurred as a result of many secondary effects associated with global and local changes in temperature due to internal and external release or absorption of energy. Global changes in the isotopic composition of water through the secondary effects associated with variation in the temperature regime are very pronounced when there is interaction between the open water basins of the Earth and the atmosphere in the form of evaporation, condensation and precipitation. The same effects are observed locally in closed systems where underground waters interact with rocks and where individual minerals and other components with water during the formation of rocks of different origin.

The temperature range within which the isotopes of water undergo detectable fractionation during the phase transition of water is −40 to +374°C. Within 220–374°C the separation factor of hydrogen isotopes becomes less than 1. At higher temperatures, the various isotopes of water in a given gas volume will have a statistically uniform distribution. The separation of hydrogen and oxygen isotopes in other compounds during their formation, and under the influence of secondary temperature effects, will occur within other temperature ranges. It is known, for example, that when oxygen interacts with the silicate phase of rocks, isotopic separation occurs up to a temperature of 1000°C (Ferronsky and Polyakov, 1982). If one knows exactly the temperature induced separation relationship between two interacting components from the actual isotopic ratios, it is possible to calculate exactly their interaction temperature. This is the principle underlying oxygen paleothermometry, which is used for determining the formation temperature of sedimentary rocks, glaciers, and other terrestrial features.

Many papers containing data on the isotopic composition of water and other natural and cosmic substances have been published. The authors of the book have analysed many of these with a view to determining the limits of variation in the concentration of the heavy isotopes deuterium and oxygen-18 in various substances. The results are presented in Table 9.2, where the isotopic values are given in units relative to the mean ocean water standard SMOW (see Ferronsky and Polyakov, 1983).

From Table 9.2 it can be seen that the isotopic composition of the surface waters of the continents and of the Earth's atmosphere, which is determined exclusively by the present temperature regime of the Earth's surface, varies within broad limits. As one would expect, the water most depleted in heavy isotopes is fixed in the polar regions (δ^2H up to −500‰, $\delta^{18}O$ up to −60‰ (Craig, 1963)), while that most enriched is in closed basins in arid zones ($\delta^2H = +129‰$, $\delta^{18}O = +31.3‰$ (Fontes and Gonfiantini, 1967)). It should be noted that the isotopic composition of surface waters, atmospheric moisture, and gases is controlled by secondary effects of the continuous, natural evaporation-condensation-precipitation cycle. All these processes, which occur at widely varying temperatures at the Earth's surface and in the atmosphere, cause fractionation of hydrogen and oxygen isotopes within broad limits.

Ocean water is uniform in its isotopic composition. Investigations carried out by many authors (Craig, 1965; Craig and Gordon, 1965) have shown that at depth of 500 m or more ocean water has a very uniform isotopic composition with small regional variation (δ^2H from −25 to +10‰ and $\delta^{18}O$ from 0.6 to +2‰) observed in the surface layer. Despite the enormous amount of evaporation from the ocean surface, which enriches the surface layer in heavy isotopes, there does not appear to be any appreciable change in the isotopic composition as a whole. This is because most of the evaporated water ultimately returns to the ocean. Such a situation can obviously

Table 9.2 Deuterium and oxygen-18 contents in natural objects

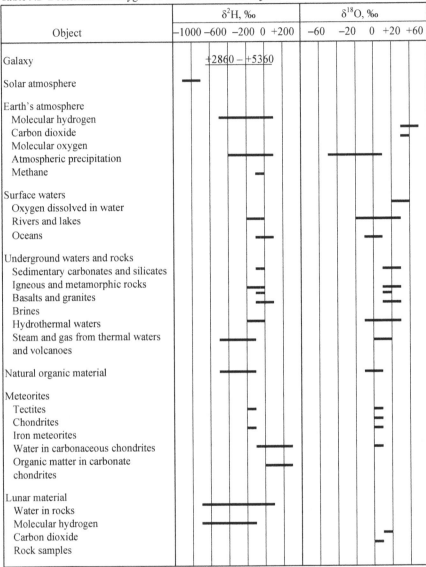

persist as long as the overall temperature balance on the Earth is maintained. If this balance is upset over a long time, the balance of isotopic composition in ocean water will break down. However, as has been shown by recent paleotemperature measurements of old ocean sediments and glaciers based on $^{18}O/^{16}O$ (Emiliani, 1970; 1978), temperature variations at the ocean surface in the equatorial zone have not exceeded 5–6°C during the past 0.7 million years, and the $^{18}O/^{16}O$ ratios in the water have not deviated by more than 0.5‰ from the present-day value. Similar results have been obtained by determining deuterium in water contained in clay minerals of marine origin and of various ages.

The isotopic and chemical composition of the ocean water may therefore be considered to have remained virtually constant over the past 250–300 million years. This is confirmed by a number of facts obtained during paleotemperature studies based on the comparison of the oxygen isotope ratios of organogenic carbonates and shells of ancient and modern molluscs (Bowen, 1966; 1991; Teys and Naydin, 1973).

Taylor (1974) came to a similar conclusion after analyzing data on the hydrogen and oxygen isotope ratios in Cambrian and Precambrian siliceous charts obtained by Knauth and Epstein (1971; 1976). Moreover, he has suggested that the isotopic composition of the ocean water has remained unchanged, at least during the majority of the upper Precambrian era.

Similar results were derived while analyzing the carbon and oxygen isotope composition of Precambrian limestones and dolomites in Africa, Canada, and Europe (Schidlowski et. al., 1975). It was found that the near-constancy of $\delta^{13}C \approx 0$ during geologic time corresponds to the constant ratio of the organic carbon to the total amount of carbon in sedimentary rocks ($C_o/C_t \approx 0.2$). On this basis it has been concluded that at least 80% of the modern oxygen isotopes were formed earlier than 3×10^9 years ago.

A very important question concerns the isotopic composition of deep underground waters and also the δ^2H and $\delta^{18}O$ composition of liquid inclusions in minerals of magmatic origin and in rocks belonging to the Earth's upper mantle. The data available concerning studies of isotopic ratios in various types of mineralized underground waters and brines in numerous regions of the United States, Canada, Japan, and the former Soviet Union with oil and gas fields, were considered in detail and shows that the relative deuterium concentration in such waters varies from +29 to −109‰, with a clear tendency to depletion in heavy isotopes relative to ocean water.

It has been shown by many researchers that the water-bearing carbonate and silicate rocks are always enriched in heavy isotopes of oxygen. Oxygen exchange occurred through the interaction of deep underground water (depleted in heavy isotopes as a result of mixing with local atmospheric precipitation in the course of their formation) with water-bearing carbonates and silicates at appropriate temperatures. This process led to some enrichment of the brines in heavy oxygen compared with local precipitation, and sometimes compared with ocean water.

It may be assumed that one of the secondary temperature effects leading to the enrichment of deep underground waters in deuterium and oxygen-18 was that of the surface evaporation of water of these basins being included in inland seas and lakes.

The range of variation in the isotopic composition of the Earth's thermal waters is from −18‰ to −207‰ for δ^2H, and from +7.5‰ to −22.5‰ for $\delta^{18}O$. The corresponding values for steam and gas from springs and volcanoes are from −40‰ to −520‰ for δ^2H and from +0.7‰ to 22.5‰ for $\delta^{18}O$. Rubey (1964), and a number of other authors, think that those places where thermal waters emerge at the Earth's surface (from crystalline rocks and in active volcanic regions) are where juvenile water emerges, and mark the sources which gave rise to our present oceans. Craig (1963) has made a detailed analysis of the isotopic composition of water from hot springs in many parts of the Earth. His investigations have shown that all hot springs discharge water having an isotopic composition identical with that of the local atmospheric precipitation. The widely observed higher enrichment is explained by oxygen substitution, which takes place when these waters interact with water-bearing carbonate and silicate rocks.

Measurement of tritium concentrations in thermal waters performed by Theodorsson (1967) in Iceland, Begemann (1963) in the United States, and other investigators, have yielded values that are absolutely identical with the tritium concentrations in local surface waters and rainfall. Because tritium is an isotope formed primarily by cosmic rays and concentrations can only be increased during nuclear bomb tests, it is impossible that any tritium could have been derived from underground sources. In any case, the circulation period of waters to hot springs fed by atmospheric and surface waters ranges from several months to several years. Thus, the above mentioned investigations suggest that hot springs do not supply juvenile water.

Let us consider the δ^2H and $\delta^{18}O$ variation in sedimentary rocks, liquid inclusions, and minerals in magmatic rocks, granites, and basalts. The limits of variations of these isotopes are shown in Table 9.2. It can be seen that for all the types of rocks in question the isotopic ratios of hydrogen are negative relative to ocean water, whereas the oxygen values for all the rock types listed are positive relative to ocean water. The oxygen isotope ratios range from 6–7‰ for basalts and 7–12‰ for granites and igneous rocks. In younger carbonate and silicate sedimentary rocks (no older than Riphean) the relative content of heavy oxygen ranges from +25‰ to +38‰, whereas in the ancient (Proterozoic, Archean, and older) carbonates, clay, and siliceous shales it is much lower: from +7 to +25‰. The global trend of the heavy oxygen content to decrease over time is interpreted in different ways.

Weber (1965), who examined more than 600 samples of ancient and modern calcites and dolomites from various geographical regions of the world, reported that the gradual increase of the oxygen-18 content in younger carbonate rocks is the result of enrichment of the ocean water in heavy oxygen due to discharges of juvenile water during the growth of the ocean.

Silverman (1951) has reported that if the primary ocean was created by the emergence of juvenile waters, its water should be enriched in oxygen-18 compared with modern ocean water. The result of the melting of rocks of the Earth's deep interior and the degassing of water in equilibrium with large masses of the mantle silicates, should be enrichment in oxygen-18 by at least +7‰.

Craig (1963) has also considered that if primary oceans were formed due to emergency of the juvenile waters, then it should exhibit a relative $\delta^{18}O$ content of about +7‰. The subsequent depletion of the ocean water in ^{18}O could have occurred as a result of the accumulation of carbonate and silicate sedimentary rocks enriched in heavy oxygen compared with the original igneous rocks. Craig has assumed that the primary calcium carbonates and silicates precipitate from the solution with relative ^{18}O contents up to $+30‰$. The relative content of heavy oxygen in sedimentary rocks decreases on average to $+20‰$ in the course of isotope exchange with fresh waters. Since the primary igneous waters had on average ^{18}O content of about $+10‰$ before their destruction, then during the formation of sedimentary rocks a continuous depletion of the ocean in heavy oxygen should occur.

Migdisov et al. (1974) have pointed to the importance of such processes as the volcanic activity of the Earth in the evolution of the isotopic composition of the ocean. Such activity was accompanied by increases in the weathering of the igneous rocks and the sedimentation of the weathering products. In addition, the volcanic activity is accompanied by the emergency of juvenile water and CO_2, leading to the enrichment of the ocean in ^{18}O. A similar effect is produced during the process of the deep metamorphism of sedimentary rocks accompanied with the emergence, to the ocean, of water and CO_2 enriched in heavy oxygen.

It should be pointed out that the volcanic activity, and related processes of rock sedimentation that may have taken place in certain geologic epochs, could not have markedly affected the ^{18}O change in the ocean because of the great differences between the total masses of oxygen in the ocean water and in sedimentary rocks formed during volcanic activity. Savin and Epstein (1970) have evaluated the global effect of the sedimentary process upon changes in the isotope composition of ocean water during the whole observed history of the Earth. They found that all the accumulated sedimentary rocks (pelagic sediments, carbonates, sandstones, and clays) could have resulted in the depletion of the hydrosphere in oxygen-18 by $3‰$ and during the whole post-Precambrian period by only $0.6‰$. As for deuterium, the ocean could have been enriched with it during the whole history in the course of sedimentation by less than $0.3‰$, i.e. less than the errors in the measurements.

Considering the possible periodic changes in the isotopic composition of the ocean water, it should be noted that there are more effective processes than the sedimentation of rocks during volcanic activities. Those include temperature variations on the Earth over time and the associated accumulation and melting of ice in glacial and non-glacial times. According to the data of Craig (1963), only in the Pleistocene have ^{18}O variation in the ocean amounted to $1‰$, and $7‰$ for deuterium; in the earlier epochs these variations could have been even greater.

The contribution of juvenile water to the oceans during volcanic activity has not as yet been confirmed experimentally. Some researchers (Arnason and Sigurgeirsson, 1967; Friedman, 1967; Kokubu et al., 1961), while studying the water isotope composition of volcanic lavas and gases, do not consider this water to be juvenile. Only Kokubu et al. (1961), analyzing liquid inclusions in samples of basalts in Japan, have concluded that they dealt with water of magmatic origin. Later, Craig showed that in terms of the hydrogen and oxygen isotope ratios this water lies on the curve for local atmospheric precipitation and, therefore, is of meteoric origin.

The deuterium content in waters of the volcanic lavas and gases varies over wide ranges. According to data obtained by Friedman (1967), who carried out investigations during volcanic eruptions on the Hawaii Islands in 1959–1960, the deuterium content in water of lava samples at the top of the volcano ranges from -55 to $-79‰$ and in samples taken from its slopes from -57 to $-91‰$.

Arnason and Sigurgeirsson (1967) studied the isotopic composition of steam and gases in the eruption of one of the Iceland volcanoes during 3.5 years (1964–1967). Their data are close to those obtained by Friedman and indicate an average deuterium content of −55.3‰.

Sheppard and Epstein (1970), while studying the oldest (to 1.14 billion years) samples of rocks and minerals (peridotite, olivine, dunite, mylonite, kimberlite, etc.), which, in their opinion, are representative of rocks of the upper mantle of the lower part of the Earth's crust, found an average value of deuterium of −58 ± 18‰. On the basis of the obtained data they concluded that the hydrogen isotope fractionation between phlogopites and water can amount to about 10‰ at 700°C. In view of this fact the $^2H/^1H$ ratio for juvenile water should be −48 ± 20‰ relative to the standard SMOW.

Craig and Lupton (1976) studied the hydrogen isotope composition of basalts sampled from the mid-oceanic ridges of the Atlantic and Pacific Oceans. It was found that the studied samples of deep water tollite exhibit $\delta^2H = -77‰$. On the basis of analysis of the obtained data together with results of neon and helium isotope composition studies, the authors have concluded that the obtained δ^2H value may possibly reflect the juvenile hydrogen isotope composition.

A wide range in the deuterium content of terrestrial materials is found in various organic compounds (from −12‰ to −430‰). The corresponding range of oxygen-18 concentrations is from +16‰ to −5‰. This can only be attributed to great ranges of hydrogen and oxygen isotope fractionation during the oxidation-reduction processes, leading to the formation and evolution of natural organic compounds (Schiegl and Vogel, 1970; Ester and Hoering, 1980).

It is worth noting that oil in deposits of different origin and age has a stable deuterium concentration of about −100‰ relative to the standard. The natural photosynthetic process, despite its importance in the geochemistry of the upper terrestrial shells, could not have had a marked effect upon the evolution of hydrogen and oxygen isotopes in the hydrosphere. This is because the mass of organic substance accumulated during the Earth's history is too small in comparison with the mass of the sedimentary rocks and especially with the water mass in the hydrosphere.

With regard to the deuterium and oxygen-18 concentration in different types of meteorites, there is a wide range of deuterium variation in carbonaceous chondrites. As investigations of the isotopic composition of water and organic compounds in carbonaceous chondrites have shown (Briggs, 1963), the upper limits for the deuterium content of these two compounds are very similar (+290 and +275‰), whereas lower limits are −154 and −15‰ respectively.

While studying the oxygen isotope ratios in minerals of stone meteorites (Taylor et al., 1965; Reuter et al., 1965; Vinogradov et al., 1960), like many common features which make them similar to their terrestrial analogues, a number of peculiarities have been found. The ^{18}O content in the meteorite piroxenes varies from −0.5 to +8.6‰ depending upon meteorite type. In piroxenes of terrestrial igneous rocks the range is from +5.5 to +6.6‰. Olivine of carbonaceous chondrites exhibits considerable ^{18}O variations, being approximately 6‰ lower than those in olivine of other stone meteorites. If the terrestrial basalts and gabbro are enriched in ^{18}O as compared with the ultrabasic rocks, than the basaltic chondrites become depleted compared with the ultramafic chondrites.

An interesting interpretation of the data on oxygen isotope composition in different types of meteorites has been given by Clayton et al. (1976); Clayton and Mayeda (1978a, b). On the basis of the measured $^{18}O/^{16}O$ and $^{17}O/^{16}O$ isotope ratios and the construction of a diagram with $\delta^{18}O$–$\delta^{17}O$ coordinates they have distinguished six types of meteoric substances. Each of these types, characterized by its own oxygen isotope composition, cannot be obtained from any other type by means of fractionation or differentiation of substance. These groups are: (1) the terrestrial group, consisting of substances of the Earth, Moon, and enstatite ordinary chondrites; (2) the L and LL types of ordinary chondrites; (3) the H type of the ordinary chondrites; (4) the nonhydrated minerals of the C2 and C4 chondrites; (5) the hydrated minerals of the base of C2 chondrites; (6) urelites (Herndon and Suss, 1977).

The authors found that the substances of the solar nebulae from which the bodies were formed were inhomogeneous in the oxygen isotope composition. They pointed out that, as a first approximation, a two-component model of this substance can be assumed. The portion of the substance whose oxygen isotope composition has been formed on the basis of mass fractionation effects provided by chemical reactions and diffusion processes, can be indicated on the $\delta^{18}O$–$\delta^{17}O$ diagram by a straight line with a slope of 0.52. The other portion of the substance, like common oxygen, contains the component enriched in ^{16}O and having independent origin. Figure 9.1 indicates on the three-isotope diagram $\delta^{18}O$–$\delta^{17}O$ the distribution of the oxygen isotopes in various types of meteorites.

In their δ^2H and $\delta^{18}O$ contents, meteorites in general are not—when considered as chemical complexes—identical with any type of terrestrial rock. At the same time, the range of isotopic ratios points to their close affinity with certain types of

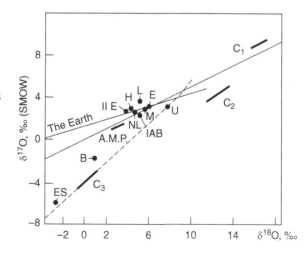

Fig. 9.1 Oxygen isotopic composition of various types of meteorites on the three-isotope diagram: carbonaceous chondrites (C_1, C_2, C_3); ordinary chondrites (*HL*); enstatites and autrites (*E*); eucrites, hovardites, diagenites, mezosiderites, pallasites (*AMP*); nakhelites and lafayettites (*NL*); uraelites (*U*); irons (*IAB*, *IIE*); Bencubbin and Wetherford (*B*); Eagle Station and Itzavisis (*ES*); Moon (*M*). (Clayton and Mayeda, 1978a)

rock. It is possible to say only that the conditions of the formation of meteorites were different from that of terrestrial rocks.

In connection with this question, there is natural interest in information on the isotopic composition of lunar material brought to Earth under the space programmes of the United States and the former Soviet Union. Published data on the isotopic composition of water in lunar dust and breccia from the Sea of Tranquillity (concentration of 81–810 ppm) and of hydrogen gas (concentration of 18–66 ppm) point to extremely low deuterium concentrations in these materials (Epstein and Taylor, 1970; Friedman et al., 1970). However the deuterium content in such small samples varies over wide ranges (from −158‰ to −870‰ for water and from −830‰ to −970‰ for hydrogen gas). This clearly demonstrates the non-terrestrial origin of water and hydrogen gas on the lunar surface. Moreover, the wide fractionation range for hydrogen isotopes suggests that the lunar rock samples probably have different temperature history.

The published data (Epstein and Taylor, 1970; Onuma et al., 1970; Friedman et al., 1970) indicate a much narrower range of oxygen-18 concentration: from +19‰ to +14‰ for carbon dioxide and from +7.2‰ to +2.8‰ for rock samples.

A very interesting oxygen isotopic composition pattern emerges from the analysis of individual minerals in lunar rocks. The isotopic composition of the oxygen in individual minerals varies appreciably; this is also evidence of the complex temperature history which these minerals experienced during formation of the corresponding compounds. The latest data on deuterium and oxygen-18 in samples from the Sea of Storms (Epstein and Taylor, 1971; Onuma et al., 1970; Friedman et al., 1971), presents another picture. While the oxygen isotopic composition lies within the same range (from +7.15‰ to +3.83‰), deuterium concentrations are much higher: in one of the crystalline samples a value of +283‰ was found (Friedman et al., 1971).

As to the problem of the existence of water in lunar rocks, there are conflicting opinions. Friedman et al. (1970; 1971) affirm that water is presented in the lunar rock samples. Epstein and Taylor (1970; 1971; 1972) consider that Friedman and co-workers found water which was brought back by astronauts or obtained during transportation and processing of samples. Analyzing the lunar rock samples brought to the Earth by Apollo-14 and 15, Epstein and Taylor (1972) noted a rather great homogeneity of rocks on the whole in oxygen-18 (the observed changes range from +5.3‰ to 6.62‰). The extracted amounts of water (about 10 mol/g) contain deuterium from −227‰ to 419‰. The $^{18}O/^{16}O$ ratio in water range from −5.9‰ to −18.2‰. The obtained values are rather close to the ^{18}O content in atmospheric moisture in the region of Los Angeles, where analyses have been carried out. This circumstance allowed Epstein and Taylor to conclude that the studied samples could have been contaminated with local moisture. The deuterium content in the gaseous hydrogen, as earlier, was low (about molecules per million), so that the $^2H/^1H$ ratios ranged from −250 to −902‰.

When we analyze the isotopic composition of water in various terrestrial and cosmic materials, the following observations are particularly striking:

1. Despite the very wide range of deuterium and oxygen-18 concentrations in various materials, the isotopic composition of ocean water remains constant over a long period of geologic time.
2. The concentration of deuterium in all terrestrial and cosmic materials tends fairly consistently towards depletion as compared with ocean water, reaching, in the gas phase of molecular hydrogen, values characteristic for the Sun's atmosphere (see Table 9.2). The only exceptions are carbonaceous chondrites and lunar rocks, where the water and molecular hydrogen present are sometimes enriched in deuterium. This may be due to secondary temperature phenomena which occurred during their formation and the subsequent appearance of compounds of more complex chemical composition.
3. The hypothesis that the present oceans were formed by the gradual emergence at the Earth's surface of juvenile waters from the depths of the Earth is not confirmed by evidence.
4. Underground waters are enriched in ^{18}O relative to rocks and minerals.
5. On the whole the hydrosphere is enriched in deuterium and depleted in oxygen-18 relative to rocks and minerals.
6. Meteorites and lunar rock samples differ in hydrogen isotope composition from terrestrial materials. At the same time the ranges of the oxygen-18 isotope variation in lunar rock samples are close to those in terrestrial rocks of similar mineralogical composition. All these facts point to a common material from which the Earth, Moon, and meteorites were formed, but this formation occurred at different temperature conditions and under different sequences of processes accompanying the formation over time and space. It has been noted that water is contained in extremely small amounts in lunar rocks and some researchers consider its presence to be doubtful.

How do the facts established above tally with current ideas about the origin of the Earth's hydrosphere (and especially the oceans) following from the 'cold' and 'hot' creation hypotheses concerning the Earth's origin?

According to the accretion hypothesis, the hydrosphere was formed through the emergence at the Earth's surface of its light components in the course of global and local melting of the initial cold homogeneous body. If that were true, the ocean should have a relative ^{18}O content up to +7‰, corresponding to the oxygen isotope ratio in basalts. The deuterium content in water of the initial oceans should have been about −80‰, estimated by the igneous rocks (Craig, 1963), about −60‰, estimated by the volcanic steam and gas (Arnason and Sigurgeirsson, 1967; Friedman, 1967), and −48 ± 20‰, using ancient rocks of the upper mantle (Sheppard and Epstein, 1970).

As one can see, the present ocean is enriched in deuterium and depleted in oxygen-18 compared with the probable initial ocean.

Among the known processes which can result in the enrichment in deuterium of the initial ocean, the process of the dissipation of free hydrogen from the Earth's gravitational field during which protium escapes preferentially is usually invoked. In fact, within a sufficient degree of accuracy one can consider that the ratio of the escape velocities of $^1H_2$ and $^1H^2H$ molecules is proportional to $(M^1_H{}^2_H/M^1_H{}^1_H)^{1/2} =$ 1.22. This value can be taken as the fractionation factor of hydrogen isotopes analogous to the fractionation factor in diffusion processes. In this case one can easily estimate that the enrichment of the ocean in the deuterium by 20‰ due to preferential losses of protium can be attained only by a decrease of the ocean mass by 11%. If free hydrogen is formed in the upper atmospheric layers during photodissociation of water molecules, then a reduction of the ocean by 11% (at its mass of 1.4×10^{21} kg) should result in the release into the upper atmosphere of 1.25×10^{20} kg of free oxygen. When the deuterium concentration in the juvenile water is lower than −20‰ (i.e. −48‰ according to Epstein and Taylor and −80‰ according to Craig), the amount of free oxygen in the atmosphere must increase proportionally. Taking into account the portion of oxygen which has been expended on oxidation of the igneous rocks (about 2.8×10^{18} kg) (Goldschmidt, 1954) and was buried together with ancient organic material (about 10^{16} kg) then one must conclude that even in this case a large amount of free oxygen could have accumulated in the atmosphere, exceeding its modern content in the atmosphere by more than two orders of magnitude. This is not what one finds in practice, and the free oxygen has been commonly assumed to be formed due to photosynthetic processes in the biosphere.

Easy theoretical estimations show that the possible losses of protium during the hydrogen escape from the Earth's gravitational field are rather low. Therefore, there is no reason to consider that the required enrichment of the ocean in the deuterium by 50–80‰ due to the dissipation of protium has taken place. Earlier we considered the possible ranges of enrichment of the ocean in deuterium due to sedimentation of the decomposed igneous rocks. This effect can be considered as insignificant, resulting in changes of about 0.3‰. Since there are no other processes known which lead to the enrichment of the hydrosphere in deuterium, such a phenomenon is doubtful.

Let us consider now ways of depletion of the initial oceans in oxygen-18. The most effective depletion process is the sedimentation of rocks. Despite the possible errors in the estimation of initial amounts of various types of sedimentary rocks required for calculations (these values differ markedly in works of various authors), the final effect of sedimentation of rocks estimated by Savin and Epstein (1970), yielding 3‰ at least, has not been reduced. In this case, when the oxygen of carbon dioxide is in isotope equilibrium with rocks during its degassing, the ocean could have become only partly depleted in oxygen-18.

Moreover, we must assume that the enrichment of acid rocks in heavy oxygen has taken place due to the water of the hydrosphere since one can hardly suggest any other source of available heavy oxygen. Assuming that the average $\delta^{18}O$ value in granites is +10‰, the ocean's juvenile water should have it to +20‰. In the case of the degassing origin of the hydrosphere this last $\delta^{18}O$ value seems very improbable.

There have been attempts to resolve the above-mentioned concentrations concerning the oxygen isotope composition with the help of the appropriate model of growth and change of the ocean isotopic composition with time. For example, it has been suggested (Chase and Perry, 1972; Ahmad and Perry, 1980) to use the model of recirculation through the mantle of the ocean water depleted in heavy oxygen, where it should become enriched in ^{18}O. But such a model has difficulty in explaining the physical reasons behind such a process occurring in nature. This model also contradicts facts already mentioned, which are evidence of the very early stage of ocean formation and the constancy of its chemical and isotopic composition over long geologic time. On the other hand, the model has been suggested to account for the low δ^2H values in mantle and igneous rocks (Taylor, 1978) for the modern ideas of plate tectonics and the spreading of the ocean floor. It is known that the hydroxyl-bearing minerals of the authigenic sea sediments have δ^2H values close ($\sim-60‰$) to those of magmatic rocks. Assuming that the sea sediments are continuously contributing to the mantle through subduction zones, the process can be accepted as the mechanism of injection of light hydrogen into mantle rocks. But in this case the same process should be accompanied by the removal of heavy oxygen from the ocean and the enrichment in heavy oxygen of the igneous rocks contained in the same sea sediments. New difficulties arise at this point.

Thus, the observed facts of the isotopic composition of water in the oceans and in underground rocks contradict the 'cold' origin hypothesis.

These same facts lead to the conclusion that juvenile water (i.e. water formed in the deep layers of the Earth and in its upper and lower mantle, which emerges at the surface) does not, and never did, exist on our planet and that the hydrosphere resulted from an atmophile process. Such a process follows from the 'hot' model of origin considered by, for example, Goldschmidt (1954).

If one assumes that during the first stage of the Earth's geochemical differentiation water remained in the gas phase and did not escape, and that there was no free oxygen, then the isotopic composition of the hydrogen and oxygen in the water falling on to the Earth's surface might well have corresponded to the concentrations now found in the ocean water. But the 'hot' hypothesis of origin of the Earth has been subjected, as already indicated, to justified criticism. Therefore, one should look for a mechanism of the Earth's formation that can satisfy the conditions of formation of the isotopic composition of the hydrosphere, following the 'hot' hypothesis. Such a mechanism should be in agreement with the geochemical conditions which were formulated, for example, by Urey (1957) and are best fitted to the 'cold' hypothesis of the Earth's origin.

9.3 Evidence from Carbon and Sulfur Isotopes

Carbon and sulfur form a number of mobile compounds in nature (CO, CO_2, CO_3^{2-}, SO_2, SO_4^{2-}, SO_3, H_2S etc.), which take part in the intensive cycling of the matter in the geosphere in close relationship with water of the hydrosphere. The study of the principles of the distribution of carbon and sulfur isotopes provides important data concerning the origin and sequence of evolution of the hydrosphere and the Earth itself.

Carbon isotopes. Solar carbon, according to the observational data available, is enriched in the light isotope. Some authors (Reghini, 1963) have considered the degree of enrichment to be rather great ($^{13}C/^{12}C \approx 10^{-5}$), others (Burnett et al., 1965) consider it to be more moderate ($^{13}C/^{12}C \approx 10^{-2}$).

Meteorites remained up to recent time the most easily available medium for studying carbon cosmic material. Comparing the $\delta^{13}C$ values obtained for iron, iron-stone and stone meteorites, it has been found that they range from -5 to $-30‰$ with the average value of $-22‰$ (relative to the PDB-1 standard). Carbon is commonly contained in insignificant amounts (hundredth %) and is present in a dispersed state. In the Canion Diablo iron meteorite, carbon has been found in the form of a graphite inclusion in troilite with $\delta^{13}C = -6.3‰$ and carbon of the carbide iron with $\delta^{13}C = -17.9‰$ (Craig, 1953). The graphite of Yardimlinsky's iron meteorite has $\delta^{13}C = -5‰$ (Vinogradov et al., 1967). Generalizing the carbon isotope data in iron meteorites (Deines and Wickman, 1975), it has been found that the observed depletion of taenite in heavy isotopes ($\delta^{13}C = -18.5‰$) compared with graphite ($\delta^{13}C = -5.5‰$) is common.

In view of the fact that the chondritic model of the Earth's origin is widespread, knowledge of the isotope composition of carbon and chemical forms of its occurrence in the most widespread type of meteorites, chondrites, is of particular interest. Carbon has been found in chondrites in the form of graphite, carbonate, carbide, diamond, and organic compounds. Ordinary chondrites have a light isotope composition of carbon with an average value $\delta^{13}C = -24‰$. The carbon content itself amounts to only 0.1–0.01% (Hayes, 1967).

Carbonaceous chondrites have been studied by Boato (1954), who reported that they can be divided into two types according to the gross content of carbon and water in them. In the first type the content of carbon ranges from 1.6 to 3.3% and $\delta^{13}C$ from $-3‰$ to $-12‰$, in the second type $\delta^{13}C$ varies from -13 to $-18‰$ with a carbon content ranging from 0.3 to 0.85%. In these meteorites the average $\delta^{13}C = -7‰$.

Thus, on the whole the following general principle is observed in meteorites: the lighter isotope composition of carbon is observed in those meteorites where its gross content is lower. Boato has suggested that an initial undifferentiated substance of the solar system is represented by the carbonaceous chondrites of the first type. In his opinion, the observed correlation between the isotope composition and the gross amount of carbon is a likely indicator of the preferential escape of ^{13}C in the process of depletion of the initial material in volatile elements.

Other viewpoints concerning the origin of the heavy isotope composition of carbonaceous chondrites were proposed. For example, Galimov (1968) has reported that the enrichment of the carbon of carbonaceous chondrites in heavy isotopes can be accounted for by the excessive yield of ^{13}C in nucleosynthesis by reaction $^{16}O(n, ^{4}He)^{13}C$ in the course of the formation of carbonaceous chondrites in a medium with a high oxygen content. Therefore, the isotope composition of carbon in the meteorites and ^{13}C cosmogenic variations in the primeval carbon, were conditioned by the nucleosynthesis of the carbon isotopes before formation of the solar system and nuclear reactions proceeding on high-energy particles in the pre-planet material during the early stage of evolution of the solar system. Later only redistribution occurred in various geospheres of the isotope concentrations inherited from the initial course.

According to Epstein and Taylor (1972), the range of $\delta^{13}C$ variations in the lunar rock samples is wide (from $-20‰$ to $-30‰$). It has been found that, as well as in the case of carbonaceous chondrites, the degree by which the samples are enriched in heavy carbon is dependent on the content of carbon itself. In the lunar basalts the $\delta^{13}C$ values range between $-30‰$ and $-18.6‰$ and in the fine-grained soils and breccia the $\delta^{13}C$ values are greater than $-3.6‰$. It is of interest to note that in lunar samples no carbon with $\delta^{13}C$ within the range from $-3.6‰$ to $-18.6‰$ has been found. At the same time, as pointed out earlier, the average value of the carbon content in chondrites is within this range. In order to explain the observed enrichment of soil and breccia with heavy carbon a number of hypotheses were suggested concerning the effect of solar wind and meteorites upon the surface of the Moon. It is assumed that the relative content of heavy carbon in the solar wind ranges from $-10‰$ to $+10‰$ (Galimov, 1968).

Comparing experimental data on the isotope ratios of carbon in meteorites and in the Moon it has been found that their isotope ratios are markedly different. This difference, as a rule, depends upon the gross content of carbon and its chemical form in the subject. Therefore, it follows that the determination of genetic relationships while studying the questions of the origin of cosmic bodies cannot be carried out without account being made of the isotope data. For example, since the isotope composition of carbon in lunar rock differs from that in chondrites, the chondritic model of the moon cannot be true.

Let us consider now the principles governing the carbon isotope composition in the main reservoirs of the Earth and the main mechanisms of the fractionation of carbon isotopes in nature (Table 9.3).

It is known that atmospheric CO_2, together with radioactive carbon dioxide, constitute an indivisible exchangeable system. The partial pressure of CO_2 in the atmosphere corresponds to the equilibrium state of this system. On the other hand, the atmospheric CO_2 is a source of carbon in biochemical systems. After the death of an organism some portion of it returns to the atmosphere and another portion is transformed in the hydrocarbon-bearing sediments and minerals. Thus, atmospheric carbon dioxide represents the common chain of organic and inorganic cycling of carbon on the Earth. The most important process of fractionation of carbon in nature corresponds to these two chains of the carbon cycle on the Earth. Biological fractionation results in a considerable enrichment of organic sediments in light isotopes of carbon system compared with atmospheric carbon dioxide. Fractionation in the atmosphere-hydrosphere leads to enrichment of the carbonate-ion heavy isotopes of carbon.

On the whole the relative content of ^{13}C in the atmosphere is $-7‰$ and sea air is subjected to smaller variations of $\delta^{13}C$ compared with land air. Daily, yearly, and other cycles of variations of the isotopic composition are observed, which correlate with the process of photosynthetic activity of the organisms. The average isotope composition of land carbon is characterized by $\delta^{13}C = -10‰$ (Galimov, 1968).

Table 9.3 Stable carbon isotope variations relative to the pdb-1 standard in natural objects (Craig, 1953; Galimov, 1968)

Object	$\delta^{13}C$, ‰						
	−80	−60	−40	−20	0	+40	+60
Oceanic Hot springs Atmosphere Wells	*Carbon dioxide*						
Bitum Coal Crude oil	*Organic matter*						
Meteorites Marine non-organic Marine organic Carbonatites Fresh water Hydrothermal Gaseous caps (eocene) Sicilian sulphurous limestone Saline domes Aragonites	*Carbonates*						
Paleozoic Mesozoic Tertary Quaternary	*Methan*						

Isotope fractionation results in variations of the carbon isotopes in living material from −7‰ to −32‰ depending on the ecological system and photosynthetic type. On average this value is −25‰ for inland plants with a Calvin photosynthetic cycle (Park and Epstein, 1960).

The process of carbon isotope fractionation in the atmosphere-ocean system results in enrichment of the seawater bicarbonate up to the average value of $\delta^{13}C = -2$‰. Here the value lies within narrow ranges from −1.3‰ to −2.9‰ which can be accounted for by a high rate of exchange of atmospheric carbon dioxide with the ocean. Variations of heavy carbon in the sea's sedimentary rocks (limestones) range from +6‰ to −9‰. But 70–80% of all the carbonates have values ranging from +2 to −2‰ with an average value close to zero. Note that theoretical ideas concerning the above-mentioned process of carbon isotope fractionation in nature do not contradict the experimental data (Galimov, 1968; Epstein, 1969).

Fractionation of carbon isotopes is also observed in the course of decomposition of organic material. Attempts were undertaken to study the variations of the carbon isotope composition in the vertical section of bottom sediments with a simultaneous determination of their age by the radiocarbon method. It was found that the organic material in sediments is enriched in light isotopes of carbon compared with living organisms (Eckelmann et al., 1962). The accumulation of ^{12}C increases from young to old ooze layers. It is assumed that enrichment in the light carbon isotope occurs as a result of the presence of a greater amount of the heavy isotope in carboxyl groups of aminoacids, which are less stable and more subject to decomposition, or due to the preservation of the lipid fractions, which contain an excessive amount of light carbon isotopes (Epstein, 1969).

Besides the above-mentioned carbon isotope fractionation process there are also some other mechanisms of its fractionation in nature, including some high-temperature processes. These mechanisms only have regional importance. Among these processes we shall note only the Fisher-Tropsh reaction (Lancet and Anders, 1970), which is as follows:

$$nCO + (2n + 1)H_2 \rightarrow C_nH_{2n+2} + nH_2O,$$
$$2nCO + (n + 1)H_2 \rightarrow C_nH_{2n+2} + nCO_2.$$

Both oxidized and reduced products are formed by this reaction. The major portion of the oxygen contained in CO is transformed into H_2O and a small proportion (several tenths of a percent) into CO_2. Among the reduced products, methane is predominant. The content of hydrocarbons that are volatile at 400°C amounts to ~30% to the end of the reaction. About 1.5% of the polymerized organic compounds remain on the catalyst at this temperature.

The carbon dioxide formed in the Fisher-Tropsh process is enriched in ^{13}C by about 30% relative to the initial carbon present in CO. The carbon present in the fraction which is volatile at temperature <400°C becomes enriched in the light isotope by approximately 20‰, whereas the high-molecular product remaining on the catalyst exhibits even more enrichment in the light isotope relative to the initial CO (~33‰). At the beginning of the reaction the carbon of the methane fraction exhibits maximum enrichment in the light isotope. At this moment the enrichment of CO_2 relative to CH_4 reaches ~100‰ and the enrichment of CO relative to CH_4 reaches ~60‰. During the course of the reaction methane becomes depleted in the light isotope so that, in the later stages, its isotope composition becomes approximately equal to that of the initial CO.

The Fisher-Tropsh process is usually considered as the most probable model of the abiogenic synthesis of organic compounds in nature. It is assumed that this process causes the formation of organic compounds, for example, in the carbonaceous chondrites (Lancet and Anders, 1970). It is thought by some researchers that the hydrocarbons and bitumens of igneous rocks are formed by the Fisher-Tropsh reaction (Galimov, 1973).

Within the scope of questions about the Earth's origin and its hydrosphere formation, estimation of the average value of the relative content of heavy carbon in the terrestrial crust as a whole, and the comparison of this value with $\delta^{13}C$ value of the other Earth's shells and meteorites, is of importance. This estimation is based on the knowledge of the amount of carbon and its isotope composition in the Earth's major reservoirs.

According to the data of Craig (1953) the $\delta^{13}C$ value in the Earth's crust ranges from $-7‰$ to $-15.6‰$ with an average of $-12‰$. Galimov (1968) has determined at range of variations of this value between $-3‰$ and $-8‰$ with an average value of $-5‰$. Epstein (1969), using data of the carbon content in sedimentary rocks, which are representative of the main reservoirs of carbon in the upper sphere of the Earth (73%, $\delta^{13}C = 0$), and data on the carbon content in shales, representing a second considerable reservoir of carbon (26%, $\delta^{13}C = -27‰$), has found that in the Earth's crust the average value of $\delta^{13}C = -7‰$.

Thus, despite some differences in estimation of the average $\delta^{13}C$ value, obtained by different authors for the Earth's crust, all agree that the Earth's crust is enriched in the heavy carbon isotope compared with meteorite and solar material.

As pointed out earlier, the isotope fractionation is reduced or even absent at high-temperature processes. Assuming the chondrite model of the Earth's origin, and that carbon has appeared in its upper sphere along with other volatile components, due to degassing from its interior during geologic time, a light isotope composition of carbon in terrestrial crust should be expected. But in fact the opposite phenomenon is observed.

In the framework of the degassing hypothesis there is an idea that the type of material of carbonaceous chondrites, whose isotope composition is similar to that of the Earth's crust, has played an important role in the formation of the isotope composition of the Earth's crust (Galimov, 1973). But it is difficult to find the sources through which the emergence of this substance to the surface occurred.

In discussing the principles of hydrogen and oxygen isotope distribution, we have already pointed out that the existing opinion of the emergence of juvenile material through hydrotherms appears to be incorrect. Although it is impossible to pinpoint the sources of the emergence of mantle carbon at present, one can assume that they existed in the past.

The carbon isotope composition of sedimentary carbonates is known to be a sensitive indicator of the general geologic paleo-situation. If, during the Earth's history, there were periods in which a considerable emergence of the mantle carbon dioxide occurred, then it should have been reflected in the isotope composition of the sedimentary carbonates. It was concluded earlier, on the basis of data on the heavy carbon content in limestones, that their isotope composition bears no relation to geologic age (Craig, 1953). Galimov (1968), generalizing the available data on the isotope composition of sea and fresh water limestones, has come to the conclusion that carbonates were extremely enriched in the light isotope of carbon during two periods, the Carboniferous and Tertiary. He pointed out the synchronous limestone isotope changes with the amount of organic substance on the globe during these times. Galimov suggested that the enrichment of limestones in the light isotope can be accounted for by the intensive contribution of carbon dioxide in the Hercynian and Alpine orogenies. The enrichment of limestones in the heavy carbon isotope in subsequent epochs occurred due to the withdrawal of the light isotope during the plant's intensive photosynthetic activity and the burial in sediments of considerable amounts of the light carbon. But this interpretation is disputable. As Schell et al. (1967) have reported, the increase in temperature of the world oceans results in a shift of equilibrium in the carbonate-calcium system in such a way that this, in turn, leads to the more active process of photosynthesis in plants. As a result the sea and fresh water limestones become enriched in the light carbon isotopes. The decrease of temperature of the world oceans in subsequent time should result in opposite effects.

The problem of carbon origin in the upper shell of the Earth requires further investigation. But on the basis of the data available at present there are reasons to assume that neither now, nor in the past, has there been any considerable emergence of the mantle carbon to the surface through degassing.

In this connection it is of interest to note that some authors (Epstein, 1969) have attempted to consider various alternatives of the possible existence of the primary form of main reservoirs of carbon on the Earth in the past. One can assume, for example, that at a certain stage in the Earth's early history the main reservoirs of carbon contained carbonate rocks with $\delta^{13}C = -7‰$. If the carbon isotope fractionation effects between the main reservoirs have remained unchanged up to the present time, then the atmospheric carbon dioxide exhibits $\delta^{13}C = -14‰$ and the organic material $\delta^{13}C = -34‰$. On the other hand, if the major portion of the carbon was represented by the organic substance with $\delta^{13}C = -7‰$, then the atmospheric carbon dioxide had $\delta^{13}C = +13‰$ and carbon of the carbonate rocks must had $\delta^{13}C = +20‰$.

Sulfur isotopes. The available data on the sulfur isotope composition of iron, iron-stone, and stone meteorites indicate that the $^{34}S/^{32}S$ ratio, characteristic of cosmic bodies, is approximately constant (Ault and Kulp, 1959). Different $\delta^{34}S$ values were discovered in the Orgueil carbonaceous chondrite for individual chemical components of sulfur (for the elementary, troilite and sulfate forms). For the meteorite as a whole the $\delta^{34}S$ value relative to the troilite standard of the iron meteorite from the Canyon Diablo appear to be zero (Monster et al., 1965).

Analyzing the sulfur isotope composition of lunar rock samples, it has been found (Kaplan and Smith, 1970) that $\delta^{34}S$ in the dust ranges from +4.4‰ to +8.2‰. In breccias this value has been found to vary from +3.3‰ to +3.6‰ and in fine-grained soil from +1.2‰ to +1.3‰. The relatively high $\delta^{34}S$ content in the dust has been accounted for by the vaporization of the sulfur light isotope by bombardment of the Moon's surface with micrometeorites and protons of the Solar wind.

The main geochemical cycle of sulfur on the Earth is related to existence of the oceans. Despite the continuous process of sulfur contribution to the ocean together with river runoff, its concentration and isotope composition remains constant in the ocean due to sulfate reduction. During the cycle the relatively light continental sulfate with $\delta^{34}S = +5‰$ becomes enriched up to the oceanic average value of $\delta^{34}S = +20‰$ and the excess of the isotope is bonded in diagenetic sulfides.

The isotope composition of sulfur in oceanic sulfate is a sensitive indicator of its dynamic equilibrium in the cycle and the constancy of ocean salt composition on the whole. This equilibrium is determined by the rate of biogenic sulfate reduction and therefore by the total amount of biomass on the Earth. On the other hand, the content of sulfur in the oceans is closely related to its total salinity.

Wide studies concerning the sulfur isotope composition of evaporates of different ages have been undertaken, aimed at elucidation of the biochemical history of the oceans.

While studying sulfates in the Phanerozoic evaporates, it has been shown (Ault and Kulp, 1959; Vinogradov, 1980) that the sulfur isotope composition has no time-ordered changes. There occur only slight variations of the $\delta^{34}S$ values from the average value corresponding to the sulfur isotope composition of the present ocean (+20‰). The only exceptions are the Permian evaporates, which always exhibit depletion down to the value of $\delta^{34}S = +10‰$. Vinogradov (1980) has explained this phenomenon in terms of paleogeographical peculiarities in the accumulation of Permian evaporates. These peculiarities consisted, in his opinion, in an increase in the role of the inland sulfates component in the recharge of salt basins in the transition from the sulfate to the haloid accumulation of salts. On the whole, the isotope composition of sulfur in sulfates in ancient evaporates during the whole Phanerozoic period has remained constant, and close to the modern composition of the oceans. This circumstance is evidence in favor of the constancy of the ocean's salinity, amount of the biomass, and concentration of oxygen on the Earth during the whole Phanerozoic period.

While studying the Precambrian metamorphic rocks from East Siberia, the Pamirs, and South Africa, which exhibit salt-forming features by a number of minerals (skapolite, apatite, lasurite, carbonatites), it has been found (Vinogradov et al., 1960; Vinogradov, 1975) that these rocks contain relatively high concentrations of sulfide, sulfate, and native sulfur. The isotope composition of sulfur in sulfate of the metasomatic minerals is characterized by high enrichment in the heavy isotope (from +13‰ to +45‰) and the identical content of $\delta^{34}S$ relative to sulfates and sulfides of the sulfur-bearing carbonate rocks. On this basis the authors have come to a reasonable conclusion regarding the sedimentary origin of the initial sulfates. These sulfates were accumulated at the steady dynamic cycle of sulfur and served as the initial source of sulfides. In the metasamotic minerals of the Archean carbonabe rocks in Aldan (East Siberia) and rocks of the Swaziland system (South Africa) up to 3.5×10^9 years old, widely developed regional scale sulfates of sedimentary origin were found, with $\delta^{34}S = +6‰$, which are not typical for sedimentary sulfates. On the basis of experimental data analysis, it has been found that the sources of sulfur, participating in the metasomatism and metamorphism processes, were the metamorphic sedimentary thicknesses containing sulfur in the form of sedimentary sulfide. From the existence in the section of sedimentary rocks of layers of dolomites enriched in sulfate-sulfur, it has been concluded that the process of the sedimentation of sulfates has taken place in the normal facial conditions in the course of the salinization of the sea basin. But the sulfur cycle in the basin has not attained dynamic equilibrium and the sulfate-sulfur in a basin has not yet been enriched in the heavy isotope. But in individual sites of a basin such enrichment of the sulfate-sulfur of metasomatic minerals has occurred and the $\delta^{34}S = +20‰$ has been found there. This has led the above authors to conclude that the establishment of the dynamic equilibrium of the sulfur cycle, at a level which approximates the modern one, coincides with the age of the studied rocks, i.e. occurred about $(3–3.5) \times 10^9$ years ago. On this basis Vinogradov (1980) has concluded that the emergence of the main mass of sulfur from the interior to the upper shell of the Earth, and also the formation of the oxygen composition of the atmosphere and the salinity of the ocean, which approximates the modern levels, ended not later than 3.5×10^9 years ago.

Let us consider now the comparative analysis of the sulfur isotope composition of the upper shell of the Earth. The balance estimations show (Grinenko and Grinenko, 1974) that the major sulfur reservoir in the Earth's crust is the platform sedimentary thickness of the continents, characterized by an average value of $\delta^{34}S$ of about +4‰. In the geosynclinal areas containing up to 30% sulfur the value of $\delta^{34}S$ is close to zero. The ocean water contains15% sulfur in the form of dissolved sulfates, enriched in the heavy isotope up to +20‰. The sulfur of the ocean sediments, amounting to about 10% of that in the sedimentary thickness, exhibits $\delta^{34}S =$ +7.7‰. The average $\delta^{34}S$ value in the abyssal clays, limestone, and siliceous sediments is equal to +17‰. The ultramafic oceanic and inland rocks are characterized by an average value of $\delta^{34}S = +1.2$‰. For basic rocks this value is +2.7‰ and for acid rocks it is equal to +5.1‰. Therefore, the outer sphere of the Earth together with the ocean is enriched in ^{34}S by 5.5‰ and the terrestrial crust on the whole by 3‰ relative to meteorite material. It is of note that as one moves from the basaltic sphere towards the granite sphere and farther to the sedimentary continental layers and the oceans, the amount of sulfur increases with a simultaneous enrichment in the heavy isotope. Some researches (Grinenko and Grinenko, 1974) have assumed that such global processes as degassing of the mantle, crystalline differentiation, and metamorphism of rocks have the same effect. This effect manifests itself as increases in the amount of sulfur in sedimentary rocks from the Archean to the Proterozoic, Phanerozoic, and Cenozoic with simultaneous increases of sulfate-sulfur enrichment in ^{34}S.

It should be pointed out that such an approach to the observed global principle of increases in the amount of sulfur with enrichment in ^{34}S from the ultramafic rocks to the acid ones and to the oceans contradicts the evidence given above in favor of the constancy of the salt and isotope composition of the oceans during the last $(3–3.5) \times 10^9$ years. On the other hand the above-mentioned principle including the relationship between the gross content of the element with its isotopic composition is typical for the other elements stated: oxygen, carbon, and hydrogen. For all these elements we have observed enrichment in heavy isotopes while moving from ultramafic rocks to acid rocks and to the ocean. The only exception is the oxygen isotopic composition of the ocean. In order to explain this general principle, while considering the isotopic composition of different elements, different researchers employ assumptions which are often contradictory when compared. In addition to the above-mentioned principles of the distribution of the isotopes of the volatile elements H, O, C, S already considered, one should bear in mind one more cosmochemical principle. This principle states that the Earth's crust is, in general, enriched in the heavy isotopes of these elements relative to meteorite substance. The available experimental data on boron isotope distribution, despite being limited, shows that for isotopes of this element the above-mentioned principle also holds.

It should be pointed out that the data on the isotope distribution of noble gases (He, Ne, Ar, Kr, Xe) in the upper sphere of the Earth and in meteorite substances indicate the applicability of the above principle to this group of elements also.

The observed regularities of the distribution of volatile elements in the upper terrestrial sphere, meteorite, and lunar material, and also the results of cosmochemical and cosmophysical studies obtained during recent years, suggest the importance of the processes of chemical differentiation of substance which took place at the pre-planetary stage of evolution of the planets. In this case the idea of chemical and isotope homogeneity of the forming planet material is doubtful. In this connection, let us consider in brief the cosmochemical results which have been obtained in recent years by studying meteorites, the Moon, and other planets. They have a direct bearing on the problem of the origin of the Earth and its shells including the hydrosphere and the atmosphere.

9.4   Chemical Differentiation of Proto-planetary Substance

The fact that the average density of the material of planets decreases with increasing distance from the Sun is of importance and was noted in the early hypotheses concerning the origin of planets in the solar system. On the basis of experimental and theoretical data it has been found that there is a sharp difference in the mass density at the transition from the planets of the terrestrial group to the major planets of the Jupiter group. The major planets were formed of low-temperature volatile elements and compounds, mainly of hydrogen, helium, methane, and ammonium. The planets of the terrestrial group, previously thought to be formed mainly of silicates and iron, exhibit decreases in iron with increasing distance from the Sun.

It has also been found that the comets moving around the Sun in eccentric orbits are constituted of light elements and compounds.

By studying different types of meteorites it has been found that they differ sharply in their chemical composition and, for some elements, difference in isotope ratios is observed.

All these facts indicate that during the formation of the planets of the solar system, processes occurred which resulted in the chemical differentiation of their material. The elucidation of the mechanism of these processes, together with that mechanism which has caused the observed principles of their motion, is one of the main problems which must be solved in the development of cosmogonic theories. The solution of this problem is closely related to the problem of the origin of the Earth and of the formation of planet shells.

Urey was one of the first researchers to point out the necessity of taking into account the observed geochemical and cosmochemical facts in developing geochemical and cosmochemical theories concerning the origin of the solar system. He formulated the boundary conditions (Urey, 1957), which have formed the basis of the chemical differentiation of initial material in the course of formation of meteorites and planets of the terrestrial group and had a great effect on the course of the development of studies in cosmogony. In this respect Urey's conclusion that the initial protoplanetary material underwent a high-temperature stage during its evolution when its temperature could have attained 2000°C, i.e. the temperature when the initial substances were in gaseous state, was of greatest importance.

Recent studies concerning the composition and structure of meteorites have confirmed this conclusion and shown that in the course of the evolution of the protoplanetary material, processes leading to its temperature condensation took place, i.e. processes leading to the transition of the substance from gaseous to liquid and solid phases. These cosmochemical conclusions have had an effect upon theoretical studies concerning development of the general cosmogonical hypotheses, in which attempts were made to determine, from the physical view-point, the probability of the occurrence of a high-temperature stage in the initial cloud (Cameron and Pine, 1973; Cameron, 1973).

Let us consider now under which conditions and in what sequence the process of condensation of chemical elements and their compounds should have proceeded during evolution of the gaseous cloud, if this process took place at all.

The question of the chemical condensation of elements and their compounds from gaseous material was studied earlier by Urey (1959), Wood (1963), Ringwood (1966; 1979), Lord (1965), Anders (1972) and others. They were mainly trying to find an explanation for the observed picture of the depletion of the planets and different types of meteorites with volatile elements. On the basis of their studies it was found that, in different types of meteorites, certain principles are observed regarding the content of the high-temperature nonvolatile and low-temperature volatile fraction of the condensed material. At the same time, while explaining the chemical composition of meteorites on the basis of the obtained calculations, they met difficulties concerning the unsteadiness of the degree of depletion of different types of meteorites by different types of volatile elements.

For estimating the condensation temperatures of elements, Larimer (1967) has employed the common equations of thermodynamics of ideal gases and the data on the elements' cosmic abundances which were obtained earlier by Cameron. For the starting point of gas condensation, Latimer has assumed that when attaining corresponding thermodynamic conditions in the gaseous cloud, the gas pressure of an element of natural abundance attains the corresponding partial pressure. Estimations were carried out at equilibrium conditions of condensation on the assumption that the process proceeded in hydrogen gas and the condensing elements were in atomic or simple molecular states.

Table 9.4 shows the theoretical condensation temperatures of elements obtained by Larimer for two values of the total gas pressure p_τ in the cloud. It follows from Table 9.4 that the temperature of condensation of the elements has little effect upon the sequence of their condensation at two different gas pressures.

The calculation of the condensation temperatures of various complex chemical compounds was carried out by Larimer on the basis of the same principle of attainment, by a given compound, of a temperature corresponding to its partial pressure obtained from the gas thermodynamics equation. But in this case it has appeared to be necessary to account for the possible distribution of each element within the

Table 9.4 Calculated temperatures of condensation of four elements (Larimer, 1967)

$p_\tau = 10^5$ Pa		$p_\tau = 6.6 \times 10^2$ Pa		$p_\tau = 10^5$ Pa		$p_\tau = 6.6 \times 10^2$ Pa	
Element	T (K)	Element	T (K)	Element	T (K)	Element	T (K)
Fe	1790	Fe	1620	Pb	655	Pb	570
V	1760	V	1500	Bi	620	Bi	530
Ni	1690	Ni	1440	Sb_2	590	Sb_2	515
Cu	1260	Sc	1090	Tl	540	Tl	475
Sc	1250	Cu	1045	Te	517	Te	460
Mn	1195	Ge	980	Zn	503	Zn	430
Ge	1150	Au	970	S_2	489	S_2	400
Au	1100	Mn	920	Se	416	Se	375
Ga	1015	Ga	880	Cd	356	Cd	318
Sn	940	Sn	806	Hg	196	Hg	181
Ag	880	Ag	780	I_2	185	I_2	169
In	765	In	670				

different compounds. This has been done by estimating the equilibrium constants using the thermodynamic constants of the reaction products.

While considering the problem concerning condensation of different elements, compounds, and alloys, the gas pressure is of principal importance, being a function of density of the medium abundances of elements and their amounts. Larimer has considered the condensation process of the compounds and alloys to be dependent on the reaction's kinetics and the diffusion rate. In view of this the condensation of pure elements and compounds could have occurred during quick gas cooling. During slow cooling the diffusion equilibrium in the gas-condensate system will be maintained, resulting in the formation of alloys and solid solutions.

The first estimates of the temperatures and condensation sequence of the elements and compounds, which were carried out by Larimer (1967), were later developed by Larimer and Anders (1967; 1970), and Grossman and Larimer (1974). These re-estimated data, presented in Fig. 9.2, indicate that, from the cooling gas cloud at pressure of 10 Pa, such refractory elements as Os, Re and Zr should condense first. Their formation temperature was to be higher than that of Al_2O_3, being a widespread compound, whose condensation temperature is 1679 K. For Ti and Ca, forming $CaTiO_3$ and $Ca_2Al_2SiO_7$ compounds, the condensation temperature is about 1500 K. The rare elements U, Pu, Th, Ta and Nb form solid solutions in $CaTiO_3$. $CaMgSi_2O_6$ is formed at T = 1387 K after about 10% of Mg and Si has been condensed. The metal iron precipitates at T = 1375 K. The latter compound, on reacting with vapor, later forms $MgSiO_3$ which absorbs the Si remaining in the gaseous phase.

The elements condensing at higher temperatures than 1250 K form the group of the refractory compounds. At lower temperatures Cu, Ge, and Ga precipitate in the form of solid solutions in metals. Then $CaAl_2Si_2O_8$ is formed at T = 1299 K and Na, K and Rb, condense in the form of solid solutions. The alkali metals precipitate completely at T = 1000 K, Ag at T = 750 K. The metal iron oxidizes at T = 750 K, troilite at T = 700 K being formed by the reaction between metallic iron and gaseous

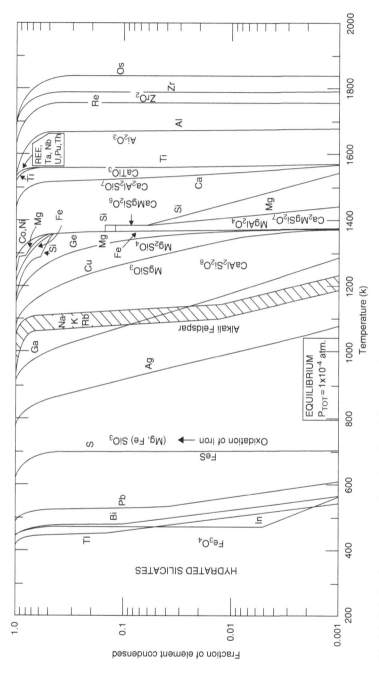

Fig. 9.2 Calculated temperatures of condensation of elements and compounds at 10 Pa (Grossman and Larimer, 1974)

H_2S. Pb, Bi, In and Tl, which are highly depleted in chondrites, condense within the temperature range 600–400 K, magnetites at T = 405 K, hydrated silicates at T = 350 K. Ar, CH_4, NH_3, H_2O, and hydrate methane condense at 200 K. The investigation of the condensational conditions of the substance over the range of pressure from 10 to 10^{-3} Pa for the high-temperature compounds, rich in Ca, Al and Ti, indicates that the sequence of their formation is preserved.

On the basis of the obtained data on the temperatures and sequences of condensation of elements and compounds, attempts were made to explain chemical fractionation in meteorites, and to reconstruct some cosmochemical events which took place in the protoplanetary cloud.

Larimer has come to a conclusion that the observed picture of fractionation of chemical elements in chondrites should be related to their fractionation in the solar nebula during the condensation of the elements and compounds. The subsequent process of heating of the meteorite bodies, being possible in his opinion, has not resulted in further fractionation except for the most volatile elements, noble gases and mercury.

The studies of Larimer and Anders (1967), concerning the observed fractionation of the thirty-one most volatile elements in the carbonaceous and ordinary chondrites, led them to conclude the following. The abundance of these elements in the carbonaceous chondrites C_1, C_2, and C_3 type and in the enstatite chondrites type I gradually decrease in order of their sequences. Taking unity for the abundance of the elements in the C_1 type chondrites, their abundances in chondrites of other types are in the ratios 1:0.5:0.3:0.7. In the ordinary chondrites and enstatite type II chondrites, nine elements (Au, Cu, F, Ca, Ga, Ge, S, Sb, Se, Sn) have approximately constant depletion coefficients ~0.25 and ~0.5. The other 18 elements (Ag, Bi, Br, C, Cd, Cl, Cs, H, Hg, I, In, Kr, N, Pb, Tl, Te, Xe, Zn) have a coefficient of 0.002.

On the basis of the experimental data given above, the authors assumed that chondrites are composed of a mixture of two types of substances: a low-temperature fraction representing the basis of condensed material, which holds the volatile elements, and a high-temperature fraction represented by individual chondrules and metallic grains, which have lost volatile elements. Further, these facts indicate the possible formation of the above-mentioned fractions directly from high-temperature nebula during cooling. They could not have been formed in any parent meteorite bodies. Conclusions were also made concerning the probable temperatures of formation of the meteorites: $T \leq 400$ K for the carbonaceous chondrites, 400–800 K for enstatite meteorites (type I), 530–680 K for ordinary and enstatite (type II) chondrites, and $T \leq 530$ K for ordinary non-equilibrium chondrites. These temperatures are considerably higher than those typical of the asteroid belt (170 K). On the basis of this hypothesis the authors have assumed that the enstatite chondrites originate from the inner part of the asteroid belt, the ordinary chondrites from its central part, and carbonaceous chondrites from the outer belt.

Larimer, Anders, Grossman, and other researchers appear to have developed a theory which successfully explains many aspects of the chemical differentiation of substances in meteorites. But this theory of the equilibrium condensation of chemical elements and compounds of the protoplanetary material has met with a number of difficulties. For example, in the carbonaceous chondrite Allende (fell in 1969, 2000 kg), the high-temperature inclusions of minerals were surrounded by incompatible low-temperature minerals. The relative ^{18}O content in the three inclusions of halenite and spinel in this meteorite range from −9.7‰ to −11.5‰, which corresponds to the equilibrium temperature of their formation, approximately 800 K, for those portions of the cloud where these inclusions were formed. This temperature appears to be too low compared with that following from studying the texture and mineralogy of the inclusion. The type I enstatite chondrites must have been formed at very high pressures so that Tl remained at T = 400 K. The presence of graphite requires high formation temperatures and the presence of water requires low temperatures. The general conditions of the formation of carbonaceous chondrites were found to be inadequate on the whole. With regard to the formation temperatures of these meteorites they should not have been hydrated, whereas the majority of them contain a great amount of water. Besides, Clayton et al. (1977) have discovered in all the C_2, C_3 and C_4 meteorites they studied as a whole, and in the spinel, olivine, and feldspar minerals, that there is excess of ^{16}O, which has been prescribed to the pre-solar component, for which the mechanism of transfer to the meteorites and larger bodies including the planets remains uncertain. The fractionation picture in the iron-silicates system also remains ambiguous. The volatile elements in which all the stone meteorites are depleted, and the existence of organic materials in the form of the high-molecular hydrocarbon compounds in meteorites, create a serious problem which is unsolved as yet. The observed amounts of the noble gases should also be explained. It has been found that the content of Ar, Kr and Xe in individual types of chondrites differs by three orders of magnitude. Helium and neon were found only in the carbonaceous chondrites and in very small amounts in the ordinary chondrites, which is in disagreement with their content in the Solar wind. Some promising ideas of application of the absorption effects and equilibrium dissolution of noble gases in the meteorite material during its formation have appeared recently for explanation of the observed data.

Grossman and Larimer (1974) have pointed out that in order to explain, from the view point of the equilibrium of a substance and its accretion, a number of chemical, mineralogical, and textural peculiarities in stone meteorites, at least four processes are required, resulting in fractionation of the condensed material:

1. The most refractory material, rich in Ca, Al and Ti, must have transferred into the internal zone of the protosun cloud. Enstatite and ordinary chondrites were preferentially depleted in this material and outer zones of the Earth and Moon were enriched in it. These formations also included the metallic portion, rich in refractory compounds.
2. Most of the metal must have been extracted from the silicate grains at T = 1000–700 K. This process could only have occurred in the initial material of chondrites, which explains the observed differences in the densities of the terrestrial planets.
3. Before the accretion started, some portion of the condensed material must have been heated in order to obtain chondrites depleted in volatile elements.

4. The chemically unseparated dust fraction should have remained in equilibrium with the gaseous phase gradually being enriched in the volatile components (In, Tl, Xe etc.). By inclusion of the chondrules in different proportions, ordinary (at T = 450 ± 50 K and pressure 1 Pa) and carbonaceous (at T = 350 ± 50 K and pressure 0.1 Pa) chondrites could have been formed.

As to conditions of the formation of iron meteorites, Kelly and Larimer (1977) have pointed out that under their apparent plain composition lies a complicated cosmochemical history. According to variations of a number of elements (Au, Co, Cu, Fe, Ga, Ge, In, Mo, Ni, Os, Pd, Pt, Re, Rh, Rn) the iron meteorites are divided into 12 groups (Scott and Wasson,1976). The cosmochemical history of their metallic phase is related to the four stages of evolution:

1. Condensation and fractionation in the protosun nebula and accretion of the substances.
2. Oxidation-reduction process in the nebula and parent bodies.
3. Melting and differentiation in the parent bodies.
4. Fractionation crystallization during solidification.

On the basis of studies of distribution of the above-mentioned elements in iron meteorites on the whole, and also in their metallic, silicate, and sulfide phase, Kelly and Larimer (1977) have found that out of 12 groups of iron meteorites five satisfy the conditions of passing through all these four stages. The seven groups have unusual histories. In order to explain the observed variations of the chemical composition in some groups, high-temperature condensation (T = 1270 K at p_τ = 1 Pa) and, therefore, high-temperature accretion is required. For the other groups there is evidence of quick metal cooling (7–200 degree/million years) and partial melting of aggregates of the parent bodies. In the third group some features are observed which indicate the occurrence of recrystallization after partial melting.

The cosmochemical history of the metallic phase in chondrites and achondrites does not seem to be a less complicated problem.

Herndon and Suess (1976) have reported that the possibility that such minerals as TiN, Si_2N_2O and CaS, which occur in enstatite chondrites and achondrites, are formed directly from the gas nebula at the accepted conditions of equilibrium condensation is doubtful. They have noted that an extremely reducing medium is needed to obtain Ti and Fe with inclusions of the elementary silica. According to Herndon and Suess the models of formation of the above-mentioned compounds can be constructed only when assuming the pressure in the gaseous cloud to be $p_\tau \geq 10^5$ Pa.

Herndon and Suess have also thrown doubt upon the possibility of iron formation in sulphide, oxide, and pure metal forms in chondrules at equilibrium conditions. They have assumed that one way of obtaining iron in the three chemical forms occurring in ordinary chondrites is described by the non-equilibrium gas condensation model, which is being developed by a number of researchers.

Blander and Katz (1967), Blander and Abdel-Gavad (1969) have considered the problem of condensation from the gas nebula of the primary formation from which planets and meteorites could later have been formed under non-equilibrium conditions of condensation. They suggested that the chemical composition of the cooling gas corresponded to the composition of elements in the solar system. Using data obtained earlier by Suess, Urey, and Cameron as a basis, the authors (Blander and Katz, 1967; Blander and Abdel-Gavad, 1969) have assumed the following sequence of ratios for the main constituents of the gas: H_2:H_2O:SiO_2:Mg:Fe as 10^{10}:10^7:$10^{5.5}$: $10^{5.4}$:$10^{5.3}$. It was assumed that the cooling of the nebula occurred, along with other possible processes, due to energy losses by radiation as a result of which the gas cooled quickly at the boundary and more slowly in the interior of the nebula. Gas condensation was determined by its pressure and the rate of cooling. In this case as the upper limit of pressure and the lower limit of the rate of cooling in the centre of the nebula are reduced, the size of the nebula is increased. The condensation process occurred under non-equilibrium conditions and displayed a threshold character. It has been shown by the above authors that under the isobaric process of gas cooling at certain conditions, iron will precipitate first and then the silicate phase is condensed. The compounds enriched in calcium can condense earlier than silicates. It has been found that the process of material cooling at conditions of restricted equilibrium results in large variation of the chemical elements in the condensation products, which can explain the high content of oxides of iron in the silicate phase and also the observations of the different content of volatile elements in chondrites. Finally, they have studied the formation mechanism of individual chondrules in meteorites, and on the basis of the obtained experimental data have shown that they are the primeval formations of the liquid droplet condensate precipitated in the overcooled condition and subsequently solidified.

On the other hand, Arrhenius and Alfven (1977), while studying the conditions of evolution of the protoplanetary material on the basis of their own evidence, assumed that the process of its condensation could have started directly from the plasma state of the cloud, being at a temperature of 10^4 K and pressure $<10^{-2}$ Pa proceeding under disequilibrium conditions. In this case the refractory elements and compounds condense at $p_\tau = 10$ Pa and T = 1250 K. The elements in which chondrites are normally depleted condense in the temperature range 1300–600 K, hydrosilicates at T < 350 K, CH_4, NH_3, H_2O, and hydrated methane at T < 200 K. Arrhenius and Alfen note that the temperature of solidification of material could have been considerably lower than the temperature of the surrounding gas. This is because the meteoric substance, according to the observation data, was not differentiated at the first stage and, therefore, it was formed under extreme disequilibrium temperature conditions with the surrounding gas.

Experimental observations and conclusions concerning the conditions of condensation and chemical fractionation of the protoplanetary material in meteorites, and also direct cosmochemical observation of the Moon and planets, are leading to a gradual revision of our ideas on the nature of the chemical differentiation of material in planets.

Analysing the theories of possible formation of the Earth with regard to the two-component model, Larimer and Anders (1967) have considered that starting from the suggested hypothesis, the high-temperature component could have formed the planet's core at T ≥ 1159 K. The low-temperature fraction should have amounted to 10%. In view of the actual abundances of Bi, In and Tl in the Earth's core and also the higher temperature at distance of 1 A.U. from the Sun, the temperature of formation of the upper shell of the planet could have been ≤400 K.

Turekian and Clark (1969) and Clark et al. (1972) have suggested that the formation of the Earth occurred in a sequence determined by the process of substance cooling, which started at T = 2000 K and pressure 10^2 Pa. According to their model the primeval Earth immediately obtained a shell structure due to a decrease of the iron-silicate ratio from the Earth's centre to the surface. In the course of the Earth's growth its gravitational energy increased, resulting in melting of the body being formed. In this case the liquid iron must descend to the planet's centre, forming its core. Finally, when the Earth cooled, 20% of its substance, rich in volatile components, were added. In Turekian and Clark's model an attempt has been made to avoid the geochemical difficulties which were noted by Ringwood (1966).

Vinogradov (1971; 1975) came to the conclusion that at the high-temperature stage of the evolution of gaseous cloud physicochemical differentiation, accompanied by emergence of iron and silicate phases, had already started. These phases, during further processes of the cloud's evolution, provided the basis for formation of the cores and mantles of the future planets.

On the basis of Larimer and Anders's two-component model, Laul et al. (1973) estimated the formation temperatures of the Earth, Moon and meteorites of different type. They took Tl as the cosmothermometer, since Tl/Rb and Tl/Cs ratios for the lunar and terrestrial rocks were found to be the same and the K/U, Rb/U, and Cs/U ratios for the same rocks approximate to constant values, where uranium was considered as a nonvolatile element. The magmatic processes are known to be inefficient in the fractionation of rocks.

In Table 9.5 the temperatures obtained by Laul et al. (1973) are shown and, for comparison, those obtained by Onuma et al. (1972) on the basis of oxygen isotope ratios for the same objects are added.

Table 9.5 Accretion temperatures of the earth, the moon, and meteorites

Object	Heliocentric distance (A.U.)	Pressure (Pa)	Low-temperature phase content (%)	T (K) Tl	$^{18}O/^{16}O$
The Earth (oceanic basalts)	1.0	10	11	458	450–470
The Earth (continental basalts)	1.0	10	11	456	450–470
The Moon	1.0	10	1.5	496	455
Chondrites type H, L, LL	2–3	1	25	420–500	445–480
Sherghotites	2–3	1	21	433	455
Nakhlites	2–3 (?)	1	40	438	460
Eucrites	2–3 (?)	1	0.5	432	475
Carbonaceous chondrites type C_1	3	0.1	>0.95	394	360
Carbonaceous chondrites type C_2	3	0.1	5.5	394	380

The data presented in Table 9.5 indicate, assuming that the initial assumptions are right, that the Earth, Moon, and all the above-mentioned types of meteorites were found within the narrow temperature range which could have existed in that part of the protoplanetary cloud where the formation of the considered bodies occurred. However, these data characterize the formation of the low-temperature phase. Assuming that the above-mentioned bodies contained the high-temperature phase in different proportions, this interpretation has certain difficulties with regard to the accretion hypothesis.

The latest studies of the Moon, the terrestrial group of planets, and Jupiter show that the processes of chemical differentiation of the protoplanetary substance have been more complicated in their character than those described in the existing models.

According to data obtained by the rubidium-strontium method (Wasserburg et al., 1972; Wood, 1974) the age of the sampled inland rocks ranges within $(4.3–4.6) \times 10^9$ years. This interpretation of the results of the age, combined with petrological studies, leads to the conclusion that the Moon underwent a stage of melting and chemical differentiation (at least in its upper shell) at the final stage of its formation. At that time the whole of the upper layer of the Moon within depth ≥ 100 km was in a melting state and during crystallization formed a low density core ($\rho = 2.9$ g/cm^3), mainly of plagioclase composition, within a short time interval (~200 million years). The surface melt was not formed as a result of the processes occurring in the interior of the Moon but more likely reflects the high-temperature conditions of its formation. Only some local areas of the lunar seas where filled by subsequent lava eruptions of a somewhat younger age and corresponding chemical composition. The formation of a crust with a lower density, as compared with the average density characteristic of the whole body, indicates that the chemical differentiation of the lunar substances had been completed up to the moment of the body formation, including the formation of a small size metallic core. On the basis of estimation of the energy losses and the core heat conductivity, the time of cooling of the surface layer and crystallization of the core was found to be 10^8 years. The calculations have shown that, in order for the energy released by radioactive decay to melt the lunar substance, such decay would have to continue for 2×10^9 years. According to direct measurements the heat flux from the lunar surface at the present time has been found to be unexpectedly large, about 3×10^{-2} J/m^2 sec.

The conclusions concerning the heat history of the Moon throw doubts upon the model of the cold accretion of substance during formation of the Moon and required a reassessment of the role being played by radiogenic energy in the thermal history of the Moon, and obviously in the history of all the planets of the terrestrial group.

Cameron's model attempts to explain the early lunar heating by the release of gravitational energy during the quick accretion of substance and meets the serious cosmochemical difficulties.

Analyzing the chemical composition of the samples of lunar soils, breccias, and igneous rocks, it has been found that they are characterized by a number of cosmochemical peculiarities which distinguish them from terrestrial rocks and meteorites. The igneous lunar rocks are enriched by 10–100 times in Ca, Zn, Hf, Ta, rare earth, and other refractory elements and depleted in alkaline, halogen, and volatile metals (Bi, Tl, Hg) compared with carbonaseous chondrites of the C_1 type. The lunar rocks as a whole are enriched in high-temperature and depleted in low-temperature elements relative to their cosmic abundance. Ganapathy and Anders (1974) have given comparative data on the abundance of elements in lunar and terrestrial rocks (Table 9.6).

Attempts were made to explain the observed distribution of chemical elements in lunar rocks on the basis of the different models. The idea that the Moon could have been formed from high-temperature compounds, which are represented in inclusions of the carbonaseous chondrites (C_1 type), meets with difficulties in that these compounds in chondrites are enriched both in high-temperature siderophylous (Re, Os, Ir) and litophylous elements. The lunar rocks are extremely depleted in the latter elements. The two-component model of the accretion of the Moon is therefore preferable. But this suggestion meets difficulties in explaining the oxygen isotope composition of the lunar rocks where $\delta^{18}O \approx +6‰$, whereas inclusions of the carbonaseous chondrite Allende indicates $\delta^{18}O \approx -10.5‰$.

Attempts directed towards explaining the oxygen isotopic composition in the framework of the two-component model of the primeval lunar substance have therefore failed.

Other important facts which have not been explained from the viewpoint of the conditions of the Moon's substance are:

1. The Moon does not have water and obviously did not have any (at least in its upper layers). In the case of its accretion from the substance which makes up the Earth, the lunar rocks should be hydrated or have features of hydration, but these are not actually observed.
2. No signs were found of granitization of rocks, which is likely to be closely related to the absence of water.
3. The lunar minerals tend to be reduced due to a deficit of oxygen, which can also be accounted for by the absence of water.
4. Iron is also in a reduced condition and indicates a phenomenon which is usually for terrestrial silicates, also obviously accounted for by the absence of water.

Table 9.6 Relative abundance of some groups of elements in terrestrial and lunar rocks

Group of elements	Temperature of condensation in the cloud (K)	Abundance relative to cosmic	
		Earth	Moon
Refractory elements (Al, Ca, Ti, Ba, Sr, U, Th, Ph, Ir)	>1400	−1	2.7
Silicates (Mg, Si)	1400–1200	1*	1*
Metals (Fe, Ni, Co, Cr, Au)	1400–1200	~1	0.25
Zvolatile elements (S, Na, K, Cu, Zn, Te)	1300–600	0.25	0.05
High volatile elements (Cl, Br, Hg, Pb, In)	<600	0.02	0.0005

*Given relative to the magnesium silicates

5. An age paradox is observed: the lunar dust, being the product of rock destruction, appeared to be older than the rocks themselves. This phenomenon is accounted for by the presence of the 'magic' component in the dust composed of potassium, rare earths, and phosphorus and also by the effect of the solar wind.
6. There are visible signs of the formation of minerals of surface sediments from the vapor phase.

The obtained data on cosmochemistry and the thermal history of the Moon have created great difficulties in justification of the accretion, chondritic model of its formation. The other models of lunar formation based on the hypothesis of separation of its substance from that of the Earth were found to be less inconsistent.

An analysis of the high-resolution photographs of Mercury obtained during the flight of Mariner-10 allowed the following conclusions concerning this planet to be drawn (Murray et al., 1974).

The surface of Mercury resembles the lunar seas and represents the hardened melt of large-scale fluxes of lava. According to the visible signs this melt is mainly constituted by silicates with density of about 3 g/cm^3. Because the mean density of the planet is equal to about 5.5 g/cm^3, the internal layers of the planet should consist of substances heavier than silicates. The planet obviously represents a body differentiated by density and chemical composition with a massive iron core.

Mercury's surface, as with the lunar surface, is covered with craters of impact origin. It is quite justifiably assumed that the chemical differentiation of Mercury, as well as that of the Moon, was finished before the formation of the body. If it were not, a sign of the intense bombardment of its surface by meteorites would have been lost. For the same reason it has been concluded that Mercury has had neither a primary nor a secondary atmosphere, in contrast to Mars where aeolian processes had considerably changed the primary landscape of the planet.

The comparison of the surface structures of Mercury, Mars, and the Moon has led to the conclusion that the mechanisms of the formation of these bodies and their chemical differentiation were common.

The studies of Jupiter, carried out with the help of spacecraft, had provided a number of important results. The most interesting of these appeared to be the data of the high-temperature state of the planet and on the powerful flux of radiation from its surface. On the basis of measurements carried out by the cosmic spacecraft Pioner-10, the temperature of the planet's core is estimated to be about 50 000 K, the temperature of the surface layer is 2000 K, and that of the atmosphere is 120 K. Jupiter radiates energy whose value is 2–3 times greater than that obtained from the Sun. The approximate chemical composition of the planet is estimated as the following: 82% hydrogen, 17% helium, and 1% other elements (Cotardiere, 1975).

The obtained data on the temperature regime and chemical composition of Jupiter does not provide any basis for the usage of any of the existing hypotheses for an explanation of the conditions of its formation. The accretion hypothesis is the most vulnerable. It is worth noting that Jupiter exhibits the greatest angular momentum of all the bodies of the Solar System. Despite this fact, the hypotheses being developed concerning the origin of the Solar System dealt mainly with planets of the terrestrial group. This is most probably the weakest point of the existing hypotheses.

The data given above on the cosmochemical studies of meteorites, planets, and the Moon, compiled in the second part of the twentieth century, leads to one general and possibly indisputable conclusion that, in both the protosolar and protoplanetary clouds, there were common processes of chemical differentiation of the initial substances. These processes have obviously led to cosmochemical differentiation between the planets and the Sun, and between the planet's shells as well. The differentiation mechanism seems to be the same. Let us apply our analytical solutions in dynamics to the above-presented cosmochemical observations.

9.5 Differentiation of the Substances with Respect to Density and Conditions for the Planet and Satellite Separation

From the point of view of planetary dynamics, chemical differentiation of substances during separation of planets and satellites from the common solar nebula and formation of body shells is the problem of differentiation of the cloud's matter with respect to atomic and molecular weights or, generally saying, according to density. The physical basis for consideration of this problem comprises the dissipative processes in the nebula or cloud which are related to gravitational and electromagnetic interaction between the constituencies. This condition results in generation of gravitational and electromagnetic energy and its loss from the surface shell in the form of radiation. This condition follows from consideration of the structure of the potential energy of a system that is non-homogeneous in its elementary content, such as considered in Sect. 6.2. The given condition for the protosolar system is satisfied a priori.

The problem of differentiation of substances with respect to the atomic and molecular weights of a cloud in its own force field was considered in Sect. 6.5, where separation of a mass with respect to density is based on Roche's dynamics and Newton's theorem about gravitational interaction of the material point and a spherical shell. The real mechanism of separation of atoms and molecules in a diffused stage of the cloud should be the same mechanism of generation of the electromagnetic (gravitational) energy considered in Chap. 8.

The process of atomic and molecular mass separation in natural systems (in nebulae and bodies) seems to be not an episode in their history. This is the continuous process of any system connected with its evolution, the essence of which is collision and scattering of the particles (molecules, atoms, nuclei and so on) accompanied by their destruction and removal of smaller, up to elementary, particles, which form the flux that we call energy and its emission. In turn, as it was shown in Chap. 8, the energy generation, i.e. the transition of the mechanical energy of a system oscillation into the energy of the electromagnetic oscillation on atomic and molecular (and nuclear for the stars) level leads to change the atomic and molecular weight and the substances themselves. Let us consider the problem of separation of the planets and satellites from the protosolar nebula. For this we come back to solution of the equation of dynamical equilibrium of a dissipative system presented in Sect. 5.2 in the form (5.36)

$$\ddot{\Phi} = -A_0\big[1 + q(t)\big] + \frac{B}{\sqrt{\Phi}}, \qquad (9.1)$$

where Φ is Jacobi's function (polar moment of inertia of the system); q(t) is the time parameter continuously ascending due to dissipation of energy at 'smooth' evolution of the system in the time interval $t \in [0, \tau]$; A and B are constant values.

Solution of Eq. (9.1) was found in the form (5.37) and (5.38)

$$\mp arccos\, W \pm arccos\, W_0 \mp \sqrt{1 - \frac{A_0\big[1 + q(t)\big]C}{2B^2}} \sqrt{1 - W^2}$$

$$- \sqrt{1 - \frac{A_0 C}{2B^2}} \sqrt{1 - W_0^2} = \sqrt{\frac{\big(2A_0\big[1 + q(t)\big]\big)^{3/2}}{4B}}\, (t - t_0).$$

$$(9.2)$$

Equations of the discriminant curves limiting the amplitude of oscillation of the Jacobi function (polar moment of inertia) are (5.39) and (5.40)

$$\sqrt{\Phi_{1,2}} = \frac{B}{A_0\big[1 + q(t)\big]}\left[1 \pm \sqrt{1 - \frac{A_0\big[1 + q(t)\big]C}{2B^2}}\right], \quad t \in [0, \tau]. \qquad (9.3)$$

Expression for the period of oscillation T_v and amplitude of the moment of inertia of the system, obtained from Eqs. (9.2) and (9.3) are (5.41) and (5.42)

$$T_v(q) = \frac{8\pi B}{\big(2A_0\big[1 + q(t)\big]\big)^{3/2}}, \qquad (9.4)$$

$$\Delta\sqrt{\Phi} = \frac{B}{A_0\big[1 + q(t)\big]}\left(1 - \frac{A_0\big[1 + q(t)\big]C}{2B^2}\right)^{1/2}. \qquad (9.5)$$

Figure 9.3 shows changes of the polar moment of inertia due to oscillating dissipation of the system's energy. The changes are confined the discriminant curves (9.3). At the point O_b the integral curve (9.2) and the discriminant curves (9.3) tend to coincide and the amplitude of the oscillation of the moment of inertia of the system decreases up to zero. In this case the system reaches the stage close to

$$\Phi = const. \qquad (9.6)$$

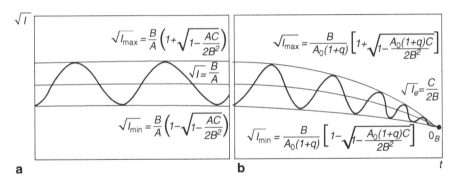

Fig. 9.3 Evolution of a dissipative system to bifurcation point

Expression (9.6) is the second solution of the non-linear differential Eq. (9.1). The point O_b is physically interpreted as the bifurcation point. Here, under action of its own force field, the mass of the upper shell of the system reaches dynamical equilibrium with the inner mass and separates into a new subsystem. Equating the radicand of Eq. (9.3) to zero one finds that

$$\frac{2B^2}{A_0(1+q_b)} = C, \tag{9.7}$$

where q_b is the value of the parameter q for the bifurcation point equal to

$$q_b = \frac{2B^2}{A_0C} - 1. \tag{9.8}$$

Then Jacobi's function (polar moment of inertia) Φ_b of the system at the bifurcation point, where the discriminant curves coincide, is

$$\phi_b = \frac{B^2}{A_0\left(1 + \frac{2B^2}{A_0C-1}\right)^2} = \frac{C^2}{4B^2}. \tag{9.9}$$

While analyzing the equation of dynamical equilibrium of a celestial body by taking into account the Coloumb interactions of the charged particles shown, we found in Chap. 8, that the relationship between the polar moment of inertia and potential energy of the Coulomb interactions, written for conservative and dissipative systems, holds. In this case solution (9.2) of Eq. (9.1) for electromagnetic interactions also holds and the system can be represented by the model of the oscillating electric dipole which generates electromagnetic oscillations (Ferronsky et al., 1982). Mechanism of that process was considered in the previous chapter from which it follows that the dissipative system cannot be electrically neutral.

The obtained solution of the equation of dynamical equilibrium of a dissipative system (9.1) represents a theoretical solution of the problem of the equilibrium or 'smooth' form of evolution of a celestial body. The condition of achievement by a dissipative system of the bifurcation point and the obtained relations for its parameters should be a qualitative solution of the problem of secondary body separation. For application to the cosmogony of the Solar System, it means that all its planets were formed from the common gaseous nebula by its separation during evolution through emission of electromagnetic energy. The evolutionary process was developed by density differentiation of the gaseous substances of the nebula and by separation of the planets and satellites from the outer shell. The normal and tangential volumetric forces of the interacting mass particles there built up permanent conditions for creation of vortexes which seems to be the main mechanism in formation of new entities. The conditions for the formation of meteorites existed during the formation of both the planets and their satellites.

Now we can define the specific physical effect of separation of bodies in the Solar System from the common nebula. It should be the inner energy (force function) of Coulomb interaction of the atomic, molecular and nuclei particles. It has been shown (Ferronsky et al., 1978) that for the planets in the Solar System, at the moment of their formation, the magnitude of the gravitational interaction energy was equal to the electromagnetic energy. In fact, the well-known expression for the energy of the Coulomb interactions U_c, written through the Madelung energy (Kittel, 1968) is:

$$U_c = n\left(e^2/R_0\right) k, \tag{9.10}$$

where $n = m/\mu$ is the quantity of interacting molecules; m is the body mass; μ is the average molecular weight; $e = 4.8 \times 10^{-10}$ is the charge of the electron; k is the Madelung coefficient; $R_0 = (V/n)^{1/3} = R_0(\mu/m)^{1/3}$ is the average radius of the Coulomb interactions of the molecules; V is the volume of the body.

Equating the expressions for the potential energy of the gravitational and Coulomb interactions, one can derive a relationship between the critical mass of the planet and its average molecular weight μ:

$$\alpha^2 \left(\frac{Gm^2}{R}\right) = n \frac{e^2}{R_0} k,$$
$$m_{kp} = \frac{e^3 k^{3/2}}{a^2 G^{3/2} \mu^2}. \tag{9.11}$$

From expression (9.11) one can find the value of the average molecular weight μ for the planets of the Solar System (see Table 9.7) (Ferronsky et al., 1978).

We consider relationship (9.11) to express the condition of formation of the condensing protoplanetary clouds, characterized by the corresponding average chemical compositions included in Table 9.7. On the basis of expression (9.11) we obtain an important cosmochemical dependence of the planet's mass from its average

Table 9.7 The average value of the molecular weight (in a.u.m) for the planets at time of their creation

Planets	Mercury	Venus	Earth	Mars	Jupiter	Saturn	Uranus	Neptune
By Eq. (9.11)	256	66	60	183	3	6	15	14
By Eq. (9.12)	467	85	109	336	6	11	28	26

chemical composition,

$$m\mu^2 = b = const, \tag{9.12}$$

where b is a constant depending in each case on the mass density distribution and the average chemical composition of the body being formed, due to the effect of the Coulomb interactions, as follows:

$$b = \frac{e^3 k^{3/2}}{a^2 G^{3/2}}.$$

Earlier, in Sect. 8.4, from solution of the Chandrasekhar-Fermi equation we obtained a rigorous solution of the problem of secondary bodies separating from the nebula at its evolution and found that the constant $b = 2 \times 10^{-16} \, g^3$ (Ferronsky et al.,1996). The corrected average values of the molecular weight for the planets calculated by (9.12) are given also in Table 9.7.

Thus, the first condition of the planet or satellite creation from the protosolar nebula at its evolution was equality of gravitational and Coulomb energy of interaction for the outer shell. Taking into account the mechanism of mass density differentiation of a cloud, the probable mechanism of a planetary body creation should be cyclonic vortexes. In this connection it is worth recalling the vortex hypotheses of creation of the Solar System proposed by Descartes which now appears to be reasonable. The argument against this theory, related to distribution of the angular momentum between the Sun and the planets, taking into account the found oscillating (volumetric) mode, which is the moment of momentum in this book, now drops out.

The second important condition of a body separation from the common nebula is equality of the normal and tangential components of the potential energy for the outer shell. This condition determines necessity of dynamical equilibrium of a new entity as a self-gravitating body and its ability to move on the orbit in the outer parent's force field.

Accounting for the Coulomb forces one may obtain a qualitative solution of the problem of chemical differentiation of substance in the course of the formation of planets. We have already pointed out that expression (9.12) is the cosmochemical criterion for the formation of planets and satellites. Coulomb forces are more effective for the elements with greater atomic numbers. In the process of dissipation of the potential energy by radiation, a differentiation of the substance occurs on the same physical basis in accordance with boiling temperatures and the relative volatilities of the elements and compounds.

To the first approximation, while estimating the chemical diffraction of the substance of the planets and satellites, one can use the following model which is similar to the two-component model of Larimer and Anders (1967).

Assume that when the protosolar nebula reaches the stage of the bifurcation point (see Fig. 9.3) it gives birth to protoplanetary condensation—the cloud, the substance of which consists of chemical compounds characterized by their average molecular weight μ in accordance with relationship (9.11). In this case all the protoplanetary cloud's elements and compounds, boiling temperatures of which are higher than that of the element and compound with molecular weight μ (refractory component), should remain in the gaseous phase, dissolved in the major component of the condensate. The quantitative estimation of abundances of the elements and compounds being present in the condensed and gaseous phases is determined by the abundances of these elements and compounds in the shell of the protosolar nebula from which the formation of the protoplanet has occurred. The separation of the high-temperature and volatile components of the protoplanetary cloud and the formation of its shells, takes place due to the Coulomb forces at the stage of formation and evolution of the secondary body.

Let us use this condensational model of planet formation to explain the observed distribution of volatile elements on the Earth and also their average isotope composition in comparison with their solar abundance (Onufriev, 1978). To a first approximation one can use for this purpose the thermodynamic laws of the ideal gases and write the separation factor η_{AB} of the low-temperature (volatile) component A and high-temperature (refractory) component и B in the gas-condensate system in the form

$$\eta_{AB} = p_A/p_B \approx exp\ [(L_B - L_A)/R_0 T]\ \ \text{or}\ \ lg\ \eta_{AB} = n(T_B - T_A),\quad (9.13)$$

where p, L and T are the vapor pressure, boiling heat of vaporization, and boiling temperature, respectively; n is the numerical factor; R_0 is the gas constant.

If the accepted model of the formation of the chemical and isotope composition of substance is true and expression (9.13) reflects, to a first approximation, the real process of such a formation, then the observed abundance of chemical elements on the Earth relative to their solar abundances can be written as follows:

$$lg\ \eta_{AB} = l_g\left(N_A^c/N_A^3\right) - l_g\left(N_B^c/N_B^3\right),\quad (9.14)$$

where N^c and N^3 are the solar and terrestrial abundances of a component respectively.

In view of the above-mentioned abundances of the noble gases He, Ne, Ar, Kr, Xe were studied on the Earth and Moon relative to their solar abundances (Onufriev, 1978). Figure 9.4 indicates the dependence of the deficit of each of the elements on Earth and the Moon $[\Delta lg(N^c/N^3)]$ upon the separation factor ($lg\ \eta$) defined by the boiling temperature of the elements relative to xenon. The observed linear dependence of the considered parameters indicates a correspondence between the experimental facts and the accepted model based on ideas about the studied process.

Besides the shallower slope of the line corresponding to the dependence on deficit of the element and the separation factor for the Moon as compared with the Earth, the line shows that the formation of the Moon and the distribution of the elements within the protoearth cloud occurred at rather high temperatures, which is also in accordance with the given theoretical and model concepts.

The probable conditions of formation of the volatile elements, some of their components, and the isotope composition of the upper shell of the Earth, were also considered (Onufriev, 1978). Assuming the maximum boiling temperature of high-temperature constituents of the Earth to be equal to 3300 K, the corresponding values of the deficit lg (N^c/N^3) depended on the separation factor lg η for the noble gases and a number of the refractory elements were plotted in Fig. 9.5. The relative volatility of the noble gases, which do not form chemical compounds, depends directly on boiling temperature. It follows from Fig. 9.5a that the dependence of lg (N^c/N^3) ≈ f(lg η) has a natural character. The group of high-temperature elements does not indicate such a dependency in the domain where the values of abundances are close to the solar ones.

The first group of elements was studied in more derail together with a number of other volatile elements and compounds at a maximum boiling temperature of 1600 K (Fig. 9.5b). It follows from the figure that the value of the deficit of each of the considered elements decreases with a decrease of their relative volatility, which corresponds to the model considerations regarding the formation of the Earth's chemical composition concerning at least the volatile elements.

From the agreement between the corresponding thermodynamic parameters and the observed data on the deficit of a number of other volatile elements and their compounds, plotted in Fig. 9.5b, the following conclusions have been drawn (Onufriev, 1978). During the formation of the Earth from the protosolar nebula the hydrogen reservoir was mainly represented by water and only a small portion of hydrogen was in the molecular form H_2. The nitrogen reservoir was represented by ammonia

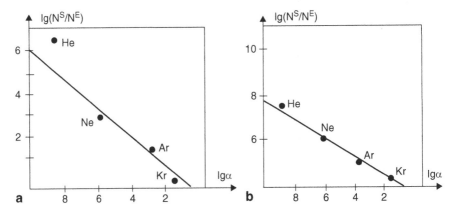

Fig. 9.4 Relationship between the value of inert gases deficit and differentiation factor for the Earth (**a**) and for the Moon (**b**)

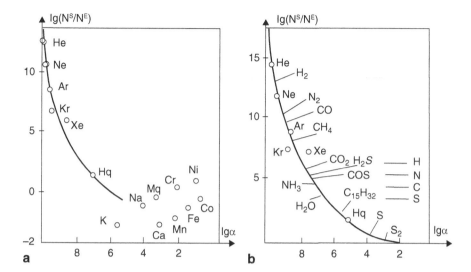

Fig. 9.5 Relationship between the value of the element and the compound deficit and differentiation coefficient for the Earth at 3300 K (**a**) and 1600 K (**b**)

NH_3 and the Earth's atmosphere as a whole was in a reduced form. Searching the possible primeval carbon compounds of the Earth (CO, CH_4, CO_2, COS), it appeared that neither one of them nor all being taken in corresponding proportions can explain from the assumed viewpoint the observed deficit of carbon. The condition can only be satisfied if assuming the existence of a higher temperature compounds of the form C_nH_{2n+2} with boiling temperature of ~400 K. Assuming the initial existence of sulphur in the form of S, S_2 and so on, the existence of the lighter low-temperature compounds such as H_2S, CaS is required.

Note that investigation of the dependence of the element's deficit upon its relative volatility, expressed through the boiling temperature, together with a consideration of the different forms of its chemical compounds, provides the key for understanding the general principle of the successive increase of the element deficit on the Earth in row Hg, S, C, N, H, Xe, Kr, Ar, Ne, He.

The considered model of formation of the chemical composition of the Earth should be completely applicable for studying the conditions of formation of their isotopes.

Without any loss of generality, one can consider that an element's isotope with a greater mass number is less volatile compared with the lighter isotope. Therefore, expressions (9.13) and (9.14) are valid for estimation of the dependence of abundance on the planets of isotopes of corresponding elements upon the separation factor expressed through the boiling temperature.

It has been shown (Onufriev, 1978) that the above-mentioned enrichment of the Earth's upper shell in heavy isotopes, compared with the cosmic abundance for the volatile elements, is natural from the viewpoint of our model. In view of this there are reasons to consider that water of the hydrosphere was initially enriched in heavy oxygen to about +20‰. The hydrosphere itself should have been in the vapour phase. With decreasing temperature the water over the Earth's surface should have been converted into the liquid phase. Precipitation of large amounts of water on the Earth's surface was the trigger mechanism for magmatic processes to start. The process of granitization of rocks on the continents should have resulted in depletion of the hydrosphere in heavy oxygen and the subsequent enrichment of granite up to the present values. Thus, starting from the considered model of the formation of the Earth, with its chemical and isotopic composition, the observed enrichment of the upper shell of the Earth in heavy isotopes of the light elements can be explained uniquely.

Note that we can now derive a conclusion regarding the meteoric origin of the hydrosphere. The condensation of water to the liquid phase occurred at the final stage of our planet's formation. One can reasonably assume that the time interval between the end of condensation of the mineral part of the Earth and the beginning of water condensation could have been large in view of the difference in their boiling temperatures. This interval could have been markedly enlarged if the greenhouse effect was provided by carbon dioxide and water vapor, similar to that now observed on Venus. Thus there are reasons for considering that the Earth's hydrosphere, in the liquid phase, is a considerably younger formation than all the other shells. Further, the observed stability of chemical and isotopic composition of the oceans in time is a result of inherited thermodynamic equilibrium which it had attained when in the gaseous phase.

9.6   Conclusion

The problem of dynamics of a celestial body in its own force field is a new task in theoretical and celestial mechanics. Its formulation and solution appears to be possible by application of volumetric (power) forces and volumetric moments. The geodetic satellites of the Earth have made it possible to study these entities closely, and thus prove this approach and demonstrate incorrectness of the hydrostatic equilibrium of the planet. By means of these satellite studies the physical meaning of the famous Jacobi equation which, in fact, is the equation of dynamical equilibrium of celestial bodies, becomes clear. We demonstrated applicability of this equation not only for solution of the problem in the framework of classical mechanics, but also in quantum mechanics and field theory. This opens the way to formulate and solve the problem of a unified field theory. Application of the dynamical approach in thermodynamics, which still remains a phenomenological branch of the science and is based on hydrostatics, make it possible to find its theoretical basis.

Finally, the problem of the carrier of the energy, generated by particle interaction, in the frame-work of the dynamical approach, is given a good background against which to discuss feasible material agents such as the elementary particles, including the infinitesimals, instead of Newton's ether. In this connection Newton, defining the inertial forces, noted: *The motion and quiescence, at their usual consideration, are distinguished like what seems to the ordinary look.* By irony of fate his words completely relate to the hydrostatic equilibrium of the Earth. The inert mass of the planet appears to be guileful. It conceals inside the forces the carrier of which is ... *a certain most subtle spirit which pervades and lies in all gross bodies; by the force and action of which spirit the particles of bodies attract one another at near distances, and cohere, if contiguous; and electric bodies operate to greater distances, as well repelling as attracting the neighboring corpuscles; and light is emitted, reflected, refracted, inflected, and heats bodies; and all sensation is excited, and the members of animal bodies move at the command of the solid filaments of the nerves, from the outward organs of sense to the brain, and from brain into the muscles. But these are things that cannot be explained in a few words, nor are we furnished with that sufficiency of experiments which is required to an accurate determination and demonstration of the laws by which this electric and elastic spirit operates.* The great thinker had a keen understanding of the real world around him, but at that time the reserve of knowledge of electromagnetism was too small. The "dark mass" which at present is a potent subject for speculation of astronomers, could turn out to be the best media for transmission of energy between interacting bodies. It physically could represent the elementary infinitesimal particles that form a background around each cosmic body, including the Universe itself.

We have no plan to study all the aspects of the problem of dynamics of the Earth based on its dynamical equilibrium. It is impossible to do all of this in one study or one book. We have tried to develop only the principal physical and analytical basis of the relevant dynamics for solution of some special tasks. We will be happy if we have succeeded.

References

Ahmad S.N. and Perry E.C. Jr.: 1980, 'Isotopic evolution of the sea', *Science Progress*, **66**, 499–511.

Anders E.: 1972, 'Physical–chemical processes in the Solar nebula, as inferred from meteorites', in: *On the origin of the Solar System*, H. Reeves (ed.), C.N.R.S., Paris, 179–201.

Arnason B. and Sigurgeirsson T.: 1967, 'Hydrogen isotopes in hydrological studies in Iceland', in: *Isotopes in hydrology, Proc. Symp. IAEA*, 35–47.

Arrhenius G. and Alfven H.: 1977, 'Fractionation and condensation in space', *Earth and Planet. Sci. Latt.*, **10**, 253–267.

Ault W.U. and Kulp H.: 1959, 'Isotopic geochemistry of sulfur', *Geochem. et Cosmochem. Acta*, **16**, 201–235.

Begemann F.: 1963, 'The tritium content of hot springs in some geothermal areas' in: *Nuclear geology on geothermal areas*, Spoleto, 55–70.

Bendat G. and Pearsol A.: 1971, *Measuring and analysis on random processes*, Mir, Moscow.

Blander J. and Abdel-Gavad M.: 1969, 'The origin of meteorites and the constrained equilibrium condensation theory', *Geochem. et Cosmochem. Acta*, **33**, 701–716.

Blander J. and Katz J.Z.: 1967, 'Condensation of primordial dust', *Geochem. et Cosmochem. Acta*, **31**, 1025–1034.

Boato G.: 1954, 'The isotopic composition of hydrogen and carbon in the carbonaceous chondrites', *Geochem. et Cosmochem. Acta*, **6**, 209–220.

Bogolubov N.N. and Mitropolsky Y.A.: 1974, *Asymptotic methods in the theory of non-linear oscillations*, Nauka, Moscow.

Bowen R.: 1966, *Paleotemperature analysis*, Elsevier, Amsterdam.

Bowen R.: 1991, *Isotopes and climat*, Elsevier, London.

Briggs M.N.: 1963, 'Evidence of an extraterrestrial origin for some organic constituents of meteorites', *Nature*, **197**, 1290.

Bullen R.E.: 1974, *Introduction to the theory of seismology*, Cambridge Univ. Press, London.

Burnett B.C., Fowler W.A. and Hoyle F.: 1965, 'Nucleosynthesis in the early history of the Solar System', *Geochem. et Cosmochem. Acta*, **29**, 1209–1241.

V. I. Ferronsky, S. V. Ferronsky, *Dynamics of the Earth,*
DOI 10.1007/978-90-481-8723-2_0, © Springer Science+Business Media B.V. 2010

Cameron A.G.W.: 1973, 'Accumulation processes in the primitive Soar nebula', *Icarus*, **18**, 407–450.

Cameron A.G.W. and Pine M.R.: 1973, 'Numerical models of the primitive Soar nebula', *Icarus*, **18**, 377–406.

Cavenaze A. (ed.): 1986, 'Earth rotation: Solved and unsolved problems', in: *Proc. NATO Advanced Research Workshop*, Reidel, Dordrecht, 320.

Chandrasekhar S. and Fermi E.: 1953, 'Problems of gravitational stability in the presence of a magnetic field', *Astrophysics J.*, **118**, 113.

Chase C.G. and Perry E.C.: 1972, 'The oceans: Growth and oxygen isotope evolution', *Science*, **177**, 992–994.

Chase C.G. and Perry E.C.: 1973, 'Oceanic growth models: Discussions', *Science*, **182**, 602–603.

Chebotarev G.A.: 1974, 'Celestial mechanics', in: *Great Soviet Encyclopedia*, 3rd ed., 386–388.

Clairaut A.K.: 1947, *Theory of the earth figure based on hydrostatics* (Transl. from French), Acad. Sci. USSR Publ. House, Moscow-Leningrad.

Clark S.P., Turekian K.K. and Grossman L.: 1972, 'Model for the early history of the Earth', in: *The nature of the solid Earth*, McGraw-Hill, New York, 3–18.

Clayton R.N. and Mayeda T.: 1978a, 'Genetic relation between iron and stony meteorites', *Earth and Planet. Sci. Lett.*, **40**, 168–174.

Clayton R.N. and Mayeda T.: 1978b, 'Multiple parent bodies of polymict brecciated meteorites', *Geochem. et Cosmochem. Acta*, **42**, 325–327.

Clayton R.N., Onuna N. and Grossman L. et al.: 1977, 'Distribution of the pre-Solar component in Allende and other carbonaceous chondrites', *Earth and Planet. Sci. Lett.*, **34**, 209–224.

Clayton R.N., Onuma N. and Mayeda T.K.: 1976, 'A classification of meteorites based on oxygen isotopes', *Earth and Planet. Sci. Latt.*, **30**, 10–18.

Cotardiere Ph.: 1975, 'Jupiter: Un piege fluide', in: *La decouverte des planets*, Paris, 71–79.

Craig H.: 1953, 'The geochemistry of stable carbon isotopes', *Geochem. et Cosmochem. Acta*, **3**, 53–92.

Craig H.: 1963, 'The isotopic geochemistry of water and carbon in geothermal areas', in: *Nuclear geology of geothermal areas*, Spoleto, 17–53.

Craig H.: 1965, 'The measurement of oxygen isotopic paleotemperatures', in: *Stable isotopes in oceanographic studies and paleotemperatures*, Spoleto, 161–182.

Craig H. and Gordon L.: 1965, 'Deuterium and oxygen-18 variation in the ocean and the marine atmosphere', in: *Stable isotopes in oceanographic studies and paleotemperatures*, Spoleto, 9–130.

Craig H. and Lupton J.E.: 1976, 'Primordial neon, helium and hydrogen in oceanic basalts', *Earth and Planet. Sci. Lett.*, **31**, 365–385.

Deines P. and Wickman F.E.: 1975, 'A contribution to the Stable isotope geochemistry of iron meteorites', *Geochim. et Cosmochim. Acta*, **39**, 547–557.

Duboshin G.N.: 1975, *Celestial mechanics: The main problems and the methods*, Nauka, Moskow.

Duboshin G.N.: 1978, *Celestial mechanics: Analytical and qualitative methods*, Nauka, Moscow.

Duboshin G.N., Rybakov A.I., Kalinina E.N. and Kholopov P.N.: 1971, *Reports of Sternberg Astron. Inst.*, Moscow State Univ. Publ., Moscow.

Eckelmann W.R., Broecker W.S., Whitlock D.W. and Allsup J.R.: 1962, 'Implications of carbon isotopic composition of total organic carbon of some recent sediments and ancient oils', *Bull. Am. Assoc. Petrol. Geol.*, **46**, 699–704.

Emiliani C.: 1970, 'Pleistocene paleotemperatures', *Science*, **168**, 822–824.

Emiliani C.: 1978, 'The cause of the ice ages', *Earth and Planet. Sci. Lett.*, **37**, 349–352.

Epstein S.: 1965, 'Distribution of carbon isotopes and their biochemical significance', in: *Proc. Symp. on CO$_2$*, Haverford, 5–14.

Epstein S.: 1969, 'Distribution of carbon isotopes and their biochemical and geochemical significance', in: *Proc. Simp. On CO$_2$: Chemical, Biochemical and Phisiological Aspects*, Haverford, USA, NASA SP-188, 5–14.

Epstein S. and Taylor H.P.: 1970, '$^{18}O/^{16}O$, $^{30}Si/^{28}Si$, $^{2}H/^{1}H$ and $^{13}C/^{12}C$ studies of lunar rocks and minerals', *Science*, **167**, 533–535.

Epstein S. and Taylor H.P.: 1971, '$^{18}O/^{16}O$, $^{30}Si/^{28}Si$, $^{2}H/^{1}H$ and $^{13}C/^{12}C$ ratios of Apollo 14 and 15 samples', in: *Proc. Second Lunar Conf.*, 1421–1441.

Epstein S. and Taylor H.P.: 1972, '$^{18}O/^{16}O$, $^{30}Si/^{28}Si$, $^{13}C/^{12}C$ and $^{2}H/^{1}H$ studies of Apollo 14 and 15 samples', *Geochem. et Cosmochem. Acta*, Suppl. 3, **2**, 1429–1454.

Ester V.F. and Hoering T.C.: 1980, 'Biogeochemistry of the stable hydrogen isotopes', *Geochem. et Cosmochem. Acta*, **44**, 1197–1206.

Ferronsky S.V.: 1983, 'The solution for seasonal variations in the Earth's rate of rotation', *Celest. Mech.*, **30**, 71–83.

Ferronsky S.V.: 1984, 'Observation of coherent pulsations in the Earth atmosphere with period about one and half hour', *Physics of Atmosphere and Oceans*, **20**, 922–928.

Ferronsky V.I.: 1974, 'Origin of the Earth hydrosphere using the data of isotopic composition of water', *Water Resources*, **4**, 21–34.

Ferronsky V.I.: 1974, 'Origin of the Earth's hydrosphere on the basis of water isotope data'. *Water Resources*, **4**, 21–34.

Ferronsky V.I.: 2005, 'Virial approach to solve the problem of global dynamics of the Earth', *Investigated in Russia*, http://zhurnal.ape.relarn/articles/2005/120.pdf, 1207–1228.

Ferronsky V.I. and Ferronsky S.V.: 2007, *Dynamics of the Earth*, Scientific World, Moscow.

Ferronsky V.I. and Polyakov V.A.: 1982, *Environmental isotopes in the hydrosphere*, Wiley, Chichester.

Ferronsky V.I. and Polyakov V.A.: 1983, *Isotopy of the hydrosphere*, Nauka, Moscow.

Ferronsky V.I., Denisik S.A. and Ferronsky S.V.: 1978, 'The solution of Jacobi's virial equation for celestial bodies', *Celest. Mech.*, **18**, 113–140.

Ferronsky V.I., Denisik S.A. and Ferronsky S.V.: 1979a, 'The virial-based solution for the velocity of gaseous sphere gravitational contraction', *Celest. Mech.*, **19**, 173–201.

Ferronsky V.I., Denisik S.A. and Ferronsky S.V.: 1979b, 'The asymptotic limit of the form-factor α and β product for celestial bodies', *Celest. Mech.*, **20**, 69–81.

Ferronsky V.I., Denisik S.A. and Ferronsky S.V.: 1979c, 'The solution of Jacobi's virial equation for nonconservative system and analysis of its dependence on parameters', *Celest. Mech.*, **20**, 143–172.

Ferronsky V.I., Denisik S.A. and Ferronsky S.V.: 1981a, 'Virial oscillations of celestial bodies: I. The effect of electromagnetic interactions', *Celest. Mech.*, **23**, 243–267.

Ferronsky V.I., Denisik S.A. and Ferronsky S.V.: 1981b, 'On the relationship between the total mass of a celestial body and the averaged mass of its constituent particles', *Physics Letters*, **84A**, 223–225.

Ferronsky V.I., Denisik S.A. and Ferronsky S.V.: 1982, 'Virial oscillations of celestial bodies: II. General approach to the solution of perturbed oscillations problem and electromagnetic effects', *Celest. Mech.*, **27**, 285–304.

Ferronsky V.I., Denisik S.A. and Ferronsky S.V.: 1984a, 'Virial approach to solution of the problem of global oscillations of the Earth atmosphere', *Physics of Atmosphere and Oceans*, **20**, 802–809.

Ferronsky V.I., Denisik S.A. and Ferronsky S.V.: 1984b, 'Virial oscillations of celestial bodies: III. The solution of the evolutionary problem in non-Newtonian time scale', *Celest. Mech.*, **32**, 173–183.

Ferronsky V.I., Denisik S.A. and Ferronsky S.V.: 1985, 'Virial oscillations of celestial bodies: IV. The Lyapunov stability of motion', *Celest. Mech.*, **35,** 23–43.

Ferronsky V.I., Denisik S.A. and Ferronsky S.V.: 1987, *Jacobi dynamics*, Reidel, Dordrecht.

Ferronsky V.I., Denisik S.A. and Ferronsky S.V.: 1996, 'Virial oscillations of celestial bodies: V. The structure of the potential and kinetic energies of a celestial body as a record of its creation history', *Celest. Mech. and Dynam. Astron.*, **64**, 167–183.

Flügge S.: 1971, *Practical quantum mechanics*, Springer, Berlin.

Fontes J.Ch. and Gonfiantini R.: 1967, 'Component isotopique au cours de l'evaporation de deux bassins Sahariens', *Earth and Planet. Sci. Latt.*, **3**, 258–266.

Friedman I.: 1967, 'Water and deuterium in pumices from the 1959–1960 eruption of Kelauea volcano, Hawaii', *U.S. Geol. Surv. Prof. Pap.*, **575B**, 120–127.

Friedman I., O'Neil J.R., Gleason J.D. and Hardcastle K.: 1970, 'Water, hydrogen, deuterium, carbon, carbon-13 and oxygen-18 content of selected lunar material', *Science*, **167**, 538–540.

Friedman I., O'Neil J.R., Gleason J.D. and Hardcastle K.: 1971, 'The carbon, hydrogen content and isotopic composition of some Apollo-12 materials', in: *Proc. 2nd Lunar Sci. Conf.*, **2**, 1407–1415.

Galimov E.M.: 1968, *Carbon stable isotope geochemistry*, Nedra, Moscow.

Galimov E.M.: 1973, *Carbon isotopes in oil and gas geology*, Nedra, Moscow.

Ganapathy R. and Anders E.: 1974, 'Bulk composition of the Moon and Earth, estimated from meteorites', *Geochim. et Cosmochim. Acta.*, **5**, 1181–1206.

Garcia Lambas D., Mosconi M.B. and Sersic J.L.: 1985, 'A global model for violent relaxation', *Astrophys. and Space Sci.*, **113**, 89–98.

Giordano C.M. and Plastino A.R.: 1999, 'Jacobi dynamics and n-body problem with variable masses', *Celest. Mech. and Dynam. Astron.*, **75**, 165–183.

Goldschmidt V.M.: 1954, *Geochemistry*, Oxford Univ. press, Oxford, UK.

Goldstein H.: 1980, *Classical Mechanics*, 2nd ed., Addison-Wesley, Reading-Massachusetts.

Grinenko V.A. and Grinenko L.N.: 1974, *Sulphur isotope geochemistry*, Nauka, Moscow.

Grossman L. and Larimer J.W.: 1974, 'Early chemical history of the Solar System', in: *Rev. Geophys. and Space Phys.*, **12**, 17–101.

Grushinsky N.P.: 1976, *Theory of the earth figure*, Nauka, Moscow.

Hayes I.M.: 1967, 'Organic constituents of meteorites: A review', *Geochim. et Cosmochim. Acta*, **31**, 1395–1440.

Herndon J.M. and Suess H.E.: 1976, 'Can enstatite meteorites form from a nebula of Solar composition?', *Geochim. et Cosmochim. Acta*, **40**, 395–399.

Herndon J.M. and Suess H.E.: 1977, 'Can the ordinary chondrites have condensed from a gas phase?', *Geochim. et Cosmochim. Acta*, **41**, 233–236.

Jacobi C.G.J.: 1884, *Vorlesungen über Dynamik*. Klebsch, Berlin.

Jeffreys H.: 1970, *The Earth: Its origin, history and physical constitution*, 5th ed., Cambridge Univ. Press, Cambridge.

Kaplan I.R. and Smith J.W.: 1970, 'Concentration and isotopic composition of carbon and sulfur in Apollo 11 lunar samples', *Science*, **167**, 61–69.

Kelly W.R. and Larimer J.W.: 1977, 'Chemical fractionation in meteorites, VIII. Iron meteorites and the cosmochemical history of the metal phase', *Geochim. et Cosmochim. Acta*, **41**, 93–111.

Kittel Ch.: 1968, *Introduction to solid state physics*, 3rd ed., Wiley, N.Y.

Kittel C., Knight W.D. and Ruderman M.W.: 1965, *Mechanics, Berkeley physics course*, Vol. 1, McGraw Hill, New York.

Klein F. and Sommerfeld A.: 1903, *Theorie des Kreisels*, Heft III, Teubner, Leipzig.

Knauth L.P. and Epstein S.: 1971, 'Oxygen and hydrogen isotope relationships in charts and implications regarding the isotope history of the hydrosphere', in: *Geol. Soc. Amer. Mts., Abstracts*, 624.

Knauth L. and Epstein S.: 1976, 'Hydrogen and oxygen isotopes retios in nodular and bedded cherts', *Geochim. et Cosmochim. Acta*, **40**, 1095–1108.

Kokubu N., Mayeda T. and Urey H.C.: 1961, 'Deuterium content of minerals, rocks and liquid inclusions from rocks', *Geochim. et Cosmochim. Acta*, **21**, 247–256.

Lancet M.S. and Anders E.: 1970, 'Carbon isotopes fractionation in the Fisher–Tropch synthesis and in meteorites', *Science*, **170**, 980–982.

Landau L.D. and Lifshitz E.M.: 1954, *Mechanics of continuous media*, Gostechizdat, Moscow.

Landau L.D. and Lifshitz E.M.: 1963, *Quantum mechanics*, Nauka, Moscow.

Landau L.D. and Lifshitz E.M.: 1969, *Mechanics, elektrodynamics*, Nauka, Moscow.

Landau L.D. and Lifshitz E.M.: 1973a, *Mechanics*, Nauka, Moscow.

Landau L.D. and Lifshitz E.M.: 1973b, *Field theory*, Nauka, Moscow.

Larimer J.W.: 1967, 'Chemical fractionation in meteorites, I. Condensation of the elements', *Geochim. et Cosmochim. Acta*, **31**, 1215–1238.

Larimer J.W. and Anders E.: 1967, 'Chemical fractionations in meteorites, II. Abundance patterns and their interpretation', *Geochim. et Cosmochim. Acta*, **31**, 1239–1270.

Larimer J.W. and Anders E.: 1970, 'Chemical fractionation in meteorites, III. Major element fractionation in chondrites', in: *Geochim. et Cosmochim. Acta*, **34**, 367–387.

Laul J.C., Ganapathy R.E. and Morgan J.W.: 1973, 'Chemical fractionation in meteorites, VI. Accretion temperatures of H-, LL-, and E–chondrites, from abundance of volatile trace elements', *Geochim. et Cosmochim. Acta*, **37**, 329–357.

Lord H.C.: 1965, 'Molecular equilibria and condensation in Solar nebula and cool stellar atmospheres', *Icarus*, **4**, 279–288.

Magnitsky V.A.: 1965, *Inner structure and physics of the Earth*, Nauka, Moscow.

Melchior P.: 1972, *Physique et dynamique planetaires*, Vander-Editeur, Bruxelles.

Migdisov A.M., Dontsova E.I. and Kusnetsova L.V. et al.: 1974, 'Probable causes of oxygen isotopic content evolution in the outer Earth shells', in: *Proc. Intern. Gepchem. Congr., Vol. 4, Sedimentary Rocks*, Moscow, 173–184.

Misner C.W., Thorne K.S. and Wheeler J.A.: 1975, *Gravitation*, Freeman, San Francisco.

Molodensky M.S. and Kramer M.V.: 1961, *The Earth's tidals and nutation of the planet*, Nauka, Moscow.

Monster J., Anders E. and Thode H.C.: 1965, '$^{34}S/^{32}S$ ratios for the different forms of sulphur in the Orgueil meteorite and their mode of formation', *Geochim. et Cosmochim. Acta*, **29**, 773–779.

Munk W. and MacDonald G.: 1960, *Rotation of the Earth* (Transl. from Engl.), Mir, Moscow.

Murray B.C., Delton J.S. and Danielson G.E., et al.: 1974, 'Mariner 10 pictures of Mercury: First results', *Science*, **184**, 459–461.

Newton I.: 1934, *Mathematical principles of natural philosophy and his system of the world* (Transl. from Latin by Andrew Mott, 1729), Cambridge Univ. Press, London.

Onufriev V.G.: 1978, 'Distribution of volatile elements in the upper Earth shell and formation of their isotopic composition', in: *Isotopy of natural waters*, V.I. Ferronsky (ed.), Nauka, Moscow, 90–118.

Onuma N., Clayton R.N. and Mayeda T.K.: 1970, 'Oxygen isotope fractionation between minerals and an estimate of the temperature of formation', *Science*, **167**, 536–537.

Onuma N., Clayton R.N. and Mayeda T.K.: 1972, 'Oxygen isotope cosmothermometer', in: *Geochim. et Cosmochim. Acta*, **36**, 169–188.

Pariysky N.N.: 1975, 'Tides', in: *Great Soviet Encyclopedia*, 3rd ed., 580–582.

Park R. and Epstein S.: 1960, 'Carbon isotope fractionation during photosynthesis', in: *Geochim. et Cosmochim. Acta*, **21**, 110–126.

Reghini G.: 1963, 'Gli isotopi nell'atmosphera solara', *Ric. Sci. Rev.*, **3**, 145.

Reuter J.H., Epstein S. and Taylor H.P.: 1965, '$^{18}O/^{16}O$ ratios of some chondritic meteorites and terrestrial ultramafic rocks', *Geochim. et Cosmochim. Acta*, **29**, 481–488.

Ringwood A.E.: 1966, 'Chemical evolution of the terrestrial planets', *Geochim. et Cosmochim. Acta*, **30**, 41–104.

Ringwood A.E.: 1979, *Origin of the Earth and Moon*, Springer, N.Y.

Rubey W.W.: 1964, 'Geological history of sea water', in: *The origin and evolution of atmospheres and oceans*, Wiley, N.Y., 1–73.

Sabadini R. and Vermeertsen B.: 2004, *Global dynamics of the Earth*, Kluwer, Dordrecht.

Savin S.M. and Epstein S.: 1970, 'The oxygen and hydrogen isotope geochemistry of ocean sediments and shales', *Geochim. et Cosmochim. Acta*, **34**, 43–63.

Schell W.R., Fairhall D.W. and Harp G.D.: 1967, 'An analytic model of carbon-14 distribution in the atmosphere', in: *Proc. Symp. IAEA: Radioactive dating and low-level counting*, Vienna, 79–92.

Schidlowski V., Eichmann R. and Junge C.E.: 1975, 'Precambrian sedimentary carbonates: Carbon and oxygen isotope geochemistry and implications for the terrestrial oxygen budget', *Precambrian Res.*, **2**, 1–69.

Schiegl W.E and Vogel J.C.: 1970, 'Deuterium content of organic matter', *Earth and Planet. Sci. Lett.*, **7**, 307–313.

Schmidt O.Y.: 1957, *Four lectures on origin of the Earth*, Ac. Sci. USSR Publ. House, Moscow.

Scott E.R.D. and Wasson J.T.: 1976, 'Chemical classification of iron meteorites, VIII: I, IIE, IIIF and 97 other irons', *Geochim. et Cosmochim. Acta*, **40**, 103–115.

Sedov L.I.: 1970, *Mechanics of continuous media*, Vol. 2., Nauka, Moscow.

Sheppard S.M.F. and Epstein S.: 1970, '$^2H/^1H$ ratios of minerals of possible mantle or lower crustal origin', *Earth and Planet. Sci. Lett.*, **9**, 232–239.

Silverman S.: 1951, 'The isotope geology of oxygen', *Geochim. et Cosmochim. Acta*, **2**, 26–42.

Spencer J.H.: 1956, 'The origin of the Solar System', in: *Physics and chemistry of the Earth*, Vol. 1, Pergamon, London, 1–16.

Spitzer L. Jr.: 1968, *Diffuse matter in space*, Interscience, New York.

Tamm I.E.: 1976, *Fundamentals on theory of electricity*, Nauka, Moscow.

Taylor H.P.: 1974, 'The application of oxygen and hydrogen isotope studies to problems of hydrothermal alteration and ore deposition'. *Econ. Geol.*, **69**, 843–883.

Taylor H.P.: 1978, 'Oxygen and hydrogen isotope studies of plutonic rocks', *Earth and Planet. Sci. Lett.*, **38**, 177–210.

Taylor H.P., Duke M.B., Silver L.T., and Epstein S.: 1965, 'Oxygen isotope studies of minerals in stone meteorites', *Geochim. et Cosmochim. Acta*, **29**, 489–512.

Teys R.V. and Naydin D.P.: 1973, *Paleothermometry and oxygen isotopic composition of organogenic carbonates*, Nauka, Moscow.

Theodorsson P.: 1967, 'Natural tritium in groundwater studies', in: *Isotopes in hydrology, Proc. Symp. IAEA*, Vienna, 371–380.

Turekian K.K. and Clark S.P. Jr.: 1969, 'Inhomogeneous accumulation of the Earth from the primitive Solar nebula', *Earth and Planet. Sci. Lett.*, **6**, 346–348.

Urey H.C.: 1957, 'Boundary conditions for theories of the origin of the Solar System', in: *Physics and chemistry of the Earth*, Vol. **2**, Pergamon, London, 46–76.

Urey H.C.: 1959, *The Planets*, 2nd ed., Yale University Press, New Haven.

Urey H.C., Brickwedde F.G. and Murphy G.M.: 1932, 'A hydrogen isotope of mass 2', *Phys. Rev.*, **39**, 1645.

Vinogradov A.P.: 1971, 'High temperature protoplanetary process', *Geokhimiya*, **11**, 1283–1296.

Vinogradov A.P.: 1975, 'The Moon differentiation', in: *Cosmochemistry of the Moon and the planets*, Nauka, Moscow, 5–28.

Vinogradov A.P., Dontsova E.I. and Chupakhin M.S.: 1960, 'Isotopic ratios of oxygen in meteorites and igneous rocks', *Geochim. et Cosmochim. Acta*, **18**, 278–293.

Vinogradov A.P., Kropotova O.I., Vdovicin G.P. and Grinenko V.A.: 1967, *Geokhimiya*, **3**, 267–273.

Vinogradov V.I.: 1980, '*Role of sedimentary cycle in geochemistry of sulphur isotopes*', Nauka, Moscow.

Wasserburg C.J., Turner G., Tera F., Podosek A., Papanastassi D.A. and Huneke J.C.: 1972, 'Comparison of Rb-Sr, K-Ar and U-Th-Pb ages: Lunar chronology and evolution', *Lunar Science III*, 788–790.

Weber J.N.: 1965, 'Extension of the carbonate paleothermometer to premesozoic rocks', in: *Stable isotopes in oceanographic studies and paleotemperatures*, Spoleto, 227–309.

Whittaker E.T.: 1937, *A treatise on the analytical dynamics of particles and rigid bodies*, Cambridge Univ. Press, Cambridge.

Wood Y.A.: 1963, 'On the origin of chondrules and chondrites', *Icarus*, **2**, 152–180.

Wood Y.A.: 1974, *Origin of the Earth's Moon*, Preprint No. 39, Center for Astrophys., Cambridge (Mass).

Zeldovich Y.B. and Novikov I.D.: 1967, *Pelativistic astrophysics*, Nauka, Moscow.

Zharkov, V.N.: 1978, *Inner structure of the Earth and planets*, Nauka, Moscow.

Subject Index

A
Accretion hypothesis, 242, 253, 273, 275
Angular momentum, 95, 115, 164, 204, 208, 233, 275, 280
Anomaly
  eccentric, 4, 5, 51, 205, 206, 209
  mean, 4, 5, 51, 205, 209
  true, 4, 5
Apsides line, 183
Archimedes' force, 52, 142, 160, 161, 166, 192
Archimedes' law, 3, 21, 189
Atmospheric pressure, 173, 213, 216–219
Atmospheric temperature, 173, 213, 219
Atom of hydrogen, 85, 115–122
Avogadro's law, 61

B
Barycentric co-ordinates, 69, 71, 98, 102
Bianchi, 230
Bifurcation point, 278, 279, 281
Bohr radius, 118
Boltzmann constant, 233, 236
Boyle-Mariotte's law, 61

C
Cartesian co-ordinates, 53, 63, 73, 82, 99, 100
Centre of inertia, 161, 164
Centre of mass, 165
Chandlers wobbling, 25, 36, 187
Chandrasekhar-Fermi equation, 237, 238, 280
Charge elementary, 226
Christoffel symbol, 92
Clairaut equation, 22, 28, 50
Clapeyron-Mendeleev's equation, 60
Clausius' virial theorem, 27, 46, 48, 51, 156
Condensing temperature, 242, 266
Confidence interval, 214–218

Conservative system, 49, 73, 97–124, 126, 135–137, 163, 164
Copernican world system, 2, 51
Coriolis' force, 52, 142, 160, 161, 166, 192
Cosmochemical events, 268
Coulomb energy, 226, 228, 237, 280
Coulomb interaction, 225–229, 237, 238, 278–280
Coulomb law, 89, 163
Covariant 4-delta, 230
Covariant differentiation, 230

D
D'Alembert operator, 231
Defect of mass, 90, 91, 193–195, 239
Degassing of volatiles, 260
Dipole electric, 229–234, 278
  oscillating, 230, 232, 278
Discrete-wave structure, 164, 235
Discriminant curves, 135, 136, 277, 278
Dissipative function
  energy, 154
Dynamical approach, 97, 98, 142, 186, 223, 284
Dynamical equilibrium, 46–53, 58–60, 62, 71–73, 88, 97, 98, 113, 120, 126, 139, 141, 142, 147, 152, 154–156, 158, 162, 163, 165, 171–173, 176, 184, 185, 187, 191, 192, 197, 198, 229, 234, 235, 238, 239, 243, 277–280, 284, 285
Dynamical equilibrium of state, 46, 53, 59, 191, 197

E
Eccentricity, 4, 5, 106, 151, 184, 205–209, 233
Eigenoscillations, 44, 142–152, 212–219
Einstein's equations, 60, 89, 124, 156, 230

V. I. Ferronsky, S. V. Ferronsky, *Dynamics of the Earth,*
DOI 10.1007/978-90-481-8723-2_0, © Springer Science+Business Media B.V. 2010